# The Aliens Among Us

# The Aliens Among Us

*How Invasive Species*
*Are Transforming*
*the Planet —*
*and Ourselves*

LESLIE ANTHONY

Yale UNIVERSITY PRESS/NEW HAVEN & LONDON

Yale University Press books may be purchased in quantity for
educational, business, or promotional use. For information,
please e-mail sales.press@yale.edu (U.S. office) or sales@yaleup.
co.uk (U.K. office).

Set in Minion type by IDS Infotech, Ltd.
Printed in the United States of America.

Library of Congress Control Number: 2017937194

ISBN 978-0-300-20890-0 (hardcover : alk. paper)

A catalogue record for this book is available from the British
Library.

This paper meets the requirements of ANSI/NISO Z39.48-1992
(Permanence of Paper).

10 9 8 7 6 5 4 3 2 1

for Petal
a lover of plants
a respecter of nature
a plucker of heartstrings

Instead of six continental realms of life, with all their minor components of mountain tops, islands and fresh waters, separated by barriers to dispersal, there will be only one world, with the remaining wild species dispersed up to the limits set by their genetic characteristics, not to the narrower limits set by mechanical barriers as well.

*Charles Elton*, The Ecology of Invasions by Animals and Plants

# Contents

# Acknowledgments

The issue of invasive species is both enormous and important. These organisms not only have been physically dislocated to new lands from the ecosystems in which they evolved, but also, as this book avers, have infiltrated both the public psyche and most institutions in their adopted homes. That's a lot to wrap your head around, meaning a project of this magnitude would never have found its way unless a critical path forward had presented itself. It was, fortunately, a human compass that offered that direction: my friend and colleague Nicholas Mandrak at the University of Toronto. Our first chat on the subject neither hinted at the Pandora's box he was opening nor revealed what his eventual commitment to shutting it would be. In the end, many hours of what likely would have been more valuable time to him was spent graciously sharing with me his hard-won knowledge on fisheries science, aquatic biodiversity, and the ecosystems of the Great Lakes. I believe I owe him an infinite amount of beer, a debt that I live in fear he will try to collect on.

At least I can blame Mandrak for the fact that Becky Cudmore, commander of the Asian Carp Program at Canada's Department of Fisheries and Oceans (DFO), suffered similarly pesky visits. Numerous phone and office consultations with another friend, Isaac Bogoch, opened a window on geographical epidemiology and invasive vector-pathogen pairs in emerging global diseases.

Beyond the considerable contributions of these three, I'll first salute those who figured most in piquing my interest on invasives: Charles Elton, author of *The Ecology of Invasions by Animals and Plants;* David Quammen, for his "Planet of Weeds" essay in *Harper's,* and the "tireless" (you'll understand this later) ecologist Daniel Simberloff for his voluminous work, much of which I've now had the pleasure to read. Informally, through Simberloff collaborations, and formally through Mandrak, I was introduced to the equally tireless Anthony Ricciardi, who provided more grist in a single phone call than any other made over the four years of this project. Editors James Little at *explore,* Clare Ogilve at *PIQUE Newsmagazine,* Dan Rubinstein at *Canadian Geographic,* Cooper Langford at *Canadian Wildlife,* and Robyn Smith at *Tyee.ca* enthusiastically published articles from which parts of this book are drawn.

My first invasive-related assignment for *explore* took me to Florida, where I hefted five-meter Burmese pythons but came away more impressed with the heft of those working to thwart their effects: Meg Lowman and Skip Snow were enormously helpful and facilitative, and George Cera and Karen Garrod both let me tag along in the field.

Next stop was Haida Gwaii, where Laurie Wein and the Parks Canada "rat patrol" of Peter Dyment, Debby Gardiner, Tysen Husband, and Clint Johnson Kendrick welcomed me into their bobbing North Pacific sanctum. I can't thank Parks' terrestrial ecologist Carita Bergman enough for the time she took to discuss ecological restoration in Haida Gwaii. Also helpful were British Columbia Conservation Officer James Hilgemann, Parks' PR specialist Heather Ramsey, Barb Wilson of the Haida Nation, and Ainsley Brow, Lexi Forbes, and Roger Packham of Coastal Conservation. On the other side of the country, Marie-Josée Légaré of Quebec National Parks (Sépaq) accompanied me on a magical trip to L'Île d'Anticosti, even tracking down mayor-cum-deer-biologist Denis Duteau for a thought-provoking dinner.

Clare Greenberg of the Sea to Sky Invasive Species Council and her team of Sam Cousins, Rob Hughes, Breanne Johnson, and Sharon Watson no doubt found me as tenacious as a knotweed—and at times harder to get rid of—yet Greenberg was never too busy to share her thoughts and expertise. Thanks also to SSISC alumnus Kristina Swerhun and, especially, SSISC and Whistler Naturalists founder Bob Brett, whose personal invasive-plant crusades are legend. Richard Beard offered insight into ecological restoration and commercial invasive plant removal when I joined his Green Admiral crew of Julia Alards-Tomalin and Wade McCleod for a day. Trevor Cox connected me with Fiona Steele at Diamond Head Consulting, who not only took time to meet, but arranged for me to accompany crews led by Jeff Hunter and Keith MacKenzie. Others who took time to converse about invasive plants include Ken Crosby at City of Surrey, Sharon Gillies at University of the Fraser Valley, and Melinda Yong at City of Burnaby.

I owe a huge debt of gratitude to my overworked but ever-helpful friend, Purnima Govindarajulu, at British Columbia's Ministry of the Environment. More aid with B.C.'s invasive bullfrogs came from Sara Ashpole, Christian Engelstoft, Natasha Lukey, and Stan Orchard. Rylee Murray deserves special thanks for a lively nighttime tour of the Fraser Valley's croaking masses.

Phone chats with Lee Frelich delivered the equivalent of a university course on invasive earthworms in Minnesota, and Erin Cameron happily extended the subject to Alberta and Finland. Ant-whisperer Sean McCann opened my eyes to the scourge of *Myrmica rubra,* and Chris Higgins of Thompson Rivers University in Kamloops filled me in on what that means to the naïve ecosystems of North America. I enjoyed a lengthy discussion on America's imported red fire ants with University of Pennsylvania doctoral student Chris Thawley. Troy Kimoto of the Canadian Food Inspection Agency schooled me on invasive insects, and University of Toronto's Sandy Smith and Robert Bourchier of Agriculture Canada further educated me on the history and current status of biocontrol in North America.

David Galbraith introduced me to lands manager Tys Theysmeyer at the Royal Botanical Gardens in Hamilton, Ontario, who groundtruthed with me the impacts of invasive common carp, *Phragmites,* irruptive cormorants, and Canada geese. Johanna Liemivuo-Lahti of the Government of Finland shared tea, biscuits, and her magnum opus on the prevention, control, and monitoring of invasive species. Sean Fox opened the gates to the Guelph Arboretum and the secrets of American elm and American chestnut restoration. A chat with Helen Kennedy of

Arlo's Honey Farm in Kelowna about bees and *Varroa* mites led to a "fruitful" discussion about invasive starlings with Jane Hatch of Tantalus Winery. Oceanwide Expeditions spokesperson Rima Deeb Granado answered emails about landings on sub-Antarctic islands, and biologist David Marson answered questions about the Asian carp lab run by DFO. Jacqueline Litzgus invited me to the mind-bending "Thinking Extinction" conference at Sudbury's Laurentian University, where Arne Moores of Simon Fraser University provided perspectives on evolutionary rates, biodiversity, and the unique vulnerability of monotypic organisms. Axel Moehrenschlager of the Calgary Zoo and University of Calgary invited me to a workshop on the vanishing sage grouse, my introduction to the world of captive breeding, reintroduction, and translocation.

Before I'd found any kind of compass, however, my stalwart researcher Penelope Lafrance gathered the buoyant mass of information required to first float this wayward ship. My agent, Carolyn Forde, worked diligently to find someone interested in a popular treatment of a burgeoning and somewhat intractable subject. That person was Jean Thomson Black of Yale University Press, who has been both patient and enthusiastic in equal measure throughout. No one, however, has been more enthusiastic than Julie Carlson, who, as editor, took on a too-often dense hedge of information with practiced eye and delicate pruner, supplanting thornier sprigs with thoughtful suggestions. As always, I was aided by the logistical and moral support of family: much thanks to my mother, Raffaelina, brother Chris, sister-in-law Karen, and daughter, Myles—who also helped blunt the modern malaise of liability by assisting in the tedious task of obtaining release forms.

And finally, my partner, Asta Kovanen, not only employed artsy skills to convert sciencey figures into concepts that could be visually grasped without number-induced trauma, but used her considerable empathy and tolerance skills to deal with four years of having invasives pointed out at every step. Through a long and often arduous journey she consistently offered the kind of encouragement and support only love—or pity—can muster.

# PART I

# Big Trouble

## *A Plague of . . . Giant Snakes?*

If you need to have a meeting about an invasive species, it's already too late.

*Unknown*

# The End?

At the millennium, Burmese pythons appeared far down the expansive list of introduced species that were keeping veteran Everglades National Park ranger Skip Snow awake at night. A gimmicky line item nestled significantly lower than the thickets of Australian melaleuca and dense Brazilian pepper that park staff had been hacking back for decades. Of less concern than an unseen plague of foreign weevils, beetles, and scale insects ravaging native vegetation. Certainly not as important as the feral pigs, tropical birds, and host of exotic fishes mocking what remained of the park's ecology. Sure, irresponsible pet owners occasionally tossed snakes they could no longer care for off the lonely highway bisecting Florida's storied River of Grass, but Snow and his fellow rangers figured these animals rarely survived. In contrast to the benevolent subtropical pasturing envisioned by the snakes' erstwhile owners, Snow's crew believed that the fifty or so pythons found over the course of a decade across the Everglades' 6,100 square kilometers—discovered more often than not decorating a backroad like some giant's discarded belt—offered proof that releasing a naïve, cage-coddled pet into an unfamiliar fray like the Everglades was a de facto death sentence.

Right up until 2003, that is, when several meter-long baby pythons turned up in a secluded area of the park lacking public access. Then, in 2004, the number of pythons discovered inexplicably spiked to sixty in a single year, prompting a stark realization: these weren't *all* discarded pets. Comparing the Burmese python's catholic tastes in prey and broad ecological tolerances across its native range in Southeast Asia to those on offer in the Everglades yielded an almost perfect fit—further cause for worry. No longer content with chance python encounters to divine what might be happening, Snow embraced a simple but ultimately instructive course of action: surgically implanting VHF radio transmitters in four captured male pythons, he and a small crew tracked the snakes through the Everglades' netherworld of hummocks, canals, and sawgrass during the spring mating season. The surprisingly wide-ranging reptiles inadvertently led researchers to twelve more pythons, which were all captured and removed. Alarm bells were already ringing loudly when, in May 2006, a five-meter female python was found coiled around a pile of recently laid eggs. Snow could now conclude only one thing: Florida's most delicate ecosystem—and everything in it—was in *big* trouble.

# PART II

# A Postcard from Florida

## Introducing Alien Invasive Species

There is no foreign land; it is the traveler only that is foreign.

*Robert Louis Stevenson*, The Silverado Squatters

# Snakes and Ladders

By the time I caught up with Snow in the fall of 2008 to write a magazine feature about Florida's growing python problem (clearly an enterprise in which puns are all but inescapable), I was but the latest on a rapidly lengthening (sorry again) list of journalists to latch onto the story (final apology). Unsurprisingly, the jungle-comes-hither tale of giant marauding serpents had generated plenty of media attention, engendering a steady stream of fear-mongering since breaking in 2005. Celebrated on a continuum of sensationalism that ran the gamut from genuine horror to undisguised glee, global news reports contained no shortage of shocking pictures or dire statistics. Photos depicting Burmese python versus American alligator smackdowns worthy of a Japanese horror flick were accompanied by estimates of up to 150,000 pythons now living and multiplying freely in the Everglades. Typically—and not altogether dishonestly—reports characterized the snakes as voracious, prolific, and capable of reaching six meters in length. It was also inferred that they were moving inexorably north through the state's reticulated waterways, as well as southward onto the Florida Keys. Big news all around, so to speak.

Snow and company had been knocking on doors in the halls of both federal and state power long before the python story blew up, but politicians had thus far refused to act, wary of reprisals from the nation's powerful $40 billion annual pet trade (the one-way-or-another source of the invaders). When a two-year-old child was constricted and killed in July 2009 by a Burmese that had escaped its cage in her Tampa home, however, officials finally moved to ban the import and possession of large snakes without a permit, launching a task force that included academic and hobby herpetologists (those who study reptiles and amphibians), invasive species experts, and even ex-wildlife smugglers. After years of inaction, an unrelated tragedy had forced legislators to connect the dots in a picture they'd long been staring at. It was lost on none that the murderous pet was the same species now proliferating in the wilds of the state's southern tier, making the specter of a small child being swallowed by a giant alien snake all too real.

As always, galvanization of political and public attention arrived far too late. Over the next few years, neither control nor eradication measures would help stem the rising python tide. Other than vehicles, the snake's only real enemy proved to be the vicissitudes of climate: a lengthy cold snap in January 2010 resulted in the documented death of numerous pythons and the assumed mortality of many more. This welcome setback, however, was short-lived, and the burgeoning python population—now consuming native, often-endangered wildlife at an alarming rate—seemed largely unaffected. From an ecological standpoint, a large population of a nonnative species had just survived its first adaptive challenge and thus its first serious pinch by natural selection. Had the fitness of Florida's Burmese pythons just ratcheted up a notch? Only time would tell, and time is always on evolution's side.

In a scenario typical for invasive species that defy easy control (which, we'll see, is most), the public's help was enlisted: in January 2013, the Florida Fish and Wildlife Commission instituted a month-long python hunt. Some 1,600 signed up to compete for two prizes in both "pro" and amateur categories: $1,000 for the longest python killed, and $1,500 for the most pythons taken. Given Florida's legally permitted arsenal of weaponry, the well-advertised Python Challenge™ (the trademark is real) seemed to guarantee a savage free-for-all that could eclipse every other invasive species bounty hunt (and there were plenty) on the planet. The additional promise of a five-star redneck gong show had mainstream media from around the continent and across the pond beelining to South Florida to variously witness, participate, and report. They would be sorrowfully disappointed on all fronts, with only sixty-eight pythons turned in and no gunfights.

Notwithstanding the absurdity of a public hunt, few reports considered the significance of Florida's python problem in a bigger picture—the rapidly growing ranks of invasive species globally, and their myriad detrimental effects. Even a nano-litany of examples (lists are as unavoidable as puns in invasion science) cast the ostensibly grievous python problem as small potatoes. As of 2010, for instance, the number of introduced species across Europe had increased 76 percent over the previous three decades to surpass 10,000, at least 1,300 of which had measurable impacts on the environment, economy, or human health. Invasive animals, plants, and microorganisms were causing an annual estimated €12–€20 billion of damage on that continent, $120–$130 billion in the United States, and tens of billions in the Antipodes—a figure that excludes the difficult-to-ascribe environmental cost of extinctions perpetrated among Australian marsupials by invasive meso-predators like foxes and housecats. Likewise, the 1950s introduction into Africa's Lake Victoria of giant Nile perch that can weigh up to two hundred kilos had led—both directly (through predation) and indirectly (through the rapacious overfishing it inspired)—to the extinction or extirpation of some two hundred indigenous freshwater fish species, precipitating an economic and social calamity that had displaced villages, facilitated trade in illegal weapons, and contributed to a regional spike in HIV/AIDS. Lastly, the globetrotting Asian tiger mosquito was vectoring a range of human tropical fevers, viruses of great medical and socioeconomic impact transplanted from Africa to its numerous new vacation homes in areas as far-flung as India, the South Pacific, Europe, and the Americas.

At this point, you might be wondering: what's the difference between an "introduced" species and an "invasive" one? Although I will expand on terminology later, I here follow the definitions of invasion-science doyen Daniel Simberloff in his recent non-technical book *Invasive Species: What Everyone Needs to Know*. That is, an introduced species is one "introduced to a new location with direct or indirect human assistance," while an invasive species is an "introduced species that has spread well beyond its arrival point and that perpetuates itself without human assistance."

Now back to that bigger picture and some background.

## Riding the Curve

In his 1998 essay "Planet of Weeds," author David Quammen made a prescient observation. In regard to the human-propelled Sixth Great Extinction currently sweeping species from the Earth at a rate unseen since the disappearance of the dinosaurs, he explains that the "fifth of the five factors contributing to our current experiment in mass extinction [is the problem of invasive species]. That factor, even more than habitat destruction and fragmentation, is a symptom of modernity. Maybe you haven't heard much about invasive species, but in coming years you will."

We have. Burmese pythons are particularly compelling in hitting so many high notes of human interest (and emotional appeal): they are snakes (phobia/hate); they are large (fear); they are abundant (loss of control); they are exotic (xenophobia); and they are of enigmatic origin (mystery/blame-seeking). A tabloid-writer's dream in an age of news-for-five-minutes, Florida's pythons claimed pole position in the Global Invasives Derby for years. Inevitably, however, the story fell fallow, succumbing both to audience fatigue and the sheer volume of new invaders making daily headlines—even in Florida, where the *vaurien du jour* at this writing is the Argentine tegu, a large, egg-gobbling lizard with the capacity to undo decades of conservation efforts supporting a range of ground-nesters from burrowing owl to sea turtles. Although news reports on invasives *might* mention disruption to native ecosystems or potential effects on biodiversity, they usually emphasize perceived hazards to human food security, health, safety, and lifestyle. Thus, tegu = a threat to your chicken coop; West Nile virus = a threat to your health; noxious giant hogweed = a threat to your children; and Asian snakehead fish = a threat to your fishing hole.

These stories range from humorous and cautionary, to disturbing and frightening. Some commentators paint these problems as Man versus Nature battles of our own creation and, increasingly, our undoing. As with other anthropogenic issues like climate change, overfishing, ocean acidification, and deforestation, however, invasive species are more accurately seen as one element of nature (us) presenting the rest with more than the usual range of challenges on an accelerated time scale. This view both widens and shifts the conversation to yield a spectrum of perceptions and perspectives that, given current human agency, might center on the idea that the problem is "inevitable." One side of this spectrum anticipates negative environmental/ecological outcomes that are "lamentable," "disastrous," or "catastrophic," while the other tracks an opposite view of "equivocal," "benign," even "beneficial" consequences. Though empirical evidence overwhelmingly supports the former camp's view, murmurings, we will see, continue from the latter.

Some of this, of course, isn't news. Historical documents, oral tradition, archaeology, paleontology, and DNA sequencing all tell us that the *Homo*-mediated introduction of nonnative species to novel and naïve environments has been a forever feature of human trade and travel; when we walked out of Africa, we were already packing invasive heat. To describe this in the terminology of invasion science, trade was a "pathway"—one of many broad routes by which invasive species are transferred from one geographic area to another. And items of trade were often "vectors"—the subroutes within pathways and the physical means by which a species is transported by humans (who can themselves be considered vectors).

Many introductions involved livestock or pets that escaped human control in their new environments, then established feral populations that subsequently spread unchecked to impact native communities of plants and animals. The fragile ecosystems of islands have been particularly vulnerable in this respect, with familiar names like Galapagos, Hawaii, and New Zealand now stand-ins for widespread ecological devastation. Other introductions involve the ubiquitous baggage accompanying human occupation (think rats, mice, cockroaches, weeds, disease); trade and commerce (weevils, beetles, and moths with fabric, foodstuffs, and wood; ants, fungi, nematodes, earthworms, and various plants with soil ballast or plant stock; mollusks, crustaceans, Cnidarians, and fishes with water ballast); the flora and fauna of a native land (British émigrés willfully seeded most continents with the nuisance of rabbits, starlings, sparrows, and an array of quaintly obnoxious garden plants); hoped-for food sources (deer, fish, bullfrog); and failed pest control measures (cane toad, Indian mongoose, mynah bird).

Over recent decades, not only has the frequency of new invasions accelerated, but they have also become more severe and geographically widespread. Both qualitative and quantitative increases have drawn attention from the content-hungry popular media, whose jeremiads have become so pervasive that the public may perceive the global proliferation of invasive species as nothing short of a reimagining of the planet's surface. Truth be told, that idea may not be as overreaching as it sounds.

Most ecosystems with which we're familiar—those we call home, imprint on and enjoy, share glowingly with others, and largely take for granted—are no longer what they seem, let alone what they once were. Many, in fact, haven't been as advertised within our lifetimes. Rather, these environments have become mere facades, artifacts of a community-reinforced eco-idealism that offer a few recognizable threads while simultaneously acting as reservoirs for a host of introduced nonnative species—many of which we can neither see nor readily identify. More significantly, research confirms the interlopers to be both growing exponentially in number and spreading at unprecedented rates. Thus, a reasonable first question is Why now?

The answer—and spoiler alert, its nuances form the core of this book—lies in the fact that our ability to both intentionally and accidentally move species around has reached critical mass at the same time that many ecosystems have reached critical vulnerability. Unable to adapt to the rapid species-shuffling precipitated by ongoing introductions, ecosystems in which nonnative species become invasive have begun falling apart like proverbial houses of cards, changes that inevitably also impact at least some form of human endeavor: indigenous crop pollinators are stymied; natural pest control is thwarted to create new pests; formerly clear waterways become unnavigable; and aquatic and terrestrial food webs are restructured in ways that threaten and endanger cherished native species. That is, ecosystem services previously relied on by the human species become the very ones being sabotaged.

Unsurprisingly, the entire scope of this emerging invasive zeitgeist is on display in Florida. With its favorable climate, mix of habitats, busy ports, generous traffic in imported organisms, and high human immigration, this subtropical peninsula has been a beachhead for lengthy checklists of foreign reptiles, amphibians, rodents, birds, fish, worms, beetles, wasps, mollusks, aquatic weeds, choking vines, and opportunistic trees. Deliberate introductions as well as escapees from private homes, ship and air cargo, wildlife attractions, and plant and animal nurseries have all contributed to significant and increasing consequences for the region. If the North American continent were a metaphorical house, by the mid-1990s Florida had become its dank basement—full of creepy crawlies no one wanted to see upstairs. As a veritable microcosm of what many perceive as a growing global cataclysm, no other place was so rife with introduced forms, nor so poised to host a genuine disaster. It was no surprise, then, that the Sunshine State was experiencing several such problems, and that the biggest catastrophe—literally and figuratively—had become its best known.

Beyond Florida, it's lost on few given to thinking about such things that as a vector for all introductions, *Homo sapiens* is the nonnative species with the greatest impact of all—yet also the only one conscious and capable of mitigating that impact. Does this suggest an ecological responsibility and imperative to act? A majority of those concerned with invasions believe so. Then how do we think about or engage with what we're reading and hearing? With what we're seeing firsthand, unconsciously creating, increasingly experiencing? This book, at least in part, looks at how you and I fit into this evolving planetary picture. It is a picture best parsed through the experiences of those most deeply embedded—researchers and managers working both in labs and on the ground. Robust as the relatively young discipline of invasion science has become, there are still so many unknowns that we might view these folks as adventurers exploring yet another dimension of the Anthropocene—the newly connoted geological era of human effects on the planet. There's enough terra incognita ahead, in fact, that it might be useful to plot our own voyage through it on a conceptual map of some kind. The waypoints on that map, if one were to exist, might be plucked from the Generalized Invasion Curve (Figure 1).

While it's unclear who first employed this simple depiction in relation to invasive species, it wasn't a great leap: the sigmoidal logistic function is frequently used in biology and ecology as a generalized model of population growth, where reproductive rate is proportional to both the existing population and available resources, all else being equal (though in natural systems all else is seldom equal, hence the use of "generalized"). The S-curve that so well describes self-limiting growth in a biological population (Figure 1a) was, in fact, derived for just this purpose by Belgian doctor and mathematician Pierre-François Verhulst, who first published the equation behind it in 1838 after reading Thomas Malthus's *An Essay on the Principle of Population* (in an 1845 paper he named the equation's solution—that is, the curve—the "logistic function"). Figures 1b and 1c are amalgams of widely circulated variations, notably a 2009 version originating with the Australian state of Victoria's Department of Primary Industries. While the intent was to visualize phases in agricultural pest and weed management, its basic tenets apply to any transplanted species that survives, establishes, and spreads to become problematic in its new home.

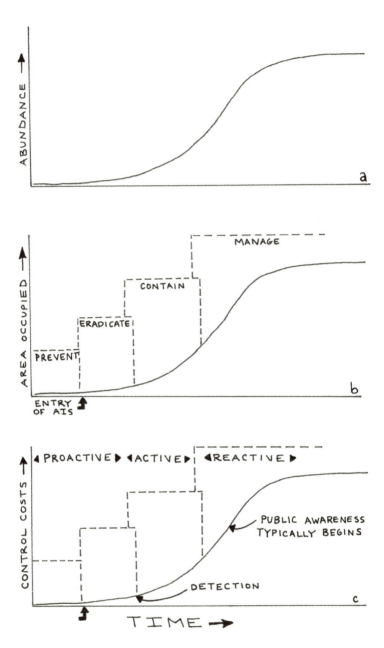

Figure 1. The Generalized Invasion Curve. The x axis is always time, but the y axis can vary depending on which aspect of an invasion you wish to highlight: (a) the basic Invasion Curve following the logistic function of self-limiting population growth, (b) idealized overlapping phases of human reaction to the entry, establishment, and geographic spread of an Alien Invasive Species (AIS) mapped onto the Invasion Curve, (c) the reality of typical actions in relation to the Invasion Curve—by the time public awareness begins, it's too late for anything but perpetual management.

In broad terms, the Invasion Curve begins after a successful introduction of a species (most introductions aren't successful, meaning an organism may have to be introduced repeatedly before it "takes"). The initial stage of population growth is approximately exponential; as the population saturates its new environment, however, limiting factors like space, food resources, and competition slow the rate of growth; then the curve flattens as the population nears maximum saturation (often referred to as "fully realized biological potential"). The horizontal axis of the graph is always time, but representations on the vertical axis can be either ecological, describing factors such as an increase in the abundance of an invasive organism or the area occupied, or economic, showing, say, the increasing costs of both damage and control. The Invasion Curve's greatest utility lies in serving as a model at any level of a jurisdiction or ecosystem—local, regional, continental, global. More importantly, an abundance of research, theory, and real-world lessons allows us to map appropriate human (societal) responses and capacities onto the Invasion Curve that suggest when, how, and why we should act at each stage of an invasion (Figure 1c). Because these will emerge, submerge, and reemerge countless times as we voyage across this conceptual sea, I offer only the briefest descriptions here:

1. **Prevention**. When a potential invader is absent from an environment to which it might pose a threat, the obvious goal is to prevent introduction. Such measures start with a scientifically defensible risk assessment that can inform public policy on creating legal, screening, monitoring, and/or physical barriers to possible vectors and pathways to introduction. Public education and outreach are also vital.

2. **EDRR/Eradication**. Formulation and implementation of Early Detection and Rapid Response (EDRR) protocols are critical to successfully dealing with introductions. When an introduction is first recognized, the initial goal is complete eradication—eliminating the invasive species while doing so remains feasible, usually when there are only one or a few small, scattered populations.

3. **Containment**. If the initial window of opportunity is missed, the game is truly afoot. Rapid increase in distribution and abundance of an invasive species will occur once populations reach a certain critical mass, making eradication unlikely and shifting the response to containing the spread in a prescribed area. While prevention and EDRR are logically grouped as proactive responses, containment signals a more active phase in which eradication at any scale requires intense effort.

4. **Management**: If containment fails—often because it requires more manpower, money, and public or political will than are available—the population growth dynamic moves any potential response into the territory of long-term control. Eradication of a now-abundant, widely established invasive species is deemed impossible and the goal is to reduce populations to the lowest possible level of impact on native communities and ecosystems. In practice, this can be achieved only locally.

There are a few other waypoints of note. First, public awareness—and thus buy-in and mobilization of resources—typically occurs late in the game, when the invasive species is experiencing exponential growth in its new ecosystem and expensive management scenarios are the only feasible option (in other words, when the presence of too many Japanese roses makes us wake up and smell them, so to speak). It thus behooves stakeholders to build as much educational outreach as possible into the prevention phase. Second, the earlier that detection and both public/political awareness can galvanize into action, the better: a scale of relative economic return on pest-management efforts in some depictions of the Invasion Curve (not shown in our Figure 1) charts a precipitous drop in returns from proactive to reactive phases, the implication being that throwing up your hands leads only to loss, while investment in long-term management is often the only way to yield dividends in protecting a resource asset.

Yes, there's a hint of techno-mumbo-jumbo here—but don't panic, you needn't digest it all immediately, and the examples to come will help the medicine go down. This book follows the Invasion Curve by roughly paralleling its accompanying human dimensions (noted in parentheses): Part 2, "A Postcard from Florida," where you currently find yourself, introduces the concept of invasive species and illustrates the circumstances around preventing and dealing with their actual introduction (proactive); Part 3, "The Python's Tale," focuses on the establishment and spread of invasive species and what can be done during this stage (active); Part 4, "Invasional Meltdown," takes a deep look at impacts (active/reactive); Part 5, "The Sisyphus Files," dwells mostly on control measures (reactive); and Part 6, "Sunrise on Homogena," considers a future beyond the Invasion Curve (coping/adapting/accepting).

But as seen in Figure 1, the logistic function is continuous, applies to short or long periods of time, and invokes overlap between phases—meaning that any invasive story is likely to reference most or all of the curve's elements. As a result, some of this volume's tales are wholly contained within one section, while others are spread across several. It's cumulative understanding we're after, so hopefully you'll find these stories interesting, occasionally entertaining, and ultimately revealing of the ever-increasing role of invasive species in our daily lives. Some may even prove personally familiar. Indeed most people can dredge up an invasive species story these days—even kids. Though a recollection like "Yeah … this one time, at band camp, I got burned by giant hogweed" might be paraphrasing a pop-cultural meme for humorous effect, it also reflects an emerging experiential trope.

Mining my own memories of growing up in Toronto in the 1960s confirms that this alien world was already coalescing around me: I remember the specter of sea lampreys dangling from swimmers exiting Lake Ontario; common carp boiling the surface of the St. Lawrence River in Montreal as my brothers and I tossed them bread from a waterside restaurant; the nonnative fire ants we learned to avoid during Florida vacations spent mostly trying to catch nonnative lizards; the annual neighborhood war waged on European dandelions and the exotic-weed-dominated vacant lots in which we played daily; the ghostly, Dutch-elm-ravaged sentinels in the invasive-earthworm-addled, understory-free forest across the street (which prompted my father to plant hardy invasive Norway maple trees that he'd been assured by landscapers wouldn't succumb to any such fate); my early frame of reference for birds being European starlings, English sparrows, and irruptive herring gulls—a spell of blue-collar invasives broken only by the exotic, equally nonnative Chinese ring-necked pheasant I spotted strutting through a field near my suburban home.

Later in life, as a trained biologist whose vocation became outdoor adventure as well as science and environment writing, I bumped harder against this milieu. Over the years, during unrelated assignments in New Zealand, Australia, Polynesia, South America, Europe, Asia, and a broad range of North American locations, I became aware of a pattern of local and regional calamities portending far broader disasters that had yet to enter the global conversation. And much like a character in a sci-fi movie, tipped to the presence of beings from another planet surreptitiously walking the Earth, I now saw aliens everywhere I looked. Intrigued, I began researching and writing about individual invasive species issues only to find that it was the people studying and working with them who drew me in. As a result, I've come to believe we all have a personal stake in the disposition of invasive species—skin in the game so to speak. I know I do.

And that's a good place to jump in.

## Anthropochory

On a bright May morning a few years ago, driving north from Vancouver on the highway that shadows the east side of Howe Sound, I was infused with the energy of the moment. As much as it can represent such a thing, my mood was that of spring incarnate. Everything that day seemed electric and, like the turning of a proverbial young man's fancy, the charge ran through me.

Far below on my left, lit by sun that already seemed mid-summer intense despite the northerly latitude, the Pacific Ocean shimmered like a clinquant carpet. To the right towered forest, mostly second-growth junk conifers hemming in remnant patches of the ecosystem that once dominated British Columbia's south coast: Douglas-fir, western hemlock, bigleaf maple, and, on sunnier bluffs, knots of rust-limbed arbutus, looking for all the world like they'd been cribbed from a Dr. Seuss book. Where you could peer far enough into the shadows below the canopy, moss glowed viridescent after a long, wet winter. That would end soon enough given the earlier, hotter, drier summers that climate change was delivering to the entire Pacific Northwest, but for now a spongy mat of hopeful jade still held sway on the forest floor. Likewise ebullient was a bushy swath of greenery occupying the open strip between timber palisade and sterile highway, interrupted here and there by Pollock-like splashes of canary yellow, patches that seemed to bloom larger each successive May to highlight the highway's bucolic revelry.

Taking this all in, my mind's eye distilled a painting entitled *Spring*—a portrayal of the natural world thrusting up, reaching out, bursting forth; a diorama of seasonal turnover and a living entreaty to join in. But while this mental canvas may have been defined by bold brushstrokes of florescence, two particulars muddled its background: first, the plant life awakening my biophilia was, for the most part, neither native to the area nor even the continent, and its ever-expanding mass was generally viewed negatively by most who lived here; second, despite my knowledge of those details, the scene was nevertheless having the same basic effect as native vegetation occupying the same space would have—that is, it contributed to the landscape-level narrative of well-being that we all experience through visual contact with nature. I'll thus forgive myself any oversight: I surely wasn't the first to ignore very real problems behind an otherwise appealing display. It was my awareness of this divide, however, that opened up a deeper line of thought, and that schism soon became a chasm.

Here began my real journey through the human paradoxes surrounding the growing ranks of invasive species—a class of transplanted organisms that we have created and enabled to thrive.

On one level, organisms falling within this bailiwick are perfectly deserving of appreciation—they are examples of relocated life unconsciously going about its habit of colonizing, adapting, growing, and reproducing in a novel locale. As the species that has proven most accomplished at this very endeavor, we humans can relate. But on another level, a stew of scientific inquiry and socioeconomic concerns casts these same phenomena in a different, darker light. With their unheralded successes in new habitats, many invasive species have come to be seen as ecological modifiers, dangerous interlopers, über-competitors, and agents of extinction; as violators of cherished aesthetics (never mind that this, too, is pure human construct); and, as some would conflate such considerations, the unwelcome, uncontrollable, consummate Other.

And here's why this dichotomy matters: for every thimbleful of rich irony posed by invasive species, gallons of practicable questions slosh in epistemological vats bearing labels that range from Biology, Ecology, Economics, Environment, and Conservation, to Transport, Food Security, Epidemiology, Social Psychology, and even Philosophy. It's a crucible so populous as to suggest the coalescing of an entire worldview, and indeed such an intellectual nexus had formed long before I located myself within it. After dwelling on this a bit as I drove, I stopped for coffee and moved on to other thoughts. Even when they weren't on my mind, however, I couldn't outrun invasive species.

An hour later I reached my home in the mountain resort town of Whistler. Before going inside, I removed my running shoes and smacked them together over a newly turned front garden, shedding the clots of earth packing the treads. I peeled off soiled socks, shaking these out, too, before piling all my sweat-stained clothing into the washer. Earlier that day, some 150 km away, I'd enjoyed a 10-km run through North Vancouver's Lynn Canyon park system, a mosaic of coastal rainforest, deep cataracts, upland fields, and lakeside wetlands. Passing mostly through its wilder tracts, my route also included short jaunts around parking lots, across hydro right-of-ways, and along shaded urban streets. I'd run similarly through each, brushing aside whatever grasses, weeds, or wildflowers edged the path, naïve to their dispositions and proclivities. I'd even passed a plot cordoned off by yellow police tape, treating the hand-scrawled sign announcing "DANGER: Giant Hogweed"—an increasingly familiar sight hereabouts—with the mental equivalent of a shrug.

Not that I didn't care, but like hundreds of hikers, sightseers, and dog-walkers that day, I was just a guy out for a relaxing run and some fresh air. Giant hogweed was a legitimately worrisome interloper, but the park's invasive plants surely weren't my responsibility. At issue for those who strung the tape, I now imagined, was that perhaps they *should* be—and that of any other user. Collective concern would surely help stanch the spread of invasive organisms, particularly those posing serious health risks like hogweed: its sap, should it contact human skin, reacts with sunlight to produce severe burns and permanent scarring. The merits of community knowledge and responsibility on such an issue can't be minimized; invasive plants often become so because they're so easily spread through otherwise innocuous human activity. As it was, I'd engaged in my own fair share that day.

To start, cleaning mud-caked shoes over disturbed soil could have let loose not only seeds of the numerous invasive plants through which I'd blithely run, but also spores, eggs, and larvae, respectively, of alien fungi, earthworms, and other soil invertebrates. Seeds in the form of tenacious burs or cleavers that cling to socks or clothing wouldn't all be dislodged or destroyed in the washing cycle, and so could be easily deposited elsewhere when next I wore them; those released into the dryer's lint trap would eventually find their way through household garbage to a landfill, where under proper conditions they might germinate. Even my drive home could have aided the incremental leapfrogging of invasive plant seeds or fragments along the highway in the vortex behind my moving car, or provided a free ride embedded in its tire treads, grill, or undercarriage. And there was more: arriving in Whistler I'd first pulled into a gas station. Opening the car door I'd watched a hitchhiking mosquito flit away, then shaken out the driver's side floormat without considering the microscopic potpourri accumulated over a winter's driving far and wide around the North American inter-mountain west. I'd likewise ignored a tiny ant crawling across the passenger-side floor, origins unknown.

Through these unwitting acts I had potentially transported, into relatively pristine mountain environs, not only giant hogweed, but also myriad threats currently causing serious ecological, socioeconomic, and health problems in the region: Scotch broom, Japanese knotweed, Himalayan blackberry, orange hawkweed, spotted knapweed, West Nile virus, European fire ants—the list goes on to the point of absurdity. Who has the bandwidth to pay attention to such things? And in the event they do, what could possibly be done to mitigate these dangers?

True, the agents of most of these invasions might find conditions unfavorable, or at least more difficult, in a valley some six hundred meters higher than where they were acquired, but it is equally true that this might only be a momentary obstacle—that is, proper conditions for germination of any particular seed may be simply more transient here than at sea level, so that the timing of an introduction is more critical for germination. But what if I were personally creating such opportunities two or three times a week, along with hundreds of other Whistlerites ranging back and forth to Vancouver, and the thousands of tourists streaming into town each day? Then "propagule pressure"—a composite measure used by invasion biologists to describe the number of individuals of a species released into a region to which they are not native—would be pretty damn high for a host of organisms here, making the eventual establishment of at least some of these a statistical certainty.

This transportation tide merits discussion, and the dialectic is an interesting one. By design, organisms moved about are often only achieving what almost all life forms engage in: dispersal, a natural process at the root of speciation. Equally natural is that individuals like myself, or more generally the species to which I belong, are co-opted as vectors of dispersal, again by design. Thus considered, what we term invasions are, on one level, equivalent to biologically successful dispersals, with their ecological impact on the landscape fundamentally linked to the ecological impact of the disperser. As I've already pointed out, the human lineage has been a natural and occasionally co-evolved aid in the dispersal of various organisms for tens of millions of years. There's even a word to describe this assistance—*anthropochory*. The main differences in this current age of globalization are the *speed* and *means* of anthropochory, that is, the increased rate and geographic range of introductions. As our movements become turbocharged, so do the number of dispersals facilitated; as our alteration of ecosystems and landscapes ramps up, successful invasions from these ongoing dispersals proceed apace.

Again, none of this is news. Invasive species have been much talked about over the past two decades, whether in the media, over backyard fences, or in official organs and communiqués of the sciences and agencies that have emerged around them. What *have* changed are the conversation and its urgency: discussions are less about ponderous study than applied research and decisive action; less about individual opinion than the collaborative creation and application of best practices in management and control; less about isolated local issues than continental and oceanic phenomena; less about *What if?* than *What now?* And certainly less about ignoring an inarguable global issue than making sure your shoes are clean before you get back in your car.

## Back in the Day

A brief history of invasion science is warranted here, and, inasmuch as this itself is mercifully brief, won't require much of a detour. Mirroring the trajectory of many an invasive plant, *The Ecology of Invasions by Animals and Plants* by British ecologist Charles S. Elton was introduced into the fertile soil of postwar science in 1958, only to lurk in the weeds for three decades while its simple wisdoms and foresights ramified slowly through the undergrowth of ecological literature. Eventually, the book blossomed into view, its subsequent exponential celebration and citation paralleling the equally explosive emergence of the field it prefaced. Elton, you see, was the first to distill the problem of nonnative organisms as one of myriad profound effects to the planet of human activity, and his specific talking points form a basis for continued investigation and debate. Certainly Elton both spaced his own goal posts and threw down a gauntlet for writers and biologists alike: "For thirty years I have read publications about this spate of invasions; and many of them preserve the atmosphere of first-hand reporting by people who have actually seen them happening, and give a feeling of urgency and scale that is absent from the drier summaries of text-books. We must make no mistake: we are seeing one of the great historical convulsions in the world's fauna and flora. We might say, with Professor Challenger, standing on Conan Doyle's 'Lost World,' with his black beard jutting out: 'We have been privileged to be present at one of the typical decisive battles of history—the battles which have determined the fate of the world.' But how will it be decisive? Will it be a Lost World? These are questions that ecologists ought to try to answer."

Eventually, someone would formally agree. An organization, actually. An arm of the International Council for Science known as SCOPE— Scientific Committee on Problems of the Environment—sponsored a series of international workshops and meetings throughout the 1980s, aiming to focus ecological expertise on a spate of rapidly growing environmental problems related to biological invasions. Attendees of "the SCOPE project"—including leading ecologists of the time like Daniel Simberloff—may not, at first, have recognized their participation in the birth of a science. Collectively squinting through a telescope of causes and consequences, however, past near-horizon noblesse oblige to the almost limitless universe of research beyond, galvanized many to launch a campaign of cooperative study around the phenomenology of invasions. From this, the modern discipline variously referred to as invasion biology, invasion ecology, and, increasingly, invasion science, emerged. As Simberloff succinctly reflects on this heady late-80s crucible, "My research path was set."

Invasive species had become so globally pervasive that they'd spawned their own field of study complete with learned societies and journals (for instance, *Biological Invasions*), website resources (like invasive.org) and textbooks (*Invasive Species in a Changing World* is just one example). Longer popular works now also sought to encapsulate the problem for the general public. In pole position was Mark Jaffe's 1994 book *And No Birds Sing: The Story of an Ecological Disaster in a Tropical Paradise*, which documented the "overrunning" of the South Pacific island of Guam by the introduced brown treesnake, resulting in the obliteration of native lizard, rodent, and bird populations. In the lyrical 2005 volume *Out of Eden: An Odyssey of Ecological Invasion*, author Alan Burdick picked up Jaffe's torch, putting in a serious shift on Guam between hands-on investigations of biomarine invasions in U.S. ports and Hawaii's struggles with invasive pigs, birds, bugs, and bushes. *Cane Toads and Other Rogue Species*, a 2010 compilation of academic essays and reference material, was a companion to the Australian documentary *Cane Toads: The Conquest*. Fans of invasive Burmese pythons even had choices: if 2012's illustrative *Invasive Pythons of the Florida Everglades*, by Michael Dorcas and John Willson, was too academic, in the same year there was Larry Perez's memoir-esque *Snake in the Grass: An Everglades Invasion*. If you fancied lighter fare, you could follow hunter-adventurer Jackson Landers as he cut a dubious swath with knife and fork through America's invasive species in his 2012 book *Eating Aliens*. There were even handbooks for the haters who bluster and bloviate about perceived undue scientific focus on the naughty deeds of invasive species, led by Ken Thompson's 2014 misfire, *Where Do Camels Belong? Why Invasive Species Aren't All Bad.*

Doubtless both popularization and polarization will increase as the invasive species gestalt percolates deeper into the artesian recesses of our culture. Already it has bubbled back to the surface as myriad theorizing. One example: the notion of invasions as environmental disease, which likens the spread of invasive species to metastatic cancer. Another example plumbs the language of military engagement: "To address the present and future ecologies of war, this paper presents the development of ecology as war. The radical reconfiguration of plant and animal communities in the 20th and 21st centuries, fuelled by economic globalization, has led to the militarization of ecosystems in biological discourse . . . this paper considers the implications of the trope of invasion to describe this novel paradigm of ecological movement. By retracing the outbreak of both rhetorical and biological invasions, this paper explores the possibility of re-thinking these biogeographic developments outside of a militaristic framework."

While this type of heavy going is best confined to the political journal it appears in, it's true that Elton, perhaps still possessed of World War II "Let's give it to the bastards" swagger, freighted his language on invasions from the get-go. This is apparent, for instance, when Elton—having a prophetic handle on the yet-to-be-articulated Invasion Curve—waxed poetic on how best to resist interlopers through human action: "To study this resistance, we have therefore to look at the other side of the battlefield and see what forces are concerned. If you want to repel invaders there are three stages at which you can try to do it. You can tackle them before they get in or while they are trying, so to speak, to pass through the guard—this is *quarantine.* You can destroy their first small population, as was done against the African malaria mosquitoes in Brazil, but this is a very rare event. Usually, if an invasion has got really going it can only be dealt with by keeping the numbers within bounds, that is by *control.*"

Like puns, metaphors abound with invasive species: indeed they are built in. Which ensures that someone, somewhere will be unhappy with any comparison. Even synonyms for introduced species in widespread use—nonnative, nonindigenous, alien, exotic—are fraught with more baggage than you can imagine (all of it handily unpacked in *American*

*Perceptions of Immigrant and Invasive Species: Strangers on the Land,* a 2006 volume by historian Peter Coates). The use of metaphor per se isn't at issue, but rather which one is most justified, appropriate, or least offensive to certain interest groups or worldviews. Some of that is unavoidable. As an example, received-view definitions for the word "alien" include space beings, foreigners, and nonnative plants and animals. The jury remains out on the existence of the first, but referring to fellow humans—now a global species—as *aliens* is clearly an anachronistic pejorative that references geopolitical and cultural differences, whereas its use in regard to animals and plants remains descriptive of true endemism, that is, evolutionary origins and divergence across both space *and* time. Another example is the invasive-action modality "Early Detection and Rapid Response," which also defines the long-held medical approach to cancer. Some see such terminology transference as cheapening the human dimension of confronting this disease, while others, in these days of political oversensitivity, cringe at anything that reflects militaristic sentiment. Both positions, however, seem iniquitous. Language is both invented and inventive: as part of their natural evolution, words, phrases, and ideas are continuously co-opted across all cultures for broader or narrower application than their initial use. With regard to militarism, human history has been defined by our "battling" various entities, and it matters little to the concept whether these were, in the moment, Neanderthals, Nazis, mosquitoes, crop pests, plague, traffic, or knotweed. Words exist to be appropriated or there'd be no such brain-wiring aids as metaphor, simile, analogy, parallel, pun, or comparison. Though skirmishes around the usage of one term or another may crop up in the course of our journey, I hope largely to avoid those frontlines in this volume by outflanking the semantics police. Regardless of how words or concepts might otherwise be freighted, the common language in use herein is that currently employed around the globe in the service of communicating about the emerging phenomena of nonnative, nonindigenous, alien, exotic, invasive species. That should cover us.

## Reality Bytes

Exercise: let's conjure up a model world. Not too unlike our own, but different enough to garner the descriptor *similar*—similarity being a prerequisite for instinctively and intuitively understanding the processes underlying our model. We'll render it as a computer simulation.

Now let's set our observation of this world to a specified distance—close enough to see elements of the surface, though far enough away that we can't make out the detail. Something like standing on a beach and looking down through a cardboard tube at the sand around our feet. With a computer, this is easy to pull off. The first thing to note about our simulated world is that the surface comprises pixels in thousands of vibrant colors—much like a kaleidoscope. Unlike a regular kaleidoscope, however, where mirrors refract a finite number of pixels into symmetrical patterns, ours is a little more free-form; we're going to imagine that each color corresponds to a different species of organism—microbe, fungi, plant, animal, what have you—limited by an irregular distribution (range) of varying numbers (populations) of pixels (individuals). Thus, species on our model world occur in discrete color patches to create numerous overlapping patterns, some discernable, others not so much. (If you want to extend the beach/cardboard tube analogy here, imagine that you can discern obvious patches of darker and lighter sand, as well as areas where these intermingle to greater or lesser degrees, but only when you stoop close enough to make out the grains do you truly see how many colors are involved.) For the sake of both simplicity and argument, let's suppose that *diversity* on our model world—that is, the number of species, as represented by the number of colors—is stable to start: when one color (species) disappears, a new one pops up via some independent mechanism that we are not privy to. This is analogous to an extinction/speciation equilibrium whose periodicity, were it to occur on our own planet based on what we know of evolution and plate tectonics, might be approximately every 10 million years, and discernable only through the geologic record. On our model world, however, we'll make that change-over happen once each day (simulations are excellent tools for compressing time). As our cosmic kaleidoscope clicks away inside the computer, new configurations appear too slowly to notice, yet every twenty-four hours, if we were able to look more closely, we'd find that one color had vanished from somewhere on the vast, pixelated landscape and another had appeared. Now remember that each color (organism) has a variable distribution (range) and abundance (population structure), such that with each passing day, the overall *pattern* also changes imperceptibly into one of billions of possibilities; again, too slow for us to track in real time, but if patterns are compared, say, from one week to another, slight shifts are indeed apparent. With these regular incremental changes, our model world remains fluid and diverse, a stable system of loss and replacement that might go on forever.

We leave our simulation running for a few weeks without checking on it, but when we do return, something odd has occurred. While the distributional boundaries of most colors are, as expected, in slow flux, one color that was at first confined to a very small area is now spreading rapidly over much of the model's surface, becoming both more widespread and abundant. Along with this major change to the model's dynamic, the kaleidoscopic mechanism also appears to have lost its equilibrium: instead of clicking once every day to lose a single color, it has accelerated and is now clicking and dropping colors about once every hour—but *without* affecting the independent variation mechanism that still generates one new color every day. When, now hourly, a color disappears, all its pixels go, leaving a spatial vacuum; and because we're only gaining one new color every twenty-four hours, these vacated spaces—no matter where they occur—immediately fill with replications of a small number of preexisting colors. The net effect of this new dynamic is that while each consecutive turn of the kaleidoscope results in noticeably fewer colors, there is an equally noticeable increase in the distribution of a select few others. Not only are we daily losing twenty-four colors and gaining only one for a net loss of twenty-three, but patterns of the remaining pixels are changing more quickly as well. Soon only a certain number of colors remain, and the patterns, having lost their previous richness, look more similar in all sectors of the model; in fact, colors and patterns once restricted to small areas appear now to form bands around the planet. Our model world is shifting from one of rich, varied, ongoing diversity to a place of increasingly widespread monotony and an end game that seems obvious long before it actually resolves: a long, slow dissolve to zonal homogenization.

Of course you know where this is going. The imaginary world is, in fact, a vastly oversimplified version of our own. The original turn of the kaleidoscope is an approximation of the effects of geologic time, the loss of color at intervals the extinction rate, and the independent-variation generator a simplified version of speciation via evolution. The sudden catastrophic spread of one color is the colonization of the planet by humans, the concomitant change to the turn rate of the kaleidoscope a replacement of geologic time by the forces of globalization, and finally, the accelerating dominance of a select few colors—mere hundreds out of potential millions—represents the spread of invasive species.

Elements of this scenario are easily observed on today's Earth, jibing with an enormous body of research and the views of a vast majority of scientists. And while a handful, as noted, pooh-pooh any notion of widespread cataclysm, abundant facts support the basic premise of an increasing number of species lost to extinction or extirpation each day, the rapid parallel rise of globalization and nonnative species introductions, the increase in demonstrable problems associated with invasive species, and the increase in GDP spent on these. So here's the simple takeaway: humans are everywhere, and so are invasive species. Which means that invasive species stories, the kind that might help illuminate the concepts of introduction, early detection, pathways, vectors, spread, establishment, impacts, and control, are now anywhere you care to look—even where problems have yet to make themselves known.

## Absence

Another exercise, this one a good deal simpler: grab a set of pointed dividers—the kind that nautical route-plotters use—and a 1:40,000,000 map of the world (that's about a meter across—the scale of most globes). Plant one of the divider ends at precisely 54°26′S, 3°24′E, set the distance between points to represent 1,600 km on your map, and trace a circle around the fixed position. If your dividers have a particularly fine point, you may be able to see the tiny piece of land on which they're planted, a mere pinhead at this scale, and that within the prescribed circle— encompassing an area of 8,148,103 square kilometers of ocean, some 80 percent the size of Europe—not a single additional piece of land is to be found. In fact, with South Africa lying more than 2,500 km to the northeast, and the Tristan da Cunha islands and South Sandwich Islands equidistant at more than 1,800 km removed to the north and west, respectively, the closest terra firma to your circle's center is Queen Maud Land in Antarctica, 1,750 km to the south. No other place on Earth, in fact, is as peculiarly isolated as the fifty-square-kilometer speck you've impaled, marked on most maps as Bouvet Island.

Both wildly remote and wildly inhospitable, Bouvet lies just east of the congruence of the South American, African, and Antarctic tectonic plates, an area known to geologists as the Bouvet Triple Junction. A dormant shield volcano that last erupted around 50 BCE, the island marks the terminus of the Mid-Atlantic Ridge—the world's longest mountain/rift system, stretching 10,000 km north to the Arctic Ocean. Thick glacial ice spilling from the volcano's remnant caldera caps 95 percent of an island otherwise rimmed by 500-meter basalt cliffs. Ferocious winds, heavy seas, near constant mist, and not a single protected cove make landing on Bouvet a risky venture. Fittingly and famously, it is also hard to find.

The island was discovered—miraculously according to nautical historians—on New Year's Day 1739 by Frenchman Jean-Baptiste Charles Bouvet de Lozier. Revealed through wavering fog as his ships *Aigle* and *Marie* approached from the north, it was the first land spotted below 50ºS in this part of the ocean, leading Bouvet de Lozier to conclude that it was a cape of Terra Australis, the long-theorized but illusory great southern continent (though Captain James Cook would be first to cross the Antarctic Circle and encounter that continent's icebergs in 1773, its existence wasn't confirmed by sight until 1820). Bouvet de Lozier's belief suggests why he attempted no circumnavigation, despite twelve days spent in dense fog trying—and failing—to land before running low on supplies and retreating with his scurvy-ridden crew north to South Africa's Cape Town. Nautical errors didn't end there: Bouvet de Lozier also plotted the island's position incorrectly in an era when navigation mostly involved dead reckoning. As a result, it was "lost" for some seventy years—eluding even Cook's search efforts in 1772. It turned up again in 1808, however, when James Lindsay, captain of a British whaler, relocated the island several hundred miles from the spot that its discoverer had logged. Also unable to land—and perhaps because it wasn't at all certain this was, in fact, the same island—Lindsay fixed the new position and took the not-unusual liberty of an eponymous appellation. Lindsay

Island, however, never stuck. In his 1832 memoir *A Narrative of Four Voyages*, American sealer and self-proclaimed explorer Benjamin Morrell claimed that he'd been first to land on "Bouvet's Island" a decade earlier. There was no mention of its all-too-obvious icecap, though, and given the apparent ease with which he'd found the scrap of land (not to mention his reputation for reporting impossible distances, erroneous latitudes, and fanciful discoveries), Morrell's claim was much disputed. In 1825, too, prior to the memoir's publication, but after Morrell's claim, while the isle still languished in anonymity, British explorer George Norris stumbled across it, claiming for the Crown the fog-shrouded, unapproachable lump he named Liverpool Island. Despite its ascension into the empire, however, celebrated British mariner James Clark Ross was twice—in 1843 and 1845—unable to find it. In fact, the island's location wasn't fixed on nautical charts until 1898, when it was finally circumnavigated by the German survey ship *Valdivia*.

First to make it ashore were Norwegians from another survey vessel, the *Norvegia*, in 1927. After clambering onto the central ice plateau, roughly mapping coast and interior, and placing a small cache for anyone marooned by shipwreck, Captain Harald Horntvedt felt justified in possessing the island in the name of King Haakon VII. The Brits must have owed something to their longtime competitors (and occasional rescuers) in polar exploration, because they relinquished their own claim within a year. The Norwegian Polar Institute, established in 1928, was made administrative authority of the island, and now operates under the Ministry of Climate and Environment.

Firmly on the map as Bouvetøya, the Norwegians nonetheless had little use for their dependency until a royal decree declared both the land and its surrounding territorial waters a nature reserve in 1971. An automated weather logger was installed in 1977 at Nyrøysa, a rubble shelf created by a 1950s landslide that wave action has subsequently eroded. A small field station comprising metal shipping containers anchored by guy wires was erected in 1996. In October 2007, researchers studying penguin and seal colonies on the ever-shrinking Nyrøysa arrived to discover the station gone, theorizing it was blown away after a 2006 earthquake had loosed the anchors. As if in confirmation, the tents they were forced to occupy were soon ripped to shreds by wind.

Although both terrestrial and airborne landings are now prohibited without permission, Bouvetøya remains sought after. Cited as one of only two places left unvisited by John Clouse, the *Guinness Book of World Records*' erstwhile "most traveled man," his successor to the title, Charles Veley, took seventy-two days to reach the island. In the 2012 documentary *Bouvetøya: The Last Place on Earth,* Canadian adventurers Jason Rodi and father Bruno plant a time capsule on its unclimbed highest point—780-meter Olavtoppen. Having climbed the highest mountain on each of the seven continents and trekked to both North and South Poles together, the film shows them traversing the planet's roughest ocean to its remotest locale bearing a titanium tube filled with messages and hopes for the future from around the world, a symbolic journey that Jason hopes will inspire humanity to do better. He also hopes his unborn daughter, Alix, will retrieve the capsule in fifty years and find Bouvetøya as free of our footprint as today. "This special journey was for you," he explains. "I went looking for the end of the world to find a new beginning."

While all of this is inherently interesting, you're entitled to wonder why I've just delivered an itemized history for a place where no one even lives. There's a reason: the history of introduced species closely follows the history of human exploration, discovery, hubris, imperialism, occupation, and commerce, and this one tale plots the halting potential—but ultimate failures—of all of these. Being remote, inaccessible, and uninhabitable means that only a few hundred souls have likely ever stood on Bouvetøya, the very reason it now appears to be the last territory on Earth untouched by alien invasive species. With no record of introduced flora or fauna, the island remains in its natural state, something no other Earthly jurisdiction—let alone an island—can claim. In addition to indigenous seal, penguin, and seabird species, there are also the familiar terrestrial occupants of other (now invaded) sub-Antarctic islands: two moss species, three liverworts, forty-nine lichens, five mites, and three springtails—the latter two groups being tiny terrestrial arthropods easily carried on wind, rafted on oceanic debris, or taxied around by birds.

The real story of Bouvetøya, then, is of its *non*-invasion—and it is noteworthy for being the only such tale we will ever know or tell.

Fortuitously, this also makes Bouvetøya a critical experiment in prevention and early-detection measures. As the sole place that these measures have been enacted prior to introductions of *any kind,* it can offer a quantifiable test of their efficacy and a singular chance to apply the lessons of past mistakes. In this respect, Bouvetøya's intact sub-Antarctic ecosystem resides at the very base of the Invasion Curve for all alien species—even as the numbers of these introduced species are rapidly growing for all other polar regions.

Along with the remotest tracts of rainforest, the driest deserts, the deepest cave systems, the abyssal ocean, and other places where humans rarely tread, the extreme environments of the poles imply that they should be among the least-invaded places on Earth. A number of studies, however, belie this assumption, suggesting that even were this true in the recent past, things have changed with lightning speed, propelled by two factors: increased visitation and climate change.

Although none has yet established a population, the numerous nonindigenous moths, flies, ants, wasps, beetles, spiders, slugs, mosses, grasses, and seeds recorded at British Antarctic Survey (BAS) stations over the years have prompted measures to reduce the risk of introductions. Thus, careful cleaning and checking of clothing, footwear, food, water, construction materials, equipment, vehicles, and vessels landing anywhere in Antarctica and its surrounding territories is de rigueur for BAS. This focus on surrounding territories is critical: many sub-Antarctic islands—occupied permanently or intermittently since the whaling and sealing era of the eighteenth and nineteenth centuries—are now heavily invaded, with nonnative plants outnumbering native species on several smaller islands. And while larger islands like South Georgia were better buffered from widespread takeover by dint of their size, other factors now appear to be tipping the balance in favor of invaders. For instance, about thirty of the seventy vascular plant species introduced on South Georgia survived only one or a few years, but of the remainder, twenty-five are now considered naturalized. According to the South Georgia Heritage Trust, though these had reproduced successfully for decades, many began spreading beyond whaling stations only in the mid-1980s, a likely response to warming summer temperatures. Nevertheless, South Georgia has larger invasive species problems:

predatory nonnative carabid beetles are decimating indigenous cara-
bids, and introduced reindeer and rats have had major impacts on
ground-nesting seabirds. (Seafaring has introduced nonnative rats to at
least 90 percent of the world's islands with devastating consequences
for native species and ecosystems; we'll revisit this nearly ubiquitous
problem in Part 5.) At BAS stations where rats exist, precautions have
been taken to avoid their spread to other sites: signage at all departure
and landing points; rat guards on all mooring lines; and poison bait
dispensers in all vessels and station buildings.

Though over the centuries rats have proved to be expert stow-
aways, they are macro-invaders for which tractable prevention measures
might be undertaken and continuously evaluated. But what of micro-
invaders? Researchers recently quantified propagule pressure for vascular plants associated with different groups of travelers to the high-Arctic
archipelago of Svalbard. Some 1,019 seeds were found on the footwear of
259 air travelers who arrived during the summer of 2008; assuming this
influx of approximately four seeds per traveler to hold true over an entire year, they estimated the annual seed load by this vector alone to be
around 270,000. Furthermore, eight of seventeen plant families represented in the sample—the majority not native to Svalbard—are among
the most invasive worldwide. Increasing the probability of successful introduction, the number of seeds was highest on footwear used in forested or alpine areas of the traveler's country of origin. Most telling, a
quarter of all seeds collected germinated under simulated Svalbard optimal soil conditions. The study demonstrated that "high propagule
transport through aviation to highly visited cold-climate regions and
isolated islands is occurring. Alien species establishment is expected to
increase with climate change, particularly in high latitude regions, making the need for regional management considerations a priority."

Regional management is more easily communicated, implemented, and shared in an occupied archipelago like Svalbard (pop. 2,500)—
where detailed screening is facilitated by the infrastructure of a seaport
and airport—than the wilds of Antarctica, making the danger posed by
propagule transport in such areas higher still. By similarly sampling
vascular-plant propagules carried by visitors to Antarctica during the
summer season of the International Polar Year (2007–2008), then

mapping the likelihood of establishment based on seed identity, origin, and spatial variation in Antarctica's climate, a group led by Steven Chown of South Africa's Stellenbosch University published the first comprehensive, continent-wide risk assessment for nonindigenous plant establishment there. Though the average seed load was an astonishing 9.5 per person in this study, scientists as well as science- and tourism-support personnel were found to carry even greater propagule loads than tourists—yet nearly 4.7 times more tourists (about 33,054) than those associated with science programs (approximately 7,085) visited annually, which somewhat evened the score. Comparing seed origins—and therefore the cold tolerance of the plants represented—to current and future climate models based on Intergovernmental Panel on Climate Change data, it was concluded that alien species establishment was most likely to occur on the Western Antarctic Peninsula, but that the risk to remaining areas of the peninsula and other coastal regions would grow substantially with climate warming. Founder populations of several aliens corroborated these findings. For example, the most widespread and dominant invasive lowland plant on the sub-Antarctic islands, annual bluegrass (*Poa annua*), had recently become established on the peninsula and was spreading as more areas remained snow-free throughout the austral summer.

The study's importance at a critical early juncture in the invasion history of an entire continent cannot be overstated: "By delivering comprehensive evaluations of human-associated propagule pressure and establishment likelihood, differentiated by spatial location and visitor category, our study . . . indicates those visitor groups and areas for which biosecurity measures should be most stringent, those where controls might be less pronounced, and how the spatial arrangement of these areas is likely to change through time. The assessment also offers guidance for planning early detection surveys and support for management decisions about whether new species occurrences are the consequence of anthropogenic transport or natural colonization [an emerging concern]. Such decision-making can be further informed by identifying natural colonization paths at appropriate spatial and temporal resolutions. These include wind trajectories for wind-dispersed species, satellite tracks of seabirds that are considered important natural dispersal agents, and genetic data on colonists and populations from elsewhere in the broader region. Thus, our study provides an evidence-based, continent-wide risk assessment for the establishment of terrestrial alien species in Antarctica and the understanding required to mitigate this risk, one of the primary conservation goals of the Antarctic Treaty System."

A follow-up paper on propagule transfer of potentially invasive plants to the Antarctic by the same researchers looked in greater detail at entrainment data—that is, the origins of the seed loads and the pathways of their human vectors—and included an analysis of bryophyte and lichen fragments (because these groups are the major components of Antarctic flora, such fragments stand a better chance to establish there than vascular plant seeds do). Seeds were found in the clothing and/or gear of 20 and 45 percent of tourist and science visitor samples, respectively, while comparative proportions for bryophyte and lichen fragments were similarly distributed at 11 and 20 percent. Unsurprisingly, the highest-risk items were the footwear, trousers, and bags of field scientists, especially those who had previously visited protected areas, parkland, botanic gardens, or alpine areas. Among tourists, those who had visited rural or agricultural areas prior to their trip, and/or traveled with national programs on smaller vessels, had the highest probability of transferring plant propagules. And in all cases, travel during the boreal or austral autumn months increased the probability that propagules would hitch a ride. This evidence, concluded the authors, made a sound case both for the self-cleaning of personal equipment, and for organization-based regulations such as issuing guidelines and holding regular inspections.

How would this translate in the real world? An easily contained area like Bouvetøya provides the best example that the overall approach is, so far, working. For instance, Houston-based Oceanwide Expeditions' twenty-nine-day cruise out of Ushuaia, Argentina, island-hops through the sub-Antarctic and South Atlantic and includes a forty-eight-hour stop at Bouvetøya. If conditions allow (there's no guarantee), passengers can land by Zodiac for a lecture on wildlife and history. "We're very aware of the potential of introducing alien species to anywhere in the Antarctic and sub-Antarctic, and dedicated to the prevention of any such introduction," Oceanwide spokesperson Rima Deeb Granado told me via email, noting that the company adheres to strict protocols set by the International Association of Antarctica Tour Operators in accordance with Antarctic Treaty regulations. "These include distributing pre-trip information for passengers to thoroughly clean their clothing and effects prior to the trip, but also that onboard staff supervise vacuuming—or physical cleaning if necessary—of [everything] that will go on shore. All landing passengers, staff and crew must be signed off as having gone through the procedure. In addition, we wash boots and other gear that go ashore in a biodegradable disinfectant such as Virkon before and after each landing in order to minimize the chances of spreading disease—for example from one penguin rookery to another."

These laudable procedures mirror those of scientists working on Bouvetøya and elsewhere in the Antarctic and sub-Antarctic. Nevertheless, the promise of halting accidental introductions—and the further prospect of reversing some of the damage in these areas through ecological restoration projects—evokes the very quote from T.S. Eliot's *Little Gidding* that Jason Rodi recites in *Bouvetøya:* "We shall not cease from exploration / And the end of all our exploring / Will be to arrive where we started / And know the place for the first time."

Rodi also quotes Carl Sagan: "For all our failings, despite our limitations, we humans are capable of greatness." When it comes to invasive species and other threats of the Anthropocene, any display of greatness on our part remains to be seen, so we'll take Bouvetøya for what it is—a small, fortuitous, but nevertheless important victory. Despite its singular wholesale embargo on nonnative organisms, however, this victory may not be as great as would be that of keeping yet more interlopers from one of the most heavily invaded ecosystems on Earth—the Great Lakes.

## Sensei

Like a massive concrete warship, the Canada Centre for Inland Waters (CCIW) docks below the Burlington Skyway—a repair-burdened, steel-girder bridge that has spanned the entrance to Hamilton Harbour from Lake Ontario since 1958. Squatting on forty acres of landfill, the bland, fortress-like building is anchored by a triad of rounded buttresses reflecting the post-modernist construction sensibilities of the 1960s—monolithic, monochromatic, utilitarian. Although I'd never set foot inside, as a doctoral student working on amphibian genetics in the Department of Ichthyology and Herpetology at Toronto's Royal Ontario Museum from 1987 to 1991, I'd been keenly aware of the reverence for CCIW held by fish-favoring fellow students. One of those was Nick Mandrak, a big guy with a thick mop of chestnut hair and a hearty laugh who channeled rugby player more than mild-mannered biologist. Mandrak and I often found ourselves side by side in the department's shared laboratory space: me tending to the numerous salamanders I was raising for my own project, he hunched over a PC enabled with the nascent mapping tools of geographic information systems. On the frequent occasions that I interrupted his work with pesky questions on biogeography, I'd found him sharp, inquisitive, funny, and generous with his great depth of knowledge about the evolution of Canada's postglacial landscapes and waterscapes. A quarter-century later, transitioning from a longtime post at CCIW with the federal Department of Fisheries and Oceans (DFO), Mandrak was a newly minted associate professor in biological sciences at his alma mater, the University of Toronto. A continued working relationship with DFO, however, meant that he maintained an office at CCIW, where we planned to meet.

I arrive on a cold November day under a leaden sky contemptuously spitting snowflakes, pulling into a half-empty parking lot that offers testament to several rounds of job cuts in the federal environmental monitoring sector—an ideologically motivated downsizing by Stephen Harper's conservative government, an anti-science cabal that made America's George W. Bush years seem like the Age of Enlightenment. (This would be only a humorous aside if, as illustrated by the Florida example, politics didn't play such a significant role in invasion trajectories.) When Mandrak descends from the concrete firmament to greet me at the security desk in a cavernous atrium, our conversation picks up right where we'd left it as wisecracking grad students.

"Can't believe it's my first time in this dump," I offer.

"Oh—you don't like the 'brutalist' architecture?"

"I had it pegged as Stalinesque."

"What's the difference?"

"Quality of concrete; Stalinesque would be a rung or two lower since communist five-year plans meant it wasn't built to last."

"Right. Still, can you imagine the *money* concrete firms were making in the '60s?"

"Exactly. That's why the mafia was into it. Which explains *everything* about Quebec."

"What are you here to talk about again?"

Well, invasion science, of course, and CCIW occupies an important place in that world. Despite its retro aesthetic and diminished personnel, CCIW remains one of the world's leading institutions for aquatic environmental research and monitoring. Managed by the eviscerated umbrella federal agency, Environment Canada, CCIW nevertheless hosts leading-edge programs for DFO, the National Water Research Institute, the National Laboratory for Environmental Testing, the United Nations' Global Environmental Monitoring System, the Canadian Hydrographic Service, CITES Canada's enforcement branch, and numerous dedicated Great Lakes programs that the HarperCons had yet to find a way to dismantle. Some 10,000 square meters of lab and office space are augmented by R&D facilities that support the production of marine and technical equipment, physical studies on sediment transport and wave dynamics, and various water treatment initiatives. With a library, auditorium, data center, and photography department, CCIW remains geek heaven for aquatic biologists and ecologists.

Stepping off the elevator several floors aloft, we stroll brightly lit hallways papered in the same occupant-chic common to universities— that is, graph-filled poster presentations salvaged from scientific meetings and public-information placards on all things Great Lakes. Pictures of fish abound, photos of mollusks weigh in at a distant second, birds and plankton appear in cameos. Entering Mandrak's corner office is like stepping onto a balcony: large windows look south over the murky waters of Hamilton Harbour, the imposing Skyway crowding the left-hand view while the remaining vista traces an industrialized horizon of smokestacks, mills, and distant dollops of coal and slag. There's a reason Hamilton was known as "Steeltown."

Mandrak was involved with invasive species long before I knew him, and he had the densely packed bookshelves to prove it. His first publication in the primary literature, in fact, was a write-up of his honors undergraduate biology project at the University of Toronto, where he'd assessed the probability of fifty-eight common, widely distributed fish species invading either the Upper (Huron, Michigan, Superior) or Lower (Erie, Ontario) Great Lakes if global warming—barely a phrase, let alone a topic, in 1985 when he'd done the work—ever came to pass. Through mathematical analyses, Mandrak compared traits of eleven species that had already successfully invaded the Great Lakes with those of the fifty-eight he'd listed, concluding that twenty-seven of these had the potential to invade under climate-warming scenarios. His was one of the earliest trait-based models employed in the service of invasion prediction, a bellwether of risk assessments to come.

For his master's degree, Mandrak lodged at a desk in the dusty bowels of the august Royal Ontario Museum, supervised by the equally august Ed "E. J." Crossman, curator emeritus of ichthyology and cross-appointed professor emeritus in the University of Toronto's Department of Zoology. Not only did Crossman stand tall as a doyen of Canadian aquatic science, but he rivaled Mandrak's own prodigious biomass by towering over him. During this stage of his career, Mandrak sought to elucidate historical influences on the distribution patterns of Ontario's fishes, both native and nonnative. Despite his focus on biogeography, during this period he also found himself giving numerous well-received invasive-species talks, whose impact was apparent when others enthusiastically published the unpublished data he presented. Mandrak's doctorate, again under Crossman, looked beyond the postglacial history of occupation to environmental influences on current biogeographic patterns—effects of factors like climate, water chemistry, and lake and stream morphology. Although the projects behind both graduate degrees had practical applicability to the growing field of invasion biology, Mandrak left for seven years to take on demanding teaching positions in the United States, which provided little time for research.

That changed when he landed back in Canada in 2001, hired by DFO to work on endangered species in the lead-up to the 2002 introduction of Canada's first Species at Risk Act (SARA was a final piece of legislation completing Canada's National Strategy for the

Protection of Species at Risk and its commitment under the UN Convention on Biological Diversity). Mandrak immediately learned how being up to your eyeballs in endangered species also meant being up to your eyeballs in invasives.

"In the Great Lakes, maybe more than anywhere else, it was clear that invasive species ranked second only to habitat loss as a major driver of species extirpations and extinctions," he begins when we finally get down to business. "But if you asked a member of the public what *they* thought the biggest threats were to Great Lakes biodiversity at the time, they might say pollution, maybe habitat loss, but *never* invasive species—despite it being the received view in the scientific community."

It's hard to imagine such a pervasive, binational issue remaining so generally invisible. Yet neither a trenchant, century-long battle with sea lamprey (we'll get to this), numerous high-profile problems related to zebra mussel (we'll get to this, too), nor even the occasional tabloid-worthy piranha fished from Lake Ontario (true story—I saw the fish) had influenced public consciousness in North America enough to add "Do Not Introduce Nonnative Species into Native Waters" to the ledger of things you taught children—and fishermen. Hence the continued release of transplanted baitfish, unwanted goldfish, and other aquarium-trade flotsam into the continent's ponds, lakes, rivers, and streams. In fact, if vast schools of tuna-sized silver carp didn't have a bizarre predilection for leaping clear of the water—and into headlines—at the sound of a boat engine (to the great peril of the craft's occupants), the current level of public knowledge concerning threats from aquatic invasive species might be even lower. Which is why the several discussions I would have with Mandrak—afternoon-long voyages that tacked back and forth over the oceanic topic of Great Lakes invasions—would both begin and end with carp.

## Apocalypse, Maybe

Silver carp (*Hypophthalmichthys molitrix*) is one of four species currently referenced by invasion scientists, managers, and policymakers under the seemingly nebulous rubric of "Asian carp(s)." The remaining trio—a bona fide taxonomic mouthful—comprise the silver's congener, bighead carp (*Hypophthalmichthys nobilis*), plus grass carp (*Ctenopharyngodon idella*), and black carp (*Mylopharyngodon piceus*). Note that the more familiar (and pronounceable) common carp (*Cyprinus carpio*)—also native to parts of Asia and introduced widely across North America in the 1800s from already-transplanted European populations—isn't included under the heading. This is not only because the common carp is now considered naturalized on this continent, but also because the others form a de facto cultural unit.

Known for millennia in China as the "four domesticated fish," silver, bighead, grass, and black carp are that country's most important freshwater resource for food and traditional medicine. Though all can attain sizes in excess of forty kilos and a meter in length, each is physically and ecologically distinct. Feeding in the divergent food guilds of, respectively, phytoplankton, zooplankton, aquatic vegetation, and benthic mollusks, these fish were originally imported to the southern United States in the 1960s and 1970s primarily for control of weeds, algae, or parasites in scenarios ranging from reservoirs to waste treatment to aquaculture. Some were also tested with varying success as farmable food sources in their own right. Lodged in outdoor ponds, hatcheries, and research facilities along the floodplains of the Mississippi River, each species escaped or was intentionally introduced into the wild on numerous occasions, particularly during the repeated heavy flooding of the early 1990s. Silver, bighead, and grass carp made their way up the Mississippi and its tributaries as far north as Minnesota; black carp, with a predilection for deeper water, was only occasionally pulled from the main channel of the central and lower Mississippi. At first the spread of silver and bighead carp appeared slow and sporadic—one netted here, another dredged up there. But once they had dispersed sufficiently northward, an environmental or ecological inflection point was apparently reached that led to radically increased spawning and survival rates: around the millennium, mid-continent populations of silver and bighead carp exploded, to the point that they became the most abundant fishes in several drainages. On a few notorious stretches of the agri-runoff-enriched Illinois River, for instance, that meant 60 percent of all individual fish and over 90 percent of fish biomass. Capable of straining up to 20 percent of their body weight in food from the water every day, these filter-feeding behemoths weren't just messing up an ecosystem, they were *becoming* it—with their abundant waste contributing to the accelerated productivity of smaller organisms that escaped the feeding dragnet, resulting in algal blooms and excessive turbidity.

Imagine, if you will, typically low-density populations of the two largest species of baleen whales—blues and fins—suddenly increasing exponentially across the world's oceans and the inevitable winnowing of all other marine vertebrates dependent on the same krill food resource, and you have a picture of what was occurring in watersheds occupied by bighead and silver carp by mid-decade. Now consider that grass carp consume daily up to 40 percent of their prodigious body weight in aquatic plants that otherwise provide spawning and nursery habitat for native fish, and that an Asian tapeworm it carries has already heavily impacted native baitfish. Finally, extrapolate to the further effects on rare and endangered snails, mussels, and other invertebrates should black carp gain more serious fin-hold, and it's clear why this quartet is construed as a potential trophic time bomb should it enter the Great Lakes in whole or in part. Not only are the lakes' heavily compromised ecosystems already especially vulnerable, but so too are their commercial and recreational fisheries, which are valued at $7 billion USD a year in the United States alone.

But wait, even casual students of geography might say, the Mississippi drains south into the Gulf of Mexico, *not* into the Great Lakes, so why such concern? Well, there wouldn't be ... were it not for the Chicago Area Waterway System (CAWS). Not only does CAWS control the flow of water around metropolitan Chicago, but as part of the 526 km Illinois Waterway, a navigable link between St. Louis and Chicago, its structures offer a critical direct pathway for faunal exchange between the Mississippi and Great Lakes basins.

It wasn't always that way, as early explorers of the continent's interior well knew. The very reason for the Windy City's existence was the bidirectional commerce that paused at the short but hugely important Chicago Portage, which crossed the Valparaiso Moraine separating the Des Plains River (Mississippi basin) and Chicago River (Great Lakes basin). As the town grew in lockstep with the volume of goods passing east and west, it was only a matter of time before commercial demand led to these waterways being artificially joined (the city's flag, in fact, is a stylized map of the portage: four red stars separating blue stripes that symbolize the two great waters meeting there). In 1848, the privately constructed 156-km Illinois and Michigan Canal was completed, connecting the Chicago River to the Illinois River. It was eventually replaced by the network of canals and locks comprising CAWS: the Chicago Sanitary and Ship Canal (or CSSC, completed in 1900); the North Shore Channel (1910); and the Cal-Sag Channel (1922). The system's key was its permanent reversal—through pumping and structural modification—of the flows of the Chicago and Calumet Rivers toward the Mississippi and away from Lake Michigan, so as to prevent the considerable sewage discharging into each from contaminating Chicago's drinking water, whose intakes lay just offshore in Lake Michigan.

The aging CAWS locks, variously owned and operated by the Army Corps of Engineers or Chicago's Metropolitan Water Reclamation District, have become focal points in the battle to prevent nuisance aquatic invasive species from moving between the Great Lakes and Mississippi. The Corps, directed to remedy the situation by Congress through Bill Clinton's Invasive Species Act of 1996, began by building a demonstration electrical dispersal barrier on the CSSC. Designed to repel multiple aquatic invasive species, the barrier's primary goal was to impede the downstream movement into the Mississippi of round goby—a highly invasive fish of Eurasian origin then ravaging the Great Lakes. Funding and construction delays combined to ensure that the effort came up short: by 1999 round goby had been found downstream of the barrier, which wouldn't become operational until 2002. Based on this early failure—and fear that any kind of power outage would offer Asian carp an opportunity to move upstream—it was decided that the barrier should be upgraded to a stronger, more permanent version, and that a second large, two-part barrier (IIA and IIB) needed to be built 240 meters downstream to provide additional protection through redundancy. Barrier IIA was up and running by 2009 at a cost of $10 million; IIB came online in 2010, at $13 million. Budget requests by the Corps put total operational costs of the barriers at around $7.25 million per year. If that seems like a lot to spend on the mere chance of stopping something, well, it is—unless you have a sense of what the cost of *not* stopping it would be.

Subsequently, the Corps began examining the most obvious fix: permanent hydrologic separation of the two basins. While future phases will look at other potential areas of Mississippi–Great Lakes connection, such as between the Wabash and Maumee Rivers during flooding of Indiana's Eagle Marsh, phase one of the Great Lakes and Mississippi River Interbasin Study (GLMRIS), completed in July 2013, focuses on CAWS and describes eight alternatives to prevent interbasin transfer of multiple aquatic nuisance species, including Asian carp. The options range from zero, or minimal, adjustments to the current approach (nonstructural methods of prevention), to major structural changes in water control that will effect complete hydrologic separation. Costs and timelines range from pocket change and short-term horizons for the former, to more than $18 billion over twenty-five years for the latter. Expensive, yes; and, unfortunately, not the only cost to account for.

CAWS also plays a significant role in the region's recreational (pleasure craft) and commercial (freight barge) navigation, and though estimates of the full economic value of these activities vary widely, the prospect of permanently shutting the locks has become a lively political football. A study commissioned by Michigan, a state engaged in litigation aimed at forcing action on the Asian carp threat, estimated that a shift from barge to overland shipping would yield additional costs of only $64–$69 million annually—a mere ten dollars per ton. As expected, the Illinois Chamber of Commerce took issue, publishing critiques of the study as well as its own report suggesting that any such action would be economically prohibitive, with a total added cost of $530–$580 million annually. No study measured its numbers against the potential economic cost of Asian carp entering the Great Lakes, but even cursory math puts that price tag at orders of magnitude higher in as short as five years. In any event, effective Asian carp prevention in CAWS will be a costly enterprise, perhaps more so given that U.S. agencies already have their hands full and treasuries drained by the creation and maintenance of barriers elsewhere in the watershed. Even if Asian carp can be kept from the Great Lakes, their continued unchecked spread throughout the Mississippi basin would result in their reaching thirty-one states and 40 percent of the continental United States—a bona fide disaster for America's freshwater ecosystems.

## Risky Business

In 2004, DFO asked Mandrak to take over Canada's risk assessment for Asian carp. The responsibility had originally been tasked to someone at headquarters in Ottawa who knew nothing about these fishes; Mandrak, by contrast, not only had broad research expertise in freshwater biodiversity, but also had actually seen and collected wild Asian carps in their native ecosystem. In 1993, he and Crossman had squeezed into ancient Aeroflot planes and hop-scotched across Siberia to far eastern Russia and Lake Khanka—"The twenty-fifth largest lake in the world no one has ever heard of," as Mandrak puts it. Located north of Vladivostok in the Amur River basin and split by China's northeast border, Khanka's seventy-five fish species comprise a startling convergence of South Asian and Siberian faunas, the greatest such diversity of any lake in Russia. Included are silver, grass, black, and common carps as well as northern snakehead, Mongolian redfin, lake skygazer, and goldfish. No single trip could make one an expert on Asian carps, but with his combination of boots-in-the-water experiences and knowledge of climate change and biogeography, Mandrak was more informed than most.

"So I do the risk assessment and they reward me by making me director of a new Centre of Expertise for Aquatic Risk Assessment," recalls Mandrak dryly. "But it actually wasn't much of a reward because they didn't bother to relieve me of any of my other duties."

Despite being further burdened by his employers' never-ending quest to find politically purposed efficiencies, Mandrak had at least circled back to where he'd started as a professional ichthyologist: modeling potential Great Lakes invasions. Given the urgency of the case at hand, prevention measures would require immediate government funding—which also meant buy-in from the electorate. Fortunately, exemplary reasons for both were only a mouse-click away.

"I think the turning point for the public was the binational Asian carp meeting in Peoria, Illinois, in 2006," Mandrak ruminates. "It got good coverage on PBS, and the hosts took the media out on boats to see these massive silver carp jumping out of the water. The film clips were all over the Internet—the YouTube stuff was a sensation."

It still is. Especially for Canadians, who'd never seen—let alone imagined—such a thing. Although Asian carps had yet to become established in any of the Great White North's multitudinous waterways, where an amphibious life is de rigueur for many people, the mere notion of innocently boating across a lake or river and being hit in the head by a flying fish the size of a small child struck genuine fear into the hearts of typically stoic Canucks—and their typically quivering politicians. No way did they want these creatures in the Great Lakes, from whence the fish might gain terrifying access to innumerable other drainages. Knowing all too well what they were dealing with, the Americans were motivated more by practical imperative to stop the spread of species that had already ravaged the upper Mississippi and Missouri basins. No way did they want these fish in the lakes either. Prevention was writ large—as was cooperation. Although aquatic invasions typically offer few chances for nations to collaborate on such a widespread front, the precedent of a half-century of binational oversight of the Great Lakes St. Lawrence Seaway System made for a determined partnership. For once, everyone on both sides of the border got the memo in time. Politicians, who generally cared little about the effects of invasive species on ecosystems they had no understanding of, were sure as hell worried about potential effects on econo-systems.

"Think about it," emphasizes Mandrak. "The Canadian government, so typically reactive and *never* proactive, actually asked for an Asian carp risk assessment *in 2004*. That's how big of a deal they already thought the problem *could* be. Now, a decade later we have a well-funded prevention program at DFO that we're spending more on than all aquatic invasive species combined outside of sea lamprey."

It was an impressive throw, and what the program entailed was undoubtedly a story unto itself. Mandrak, however, would leave that explanation to a colleague.

## Flipside

As senior science adviser on aquatic invasive species in DFO's Great Lakes Laboratory for Fisheries and Aquatic Sciences and head of the Asian Carp Program, it was no surprise that Becky Cudmore, like Mandrak, also had a longstanding interest in invasives. During her third year as a freshwater ecology major at Peterborough's Trent University, she'd hoped to take a course in invasion ecology but found none on offer. Undaunted, she petitioned to fill the gap, to which the famously liberal institution responded in typical fashion. "They let me develop my own course on the topic," she recalls. "Though . . . I was the only one who took it."

Also like Mandrak, Cudmore went on to complete a master's degree under E. J. Crossman. Her 1999 thesis used the Great Lakes' extensive history of lost fish species, population declines, and introduction of nonnative species as a case study in the relationship between diversity and "invasion resistance"—the widely held theory that changes in an ecological community's ability to resist invasions are related to changes in its diversity. Analyzing shifting patterns in the lakes' fish fauna over time, Cudmore indeed found that successful invasions generally followed a change in diversity. Her evidence provided insight into the invasibility of communities in general, and specific vulnerabilities of the Great Lakes, a topic since taken up by other researchers. This achievement put her not only on a path to eventual collaboration with Mandrak, but also on a job track that led directly down the highway from Toronto to Burlington.

"I came here in the late nineties when DFO wasn't really doing invasive work yet, so at first I worked on habitat," she tells me when we sit down in her office, located at the opposite end of the same corridor as Mandrak's. Tidier than those in Mandrak's office, the shelves above Cudmore's computer are no less stocked with information on invasives. "Of course you couldn't work on *anything* in the Great Lakes without coming up against invasive issues, but it wasn't until 2005 that DFO actually put together a formal AIS [aquatic invasive species] program with funds, vision and direction."

That development, driven by international commitments, was fortuitous for Cudmore. Canada, as one of 191 nations to ratify the International Convention on Biological Diversity (only Andorra, Iraq, Somalia, and the United States would subsequently fail to sign on or ratify), was obligated to draft a strategy on preserving biodiversity that included an action plan for invasive species. Cudmore was well positioned for Canada's sudden demand for expertise in the area, and has since been deeply immersed in the lakes' myriad invasive problems.

"Now I'm doing management, so a lot of my interactions are at the science-policy level," she notes. "But I find it equally fascinating. After we got the AIS program going, and after the Asian Carp Program came online, our unit saw big changes, mostly linked to a shift from dealing with tractable invasive problems that were already here, to being proactive about their causes and other potential invasions."

One not-so-intuitive change was suddenly having to find ways to keep governments' attention and funds flowing. "The concept of 'prevention' is a tough sell for governments because they run on performance measures and outcomes," says Cudmore. "It's hard to quantify what you've accomplished by writing reports about something that *hasn't* happened. You can't say 'OK, this is how many species we *prevented* from being introduced over this or that period.'"

Even more so when you're swimming against the current of a neo-conservative administration openly hostile to expenditures on research and science-based policymaking. Long-term trends, however, showcased deliverable outcomes that made even the most Luddite politicians warm to the value of funding prevention programs. The most relevant example for the Great Lakes was the issue of ballast water carried by transoceanic freighters.

While rock and dirt ballast was a historic route of introduction for insects and soil organisms to North America from Europe (we'll touch on this later), worldwide use of water for ballast in modern times quickly became a pathway for hundreds of aquatic invasions. As with the passive transport of hull-fouling organisms like barnacles, most ballast transplants are from one saltwater port to another. The Great Lakes, however, are unique in being a chain of mid-continent freshwater basins that could be invaded only by organisms transported from other major freshwater shipping areas. And this was exactly what happened, with the Great Lakes suffering several problematic ballast-mediated introductions of fishes, crustaceans, and mollusks originating in the Ponto-Caspian freshwater ecosystems of Eurasia. On the heels of debacles involving organisms as diverse as zebra mussel, round goby, and red mysid shrimp, ballast-water management requirements in the joint U.S.-Canada-managed Great Lakes St. Lawrence Seaway System are today the most stringent in the world: regulations stipulate that all ships destined for Great Lakes ports must exchange ballast in the open Atlantic Ocean before entering tidewater on the St. Lawrence River; if a ship is found not to have complied, it's then required to either retain its ballast on board, pump it ashore, treat it *in situ* in an environmentally sound manner, or return to sea to conduct an open-water exchange—all options costly enough to deter noncompliance.

"The ballast initiative has been a huge success. With focus, monitoring, and regulations we haven't seen an AIS introduced to the Great Lakes through ballast discharge since 2006, and no new fish since 1989," says Cudmore, adding, "Well, at least that we *know* of—the main reason prevention measures need to continue."

Cudmore should know, overseeing as she does one of the most extensive invasive-prevention umbrellas in the world in one of the most invaded aquatic ecosystems—particularly when it comes to fish (Figure 2). With the Asian Carp Program, she is a coauthor of detailed national and binational risk assessments with Mandrak and American counterparts like John Dettmers of the Michigan-based Great Lakes Fishery Commission and U.S. Geological Survey researchers Duane Chapman and Cynthia Kolar. She also implements risk-management strategies that include prevention measures like public outreach and pathway monitoring, as well as a research-based Early Detection and Rapid Response program (Figure 3) that just might prove to be the best investment DFO has ever made. That's because all four Asian carp are broadcast spawners attracted to similar areas in their native ranges for reproduction—a set length of river featuring certain temperature and flow dynamics with a wetland at the mouth. Tasked to identify qualifying watercourses along the Canadian shores of Lakes Superior, Huron, Erie, and Ontario, Mandrak used maps and available hydrological data to rate forty-nine sites as being of high, medium, or low value for Asian carp spawning. Crews sent out to ground-truth those sites that were rated medium to high found most were good, and a monitoring plan was drawn up. With a massive invasion under way in the adjacent Mississippi basin and a virtual army of scientists and engineers deployed to manage it, the Canadians also benefited from proximity to the best (and worst) virtual laboratory imaginable to refine their plan. "The science coming out of the U.S. was showing these fish to be *very* adaptable," Cudmore notes by way of example. "They didn't seem to need as long or as wide a river or as much flow as we'd originally thought." This kind of ecological plasticity characterizes most successful invaders.

Both Cudmore and aquatic science biologist David Marson journeyed to Illinois to learn firsthand about specialized gear and techniques used for sampling Asian carps. These included eDNA surveillance, electrofishing, and various types, combinations, and placements of nets, as well as exclusion devices (to keep turtles and other fishes from the nets). In one installment of a 2014 DFO webinar series on Asian carps, Marson reflected on the importance of this experience prior to beginning DFO's own monitoring. With such high value placed on early detection and removal capabilities, sampling Asian carps *in situ* was critical not only for properly identifying areas of suitable habitat, but also for ensuring that these areas were as thoroughly checked as possible. Marson also pointed to the value of synchronous sampling by partner agencies, as well as commercial and recreational anglers. "Information sharing is key," he said. "We want to slam the door shut."

I meet Marson when Cudmore steps me across the hall through a sanitary-looking door, behind which lies DFO's new, specially constructed Asian Carp Lab. Chrome surfaces and taps glint under fluorescent light in a room lined with steel boxes and freezers in mint condition, with a special glassed-in area for handling DNA sealed by a contaminant-foiling, pressure-lock door. Marson points to the large wall map depicting his monitoring sites, but I'm distracted by what hangs beside it: a fish of squat enormity, sinister eyes riding so low on the head that I assume it's a giant catfish—until I realize it lacks characteristic barbells around the mouth. The meter-long monster is a thirty-seven kilo bighead carp from the Illinois River. On the other side of the room, above lab benches and a countertop flow cytometer—the laser-equipped instrument used to measure DNA content in blood cells—schools a further rogue's gallery: replicas of monstrous grass and black carp and, of course, high-flying silver carp, the slapstick saint of every fourteen-year-old with an Internet connection. Impressive as it is, the space's sterile ambiance makes it seem more like a manicured set for *CSI: Asian Carp Investigations Unit* than a functional lab. "Hopefully it stays that way," chirps Marson. Cudmore smiled wanly. She knows it won't.

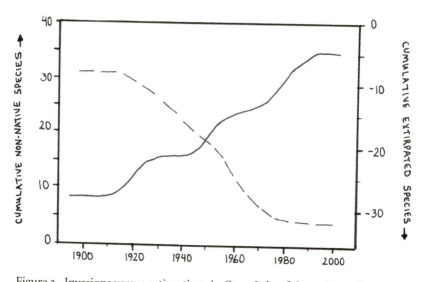

Figure 2. Invasions versus extirpations in Great Lakes fishes. Although no clear example exists of a species going extinct in the Great Lakes solely because of an aquatic invasive species, they have nevertheless contributed to extirpations. The increase in nonnative species and decrease in native species also correlates with increases in pollution, habitat destruction, and climate change—that is, ecosystem vulnerability. The pattern of increased invasion and increased extirpations crosses just after World War II. Historically, the main vectors of introduction were stocking and canals; currently these comprise recreational boating, live bait, and shipping. (Adapted from Mandrak and Cudmore 2010.)

| INVASION PROCESS | RISK ASSESSMENT | RISK MANAGEMENT |
|---|---|---|
| ENTERS PATHWAY | PROBABILITY OF INTRO | PREVENTION |
| TRANSPORT | ARRIVAL | |
| RELEASED | | EARLY DETECTION |
| SURVIVAL | SURVIVAL | RAPID RESPONSE |
| REPRODUCTION | ESTABLISHMENT | CONTROL |
| SPREAD | SPREAD | |
| IMPACT | MAGNITUDE OF IMPACT | ADAPT |

Figure 3. Invasion stages aligned with phases of risk assessment and risk management. Note that arrival, survival, establishment, and spread are components of the probability of introduction. (Adapted from Mandrak and Cudmore 2015.)

"We're the first to use trammel nets in Canada," she says, steering the conversation back to monitoring. "Asian carp might look dopey, but they're smart enough to avoid nets, and a trammel doesn't look like a straight curtain so they don't see its bags and pouches. We drive them in with rooster-tailing [boat wakes] or percussive noises: we put plungers on sticks and hit the water or bang on the side of the boat; anything it takes to spook them. It's a bit crazy, but you don't get bored."

While Cudmore seems to relish the often odd machinations of holding the line for the world's greatest freshwater ecosystem, she is even more in thrall with the practicalities underlying that task. "Risk assessment lets us think proactively based on the best scientific study and thought," she tells me, adding that some of even that expertise comes from the United States, where they're working to prevent establishment in new areas. "We're still working to prevent *introduction,* which is far cheaper and delivers other benefits—like catching invasive species not yet on our radar."

Excluding Asian carps, thirteen invasive fish currently established in the Mississippi basin were knocking on the Great Lakes' doors. And early warning risk management was still being implemented with black carp, the least known of the Asian tetrad. While only seventeen scattered individuals had been captured with no evidence of reproduction, four recent finds had been clustered closer to the lakes, prompting DFO to undertake a delayed black carp risk assessment. Like Mandrak and Cudmore's previous assessments, it would be a wise investment.

"Each year, Canada and the U.S. currently spend around $8 million and $22 million respectively just to keep sea lamprey at a livable level," notes Cudmore of the single invasive species that DFO has controlled for some seventy-five years and will, quite likely, continue to control into perpetuity. "But with the Asian Carp Program, we're only spending $3.5 million a year to prevent four species that are currently costing the Americans $75 million annually to manage."

Cudmore thought it impossible to overstate the opportunity that such an anomaly offered for invasive species management. "We've learned our lesson: it's more economical to throw up roadblocks. Our history with invasives in the Great Lakes has been so reactive, so expense-heavy. But working from the proactive, risk-assessment perspective flips everything around."

## Details, Details

Cost-benefit analysis aside, there's ultimately only one question begged by a conversation about Asian carp prevention in the Great Lakes: what would the scientists do if they actually found one?

As it turns out, Cudmore knew *exactly* what they'd do. Almost miraculously, her team had deployed sampling gear in the wild, caught Asian carp, and tested response protocols—all without any danger of a real invasion. Of course, they couldn't have known then that there was no actual risk involved. Like an unannounced fire drill, they had reacted to what could easily have been the real thing.

Carp monitoring began in 2013 with two crews based out of Burlington and one in Sault Ste. Marie. The areas sampled by the Burlington teams included 195 sites that yielded 15,849 fishes representing 73 species. The program was expanded in 2014 so that three boats and crews out of Burlington covered 560 sites and hauled in 32,000 fishes representing 84 species—a huge effort involving hundreds of man-hours. Zero Asian carps would have been the best outcome, but instead the victory was pyrrhic: three grass carp turned up—one angled, the other two captured by trammel net in the Grand River on Lake Erie's north shore. All were between fourteen and twenty kilograms, and each proved to be "triploid"—a lab-induced form that cannot reproduce.

Because sterile triploid grass carp are often used for macrophyte control in aquaculture, it's these dead-end, non-breeding forms that most often escape into the wild. But they couldn't be considered safe: the process employed to create triploids—a shock to eggs during cell division that results in three sets of chromosomes instead of the usual two, which leads to faster growth—wasn't 100 percent efficacious. Generally, any large sample of "certified" triploid stock was likely to contain a few diploids. And diploids can reproduce.

"We followed pre-planned response protocols for all three grass carp," Cudmore tells me of how the fire drill began. "We immediately undertook further sampling in the area while someone drove the fish back to the lab to test blood cells for ploidy; until those results were in, we assumed the fish was diploid and had lots of friends out there."

The operation seemed unabashedly militaristic, with Cudmore the commander in chief. A biologist was dispatched into the bowels of CCIW to obtain a blood sample standard—to be run against the carp—from live *Tilapia* maintained for just this reason. The group then readied the lab for the carp's arrival; here blood would be drawn, measurements taken, gonads and stomach contents checked, otoliths removed for analysis (these small ear bones can tell you whether the fish was grown in captivity or has been in the wild all its life). The sparkling lab I saw would have indeed gotten deservedly messy. Meanwhile, a field tech oversaw additional sampling at the capture site, checking in every hour. Each step and communication was recorded. Blood was run through the cytometer within two to three hours of capture, in each case confirming a triploid animal. Cudmore then issued a stand-down order.

But what if *no* stand-down had been issued? What if the threat had been real?

"If a carp turned out to be diploid, we'd deploy more gear specifically aimed at looking for young; if it were another Asian carp species the gear would be different; if we caught more than one carp in an area we'd bring in provincial reps and mount a much bigger response. Details would depend on species and habitat but in *every* case the goal would be to eradicate each and every individual through netting, electrofishing, chemical means, or even drawing down the water if it was in a river," Cudmore rhymes off in what seems a single breath. "And for some of those we'd have to take into account public outcry, which is why we do outreach to let people know how serious the threat is."

Serious, and elsewhere very real. Cudmore mentions that four yearling diploid grass carp were found in the fall of 2014 near Sandusky, on the south shore of Lake Erie. "The state of Ohio is currently looking into the status of this population to come up with an action plan. A bi-national committee oversees these kind of activities. Our best bet under these circumstances is, in fact, the early detection program; ideally we'd find any that strayed to this side of the lake before they had a chance to establish."

While established invasives are being managed to control their *populations*, preventing introductions requires the management of *pathways*, and there are several beyond CAWS through which carp could enter the Great Lakes, including the aquarium, bait, and live food trades. This is why it's illegal to possess or sell live Asian carp in the provinces of Ontario, Quebec, Manitoba, or British Columbia (you can, of course, possess and sell dead ones). Risk assessments show that the potential for introduction from these sources is significant. Indeed, vigilant Ontario Ministry of Natural Resources patrols of food fish importers, live-transport vehicles, and retail live markets have turned up no withering trickle: during 2011 and 2012, officers spent a combined four thousand hours inspecting dozens of wholesale and import companies in over a thousand locations, stopping six live-fish haulers carrying over 13,000 kilos of Asian carps that netted several convictions and over $100,000 CAD in fines. But while the illegality of *possession* in Ontario helped erect a barrier around this pathway, there was still a chink in that wall: without federal legislation, *importing* live Asian carp was still legal.

"That's why part of our strategy is to conduct gap analyses," explains Cudmore. "List all the activities required to stop or manage an invasive species, and list all the regulations that could help with that and see where the gaps are." Because triggers are important for a rapid response, it's incumbent on the feds to close these gaps in such a fashion that the mere *importation* of one of these species causes an alert. "We're anticipating legislation in the near future and we've already done work with border officers in helping them identify the fish involved."

In the meantime, concerned groups like Cudmore's do what they can to paper over the gaps. The result is a souvenir-shop's worth of fridge magnets, buttons, stickers, and key fobs, as well as brochures, posters, and a well-advertised hotline. With an army of recreational anglers out there, the frontline against Asian carp invasion has been at least long and energetically fortified, a backup that people working to stanch the spread of other invasives wish they had. After all, when you haul in a fish you don't recognize, a mental car-alarm goes off; whereas when you walk past a plant you don't recognize, well, you mostly figure that you never noticed it before.

## Crusader

With the toe of his boot, Bob Brett kicks hard at a knee-high stalk of Scotch broom, hoping to loosen its taproot. But when he bends to haul on the plant with both hands—vigorously, feet planted firmly—it's clear that his quest to dislodge it has failed, despite soil softened by a week of torrential rain. "These things can be pretty tenacious," he says, waving a hand dismissively. "I'll come back later."

Scotch broom (*Cytisus scoparius*) has proven to be one of the Pacific Northwest's worst invaders, a bully plant that has already had its way in Australia, New Zealand, the Iberian peninsula, Brazil, and other places. The tough, woody shrub is outcompeting native flora, especially in the endangered Garry oak ecosystems of northwest Washington State and southwestern British Columbia, where dense, fast-growing stands occupy fields, pastures, hillsides, and other open habitats like forest clearcuts, roadsides, and hydro right-of-ways. This it accomplishes through allelopathy—the release of chemicals that inhibit native plants—and by fixing nitrogen in ways that favor other nonnative plants (such as Himalayan blackberry) over native forms. It's also a significant fire hazard—with the region's changing climate making it more so. Scotch broom's pleasing springtime explosion of bright-yellow flowers (referenced earlier viz my drive along Howe Sound) is one reason that it could insinuate itself in the region before anyone realized it was a problem. But problem it is, with a single large shrub annually producing up to 20,000 seeds distributed both passively by gravity, vehicles, and animals, and actively via the plant's own exploding-pod broadcast mechanism that shoots seeds as far as five meters. Those seeds, which remain viable in the soil for some eighty years, are generally induced to germinate by a disturbance—like earthmoving or fire. The shrub can also regrow vegetatively from cut stems. It's more or less a perfect invader that can multiply rapidly to change entire ecosystems. A land carp.

As Brett turns to walk up the road, I gaze down at a smaller broom that he'd pulled up easily when we first arrived, then tossed on the ground; its prone form lies at my feet, a chthonian challenge to my own resolve. *OK, fine.* I stoop, grasp the larger stalk of the plant Brett just gave up on, bend my knees, and lean into it. Nothing. I strike another pose that delivers more leverage. Ditto. If this plant were stuck in a ceiling I could hang from it.

"They say anything thicker than your little finger is too big to pull," offers Brett over a shoulder as he shambles ahead in his rumpled rain jacket.

Not that axioms have ever stopped Brett's own efforts as a one-man wrecking crew bent on foiling the establishment of some twenty invasive species in the Whistler Valley. As long as the amount of broom remains manageable, Brett will continue pulling it out before it flowers and/or seeds. He returns, very much religiously, to some seventy-odd sites year after year to remove whatever springs forth, under the not-entirely-insane premise that, eventually, he will exhaust the limited seed bank. In fact, he'd already been to this particular spot in July. Now it's late October, and the plants we're attempting to pull have appeared since.

Like other interlopers, broom may have arrived here in fill used to shore up embankments for a newly widened highway; it might also have taken a car, truck, or train north from Squamish. Either way, Brett re-calls first pulling broom here around 1997, and because he clocks each site at least once a year while the plants are flowering, he's confident that no broom has set seed in Whistler for at least fifteen years, and that any-thing germinating anew is merely demonstrating impressive shelf life—like a site inactive for six years that suddenly sprang to life again this summer. Was it due to soil disturbance of some kind? Possibly. Different climatic conditions at a specific critical time? Equally possible. There was no way of knowing, but this much was clear: it meant more work for a lone anti-invasive fanatic.

"Can't you just remove the surface soil within a few meters of a site and potentially be done with it?" I ask, questioning the time-efficiency of pulling individual plants as they appear.

"I can easily visit all these sites on my bike," Brett answers, "and the time it takes me probably totals sixteen hours a year. That's as much as it might take just to dig out one site."

Every worthwhile crusade is backed by solid rationale and a determined general.

At the terminus of our walk, where a new overpass and previous highway route part ways, we step over bent grasses and onto the old road bed, which several years back, to Brett's chagrin, underwent a crude and disastrous attempt at ecological restoration: mulch of unknown origin was dumped atop the old asphalt, spread into a bed about thirty centimeters deep, and small conifers planted. From this witch's brew of bark bits and god-knows-what sprouted an entirely new crop of huge, swaying broom that quickly outpaced the young trees. Wherever the heavily contaminated mulch lay, broom grew. Brett's now-bleached July pullings lie strewn about like some kind of accident scene—chalk out-lines between struggling cedar, pine, and spruce saplings. Stretching skyward above a small spruce, a neck-high broom has materialized in the interim; if you discount October, when it likely hasn't grown at all, that's two months from breaking the surface to over a meter high.

The take-home—as with many invasive plants—is that broom is a champion hitchhiker, making this a main pathway to block. Until recently, however, there'd been no budget from any level of government for controlling invasive plants along roads and other corridors unless they caused problems to infrastructure. Consequently, EDRR for a new invasion, or implementing any kind of control policy that might prevent these species' further spread, fell mostly to concerned individuals like Brett. And once broom became established, it was there for good. Which explained the literally dozens of dedicated grassroots organizations like BroomBusters (broombusters.org), whose members' shared purpose has inspired a camaraderie apparent in annual meetings, potluck din-ners, and endless work details that turn on the seed-stopping motto "Cut Broom in Bloom." In some areas on Vancouver Island, annual broom-bashes are *the* community event of the year.

"Who the hell would plant a ponderosa pine *here?*" erupts the mild-mannered and typically less-salty Brett, hands flailing the air in frustration at the appearance of a rain-shadow ranchland species in a coastal rain forest. "Someone out there has a sense of humor," he says, brushing past the offending arbor.

Walking back along the highway toward the car, Brett calms himself by enumerating the no-less-odious but more familiar invasives: red-top grass, tansy, dock, ox-eye daisy, dandelion, even seemingly ubiquitous aspen, which is native to Vancouver Island, the Gulf Islands, and, like the offending ponderosa, drier reaches of British Columbia's interior. But not here, where cottonwood is the king colonizer. As we pass the scrappy broom we'd both yanked on earlier, Brett can't help himself. Striking a new point of balance, he leans in and pulls. It doesn't budge. "See you next July," he says.

## Organizer

Compact, energetic, somewhere between wiry and wired, Brett's comportment is like that of many of the keepers of mankind's knowledge vaults whom I've met over the years. Speaking in short bursts peppered with context-expanding tangents, connecting dots of information and recollection in stream-of-consciousness updates, he can sound scattered in the moment, but the big picture, when it emerges, is always bigger and more nuanced than most. His is the kind of grasp of a subject you find yourself wishing *you* had, for Brett always has a critical tidbit to donate to your cause—or the many he himself has spearheaded. His story is unique, even as it adds to thousands of similar tales in the invasive species gestalt forming around the globe.

Having abandoned advertising work in Montreal for the outdoor life, Brett moved to Whistler to teach skiing, combining it with a back-to-school master's degree in forest ecology from the University of British Columbia in 1997. While riding the seasonal roller-coaster of professional biologist and ski instructor, Brett founded the Whistler Naturalists, an active lot that averages a hundred members annually. Eschewing regular indoor meetings in favor of outdoor events, the group runs a Christmas bird count, a breeding bird survey in June, a popular "Fungus Among Us" weekend each October, and a flagship annual Whistler Bioblitz where teams of visiting scientists and public participants fan out—and up—through valley and mountain ecosystems to document as many species as possible in twenty-four hours.

The Whistler Bioblitz idea dates to 2004, during the Brett-administered Whistler Biodiversity Project and a conversation with amphibian biologist Elke Wind about the growing Bioblitz trend. Brett loved the concept—"To know nature and to keep it worth knowing" as he puts it—launching Whistler's first in 2007. With the help of dedicated co-organizers Kristina Swerhun and Julie Burrows, it's now the longest-running Bioblitz in Canada, a point of no small pride.

Generating species lists and identifying previously unknown or particularly rare species can jump-start public understanding about an area. Such documentation has historically been obtained only sporadically, scattered throughout the country in often inaccessible repositories.

In fact, little inventory work is done outside of commercial development proposals that generate "proprietary" information that's often incomplete, inaccurate, unavailable to the public, and obsolete when said development, inevitably, goes through. Efforts like those by the Whistler Naturalists, then, are more important now than ever.

"When more is known publicly about a property than a proponent knows, wiser decisions on preserving species and habitat can be made," notes Brett. "The whole idea behind both the Whistler Biodiversity Project and Bioblitz was to generate basic information about ecological and wildlife values that could help guide decision-makers."

For the long-running Whistler Biodiversity Project, Brett at first collated all available species records for the area, logging perhaps five hundred known species with some accuracy—mostly trees, birds, mammals, and fish; lists for fungi, lichens, plants, mollusks, insects, arachnids, reptiles, and amphibians were nonexistent. A decade down the road and Brett's efforts had added some 3,000 organisms to the tally; 1,200 of these from Bioblitzes alone, which essentially represented only eight days of concerted effort and so indicated the diversity to be found if one only went looking. Several of these were of special interest, including spiders previously undocumented in North America and many range extensions. Unfortunately, a number of these discoveries were also unwelcome—hence Brett's personal battle with invasive species.

"After five years of the Biodiversity Project we were finding invasives in almost every species group," recalls Brett. "Things transported with nursery stock like slugs, snails, worms, mosses, fungi, and ornamentals. Plus plants like Scotch broom that found their way here via another route. So it made sense to do something in the area of prevention."

In 2009, Brett co-founded the Sea to Sky Invasive Species Council (SSISC) with Stewardship Pemberton's Dawn Johnson (Pemberton is a bottomland agricultural community a half-hour to the north). The next year, Swerhun, with a master's degree in climate-change impacts on alpine biodiversity, took over coordination of the SSISC. As director for four years, Swerhun raised awareness of invasive species with the public, government, and industry through a successful "spotters" program; she improved communication among a diverse set of stakeholders and secured improved annual funding; and she led successful fieldwork highlighted by treatment of all known giant hogweed sites. More importantly, the SSISC became an official source of expertise that supported landowners and land managers in their identification, prevention, and control efforts. The group provided comprehensive mapping of invasive plants and prioritization of control sites; it also consulted on Whistler's pending municipal invasive plant bylaw as well as similar legislation being looked at by other governments.

A key tenet for Swerhun was building stakeholder capacity so that best practices for invasive species were part of everyday operations, stretching limited budgets further. "It was discouraging at times, especially when there was so much you wanted to do but didn't have capacity for, so picking battles was key; you had to accept that you couldn't do everything and focus on programs with a high probability of success," recalls Swerhun. "It was helpful that our directors met regularly to set priorities. Before I left, the SSISC received charitable status and found a great new executive director in Clare Greenberg."

I'd end up spending plenty of time with Greenberg, impressed by her consummate professionalism, something that typified the emerging breed of skilled, mercenary invasive fighter. As a biologist specializing in predator-prey interactions, she'd previously carried out predictive range mapping for invasives with the parks department of the state of New South Wales in Australia—fox and feral cat eradication, goat and pig control, a smattering of plant distribution. After moving to British Columbia and picking up geographic information systems (GIS) work with Environment Canada in Vancouver, she'd just launched her own contract-based mapping company when the SSISC directorship opened up. With the organization's broad, multi-taxa mandate, and the high-tech mapping of invasives fast becoming key to their control, Greenberg was a perfect fit.

Beginning in September 2013, support from seventeen other invasive species groups in British Columbia helped get Greenberg up to speed with the SSISC, but the challenges inherent to such multifunded organizations started early. "There was a lot of reporting to funders, so I had to review all the sites we'd done that summer—you know, before and after photos—and get a handle on how we'd been recording field data," she'd told me during one of our many discussions. "Now we use a

GIS spatial grid to mark the location and extent of invasions, for which we're also able to record and store photos. So instead of a spreadsheet with latitude-longitude etc., we have a geo-database—essentially a table with a spatial component that you can view on a map."

There was also the preparation of risk assessments and printed manuals detailing existing and potential invasives and their effects, the tedium of populating a website with this same information, and planning how it would all be used in outreach. Despite the coup of attracting enough funding to run two field crews, the additional challenges of regional buy-in and not-quite-comprehensive partnerships were apparent. While the Resort Municipality of Whistler (RMOW), district of Squamish, and other corridor partners provided the SSISC with funding, others brought no money to the table, despite the importance of linking all such entities across a region. "Because funding is tied to what funders want to achieve short-term with specific taxa, long-term planning for multiple species is difficult. More general funding would let us strategize better, then source specific funding for various priorities. Right now it's the other way around," Greenberg noted. "In order for invasive plant management to work across jurisdictions, each one has to be involved in a similar functional capacity, leaving private land as the only wild card."

The money. The time. The effort. The futility. Whether battling jurisdictional issues or an aggressive Japanese knotweed infestation that might take a week to knock down and five years to monitor, it all sounded so David versus Goliath. "Sure," agreed Greenberg. "Absolutely it's a huge, non-stop battle. But you can't just do *nothing*. So we're trying to use our resources in the best ways possible, focusing on EDRR for new infestations as opposed to over-concern with longstanding ones, which is pretty much the invasive-species-management norm."

But if putting out brushfires while Rome burns is the norm, why care at all?

"Because costs go up," answered Greenberg, evoking the Invasion Curve. "The costs of damage to ecosystems and infrastructure skyrocket once prevention fails, so the cost of doing nothing is far greater than the cost of doing even a little. Prevention is best, but even strategic control can retain values. You might not bother stopping burdock or orange hawkweed along a highway, for instance, but if those plants appear in the alpine you're going to move immediately to eradicate them to protect native wildflowers and the alpine ecosystem."

Effecting early detections required a tight partnership between the RMOW and the SSISC. Paul Beswetherick, landscape maintenance supervisor for the RMOW, as well as an SSISC board member and current chair of the SSISC, explained: "The RMOW has always tracked what was happening elsewhere in the corridor and how it might threaten both natural and maintained landscapes. But by focusing on the operational side and being action-oriented, the SSISC has been tremendously efficient with limited funds, effectively impacting the spread of new invasives in a very short time."

Beswetherick also sees the RMOW's recently introduced invasive bylaws—which require removal by the owner of any prohibited invasive—as a significant step. "They're forward thinking, proactive, and a legal requirement to deal with invasives will push those on the fence to say 'it's the right thing to do.' A fine is a last resort; swinging a hammer never wins you any friends so we'd rather educate landowners to act. That said, making something *illegal* creates impetus to take it seriously. Bylaws mean buy-in. In that sense Whistler is ahead of the game."

In today's world of turbo-invasions, and as with Cudmore's carp program, even the *opportunity* for prevention is a victory. To increase eyes on the ground for everything from periwinkle to opossums, the SSISC provides workshops and certification for landscapers, earthmoving companies, and environmental consultants in which they disseminate skills, knowledge, and the standard bounty of brochures, hotlines, invasive kitsch, and collateral. The hope is that more public and professionals will pitch in, report invasives, and where appropriate, remove them. This is particularly important for private land: contractors are key to convincing clients that removing an invasive from their property is in their best interest. And while Whistler might be out front when it comes to knotweed (a few occurrences) and broom (the SSISC maintains a zero-tolerance control line south of Whistler and Brett manages the rest), it's too late for other Sea to Sky communities like West Vancouver, Squamish, and Britannia Beach, which means that Whistler will be eternally vulnerable to invasion from those downstream source populations.

Mea culpa.

## More Crusading

After I wrote a feature in a regional newsmagazine about the SSISC, a well-meaning woman wrote a letter to the editor decrying the villainization of invasive plants. She maintained that they were just innocent organisms occupying space that otherwise wouldn't be filled. There was some truth to that, certainly in urban landscapes, though she probably wouldn't have been as charitable if a truculent knotweed was pushing through her basement walls, or she was strolling with Bob Brett, as I am, while he points to the distinct line where roadside invasives end and native forest begins.

"This line is *extremely* important," he asserts. "Most stuff on the side of the road is fine as long as it doesn't go into undisturbed forest and wetlands. Plants that get into shaded forest areas are a huge problem: *Lamium* groundcover is the most common but not the worst; daphne, for instance, replaces native salal in the understory and is very hard to get rid of; in addition, it has poisonous berries and the leaves have a phytotoxin."

While certain plants pose very real dangers to human and animal health, existing invasive legislation in North America generally relates only to agri-pests. Canada thistle (*Cirsium arvense*—ironically European) was designated the country's first "noxious weed" in the late 1800s because it was impeding agriculture. Ditto ox-eye daisy (*Leucanthemum vulgare*), a widespread flowering plant native to Eurasia now introduced to North America, Australia, and New Zealand. Forming dense colonies that displace native plants and modify existing communities, ox-eye daisy is difficult to control or eradicate, because it regenerates from rhizome fragments. This creates issues in cattle pastures where the animals refuse to eat it, but handily kick it around, enabling its spread. In addition, ox-eye daisy hosts several viral diseases affecting crops. In Whistler, Birdsfoot trefoil (*Lotus corniculatus*), once planted on the mountain's ski runs to function like clover in turning bare mineral soil into an erosion-resistant surface, has descended to the valley and is now turning up along roadsides north to Pemberton. Plentiful ditch-loving orange hawkweed (*Pilosella aurantiaca*) was never a concern here until it marched up into the alpine in Oregon with hikers. Spotted knapweed

(*Centaurea maculosa*), another allelopathic taproot species, regularly gets yanked by the roving Brett, who explains: "It's prolific, generates a huge seed bank, and has gotten into rangeland in the Interior. It travels in tires and along rail lines."

Speaking of which, as Brett and I walk a short stretch of rail line virtually everything we see is nonnative. Like most railway companies, the tracks' owner, Canadian National (CN), sprays annually for weeds, killing indiscriminately to the edge of the ballast supporting the tracks with the net effect of inhibiting the regeneration of native species and selecting *for* invasives that are almost always the first to colonize.

After we drive back to Whistler for a walk around my neighborhood, an invasive, bright-yellow lichen on a nonnative aspen trunk triggers a bout of free-association by Brett. Low pH forests like ours can be changed from fungal to faunal decomposition by earthworms that raise the pH, he says, and that process turns coniferous forests into deciduous forests. White pine blistering rust is invasive, he notes, tugging at a branch, but so are bedbugs . . . and Ebola. I ask him if he's read David Quammen's *Spillover*, which eerily presaged West Africa's 2014 Ebola epidemic. "No, but I referred to his 'Planet of Weeds' essay in my very first invasive talk."

That makes two of us. "Planet of Weeds" is a seminal piece because it situates this quiet crisis within the pantheon of causative factors in the Sixth Great Extinction. Eyeing up some invasive roses, Brett continues. "Hybridization between native and nonnative promiscuous plants like these means losing species by stealth. And since we don't have DNA kits to prove which plants are which, we lose them *legally* as well if someone challenges us on what's what."

I'm about to ask for clarification when I notice a cluster of broad green leaves close to the ground. "Burdock, right?"

"Yes," says Brett, stooping. "We got a photo from Bioblitz one year of a bat caught in burdock."

Hyper-aware of their physical environment, bats will typically brush by plants without consequence while swooping for insects. But they're clearly no match for a nonnative trap they don't know is there. The Bioblitzers had freed the bat, so it wasn't as bad as what I'd recently seen on a Facebook page devoted to nature photography—a trio of bats that had been caught on the Velcro-like heads of burdock and starved to death, their bodies mummifying in the Texas sun.

It's as hard keeping up with Brett as it is following his thoughts. He moves as fast as his patter roams the landscape—up mountainsides, through forests, into gardens, along the road, down ravines. "Yellow-flag iris moved downstream from a highway ditch to just above Nita Lake," he says, as we make our way into the bed of Whistler Creek where Brett believes he saw the invasive algae known as rock snot (*Didymosphenia geminata*). "Did you know yellow-flag iris was actually *planted* by the RMOW in 2005?"

I did not. We can't find any rock snot, perhaps because it's now the rainy season, when flash-flooding scours these mountain creeks after every remnant of a Pacific typhoon makes landfall. I was, however, familiar with rock snot. The brown, beige, or white algae looks as slimy as it sounds, though it feels more like wet wool than nose goo. Moved around by wading anglers on felt-soled water boots (which are banned in jurisdictions like Alaska, Colorado, Ontario, and Vermont for just this reason), it forms a thick blanket in freshwater streams that can extend for kilometers, smothering plants, altering species composition and invertebrate populations, diminishing salmon-spawning habitat, and reducing dissolved oxygen through decomposition. Having invaded most of the Lower Mainland, rock snot may, Brett worries, affect British Columbia's threatened coastal tailed frog, whose tadpoles feed by rasping diatoms off underwater rocks in these same streams. There was no known method of eradication.

"Let me tell you about a victory and a loss," says Brett, launching into the story of a single purple loosestrife plant that he and Beswetherick removed every year for a decade. "This is the first year it didn't return," he says, hoisting an aposiopetic eyebrow like a trophy. He moves on to the loss. "Someone spotted small-flower forget-me-not on Whistler Creek," he says, pointing uphill toward a debris-flow barrier. "So we had a six-person crew spend two days pulling it out. But since then, it's been popping up everywhere on the periphery. We thought we'd treat it like terrorists and fires—early detection, hit 'em hard. But it didn't work because it was clearly too well established, likely because it can grow in multiple niches from wet to dry to forest, and reproduce multiple times a season. Like dandelions . . . starlings . . . robins."

"And us," I add, citing the phenology of the monthly human re-productive cycle that allows for our perpetual breeding. Brett nods. Just then an indigenous red squirrel pops an inquisitive head from a hole in the rocks on which we stand. It looks around, moves forward, then spies our feet looming above and bolts across the rocks to disappear into the brush. "Did I mention we have a kill order out on invasive Eastern gray squirrels?" Bob finishes. "They're already in Squamish."

Though the following summer I would see my first gray squirrel in Whistler, for some reason the first thing that pops into my mind at this moment isn't the familiar visage of a fluffy rodent sitting on its haunch-es grasping an acorn, perhaps on the oak-pocked expanse surrounding the ivy-clad University of Toronto where I had once seen them daily. Nor is it the macabre image of a squirrel being shot, poisoned, or caught in a trap. Instead, it's a cartoon vignette worthy of Gary Larson's *Far Side* canon. In it, I see two mammoths grazing side by side on a Pleistocene landscape somewhere in western North America. One says to the other: "Did I mention that there's a kill order out on humans? Apparently they've already crossed the Bering Strait."

By that point, humans had indeed already done so, likely carrying a few friends. And ever since we've been telling tales, painting pictures, and sharing photos of our global conquests. Smoke signals from a disaster.

## Beetlemania

Everything you need to know about the risks to North American forests from alien insects and companion pathogens can be found in Troy Kimoto's online photo gallery. There you'll see the telltale, meticulously spaced drill holes of the Asian long-horned beetle; close-ups of the under-bark vermiculations of the emerald ash borer; and the one-two depredations of *Raffaela* fungus and its delivery vehicle, the oak ambrosia beetle.

A Vancouver-based survey biologist with the Plant Health Risk Assessment Unit of the Canadian Food Inspection Agency (CFIA), Kimoto works on nonnative forest pests, focusing primarily on sap feeders and defoliators. With dozens of these species already established in North America and more knocking on doors around the continent, Kimoto's job seems unfathomably expansive.

"I break it into manageable pieces. We started out trapping to locate nonnative insects, then implemented rearing programs in main entry points like Vancouver, Toronto, Montreal, and Halifax. I look at interception data for each target pest to prioritize research into developing the right traps and detection methods. A lot of what I do basically involves coming up with new mousetraps," notes Kimoto. "But I also try to identify gaps. For instance, historically in North America we've dealt mostly with coniferous bark beetles, so we have a good understanding of what attacks softwood and which tree volatiles [aromatic chemicals] attract them, but we don't have the same expertise with hardwood-boring jewel beetles."

How could you? With some 15,000 species in 450 genera—double the number of bark beetles—the jewel beetle family Buprestidae is one of the largest in all of beetledom. With that many critters in play, Kimoto's concerns can be individual species, genera, or entire families.

The CFIA tracks the spread of established nonnative insects domestically, but a second survey type aims to detect new international arrivals. In both cases, Kimoto deploys paper-and-plastic traps baited with insect pheromones (sexual attractants) or complex host-plant volatiles (think of the familiar "pine scent" of a car deodorizer, which can actually represent dozens of appealing chemical odors to an insect). He spends time searching forests close to port areas, and the industrial zones of ports themselves. "We can't confiscate material," he says of the areas where high-risk wood is stored or uncrated, "but if a company has insects emerge from packaging, they'll usually call us."

Oceanic shipping and air freight remain *the* major pathways for insect introductions worldwide, and wooden frames, crates, and palettes employed to protect everything from glass windows to electronics are the number one vector for nonnative forest pests. Wood packaging is supposed to be pretreated (typically fumigated) to international standards, but often isn't, so Kimoto still sees plenty of foreign insects applying for Canadian citizenship.

The main risk from insects entering via this route is to the forestry industry, for which the Asian long-horned beetle (*Anoplophora glabripennis*) now serves as poster bug. Throughout the 1990s and into the new millennium, alert warehouse staff in various states and provinces intercepted larvae and emergent adults of this large buprestid on packages arriving from Asia, but the beetle nevertheless managed to become established in numerous places. Tens of thousands of trees in the northeastern United States were cut down in an attempt to stop its spread, but with little success and many economic, environmental, and aesthetic costs. Canada managed to contain its smaller outbreaks to declare itself ALHB-free in 2007—now a joke among the cognoscenti. Which is why foreign wood continues to be the biggest concern for prevention scientists like Kimoto, even if its ingress is concentrated in only a few places.

Over 50 percent of all foreign containers bound for Canada end up in the greater Toronto area, arriving either first to that port, or to more distant docks in Vancouver (3,360 km away), Halifax (1,800 km), and Montreal (550 km). They are then shipped to Toronto by train or truck. Additional worry surrounds the opportunity for introductions to occur along these transport routes or at remote endpoints. "Look at all the heavy equipment and overseas resources being shipped to the Alberta tar sands, for example. The Canadian Border Services Agency is responsible for inspections on dunnage and wood packaging with that stuff but it can only do so much," says Kimoto. "Even if CBSA *doubled* human resources, that only means increasing inspections from 1 percent of shipments to 2 percent."

(In a telling demonstration of how much can slip through without rigorous screening, a twenty-week trial of comprehensive air cargo inspections at Maui's Kahului Airport by the Hawaii Department of Agriculture in 2001 turned up 279 insect species, 125 previously unrecorded there, and forty-seven plant pathogens that included sixteen novelties.)

Concerns surrounding seaports like Seattle and Vancouver aren't limited to the idea that invasive species might be unloaded onto land. Attracted to offshore lights, the Asian gypsy moth lays eggs on ships docked in its native range that subsequently cross the Pacific to North America's west coast. In the destination harbor, often while ships wait to unload, the eggs hatch and, much like baby spiders, the tiny moth larvae shoot out a thread that lets them be picked up by the wind and blown ashore. With species like this to track, it makes sense to enlist the public's help.

"Canada is a huge country with trees from coast to coast, and obviously we can't put out traps everywhere. So we're trying more outreach. We have a set of tearproof, waterproof, credit-card-sized 'Insects Not Wanted in Canada' that we give to foresters, orchardists, arborists. We've even made removable tattoos of nonnative forest pests for kids."

Those are great tools, but pushing out the prevention message remains a challenge for Kimoto's skeleton crew, who do what they can despite a raft of other responsibilities. For every person in the CFIA addressing invasive problems, the USDA and other federal agencies stateside have twelve to fifteen bodies working on the same things, many of them dedicated specifically to outreach.

Given the CFIA's thin frontlines, I wondered about imminent threats like the hemlock woolly adelgid (*Adelges tsugae*). This aphid-like insect has virtually wiped out Eastern hemlock along the U.S. eastern seaboard since being introduced from Japan in the early 1950s, and it is moving steadily north into Canada. The nano-sized adelgid doesn't fly, moves slowly, and is hard to search for: its presence is most often revealed by cottony egg sacs that resemble a tiny Q-tip (hence "woolly"). Kimoto explains how a buddy in the Canadian Forest Service spends his time brainstorming novel ways to locate adelgids. One involves using a slingshot to propel a series of ten racquetballs swathed in Velcro into the crown of a hemlock, targeting different aspects and different heights to catch egg sacs (there's empirical accuracy behind this; tests on a known infestation in New York State showed that ten balls could detect even the smallest population). Another is a riff on sticky traps: a researcher places twenty-square-centimeter sheets of corrugated plastic with a sticky surface under hemlocks on the leeside of prevailing winds; when the breeze picks up, branchlets fall onto the sheets, sometimes bearing adelgids.

While the CFIA has plenty on its plate, it is, thankfully, not responsible for everything. But even sharing the job has been fraught, explained Kimoto. The European gypsy moth, for example, has become well established all the way from Sault Ste. Marie—at the junction of Lakes Huron, Michigan, and Superior—east to the Atlantic. It is within the

CFIA's aegis to survey for any westward expansion, but where the species is already established—or would be—the Canadian Forest Service is responsible for treatment. With the related Asian gypsy moth, which is still riding the EDRR end of the Invasion Curve, the CFIA is responsible for both detection *and* treatment. If this sounds complicated, it is. Enough that some aspects of responsibility for invasive forest insects are actually decided in the courts.

Efforts to get in front of invasives can be hampered not only by jurisdictional jigsaws, but also by funding priorities that short-sell important aspects of basic science in favor of flashier gewgaws. Kimoto relates a now storied example.

A few years back, a single specimen of a previously undetected, elm-attacking beetle was found in a trap in Medicine Hat, Alberta. More traps were deployed but no further specimens were discovered. Meanwhile, the beetle turned up in Colorado and other states. With it now on the continent's collective radar, American researchers began combing the West only to find it everywhere. This suggested that the beetle had been there a while but gone unnoticed. Sure enough, entomologists searching university collections found it present as far back as 1994, but improperly identified. Kimoto felt that the mistake was due to a lack of alpha taxonomists—people who can identify the beetle by sight. "These days all the money in systematics goes into molecular techniques," notes Kimoto. "These are potent and useful, but there's still a need for people who can do morphological taxonomy on the spot."

Kimoto wraps up with a hopeful screed on various research initiatives and international collaborations of which I catch only snippets . . . Slovakia as the model for jewel-beetle detection in Central Europe . . . Chinese researchers putting electrodes into beetle antennae to see if they can divine which chemicals are attractive or repellent to it . . . "Maybe in 10 years we'll have the tools to detect jewel beetles," he finishes.

Kimoto's examples drive home an important point: while the ever-inventive *Homo sapiens* that once skipped across the Bering Strait excels in crafting post-hoc solutions like Velcro-covered racquetballs to solve self-inflicted biological problems, invasive species have forced us into playing the long game—historically our worst skill.

# Air Mail

Clearly an invasive species zeitgeist has infiltrated and influenced everything from ecology, biology, and conservation to global transportation and commerce, community development, and culture. Whether delivered in news, popular writing, scientific literature, regulations, public warnings, classrooms, or art (I'll leave you to Google that), we live in a world struggling to comprehend, describe, and adapt to the magnitude and trajectory of invasive species problems.

Public stirrings over alien incursions are big, in Florida perhaps biggest of all, yet Florida will always resonate most strongly for me as a place I vacationed as a kid, dragged along by parents looking to escape Canada's ferocious winters: ocean beaches, seashells, scampering lizards, strange insects, botanic gardens, and everywhere racks and racks of postcards documenting both the state's wonders and weirdness. Are postcards still a thing? I don't know. But back in the day, they were *the* go-to souvenir, functioning foremost as personal greetings from a foreign destination, but also as an advertisement to visit these same idylls. The cliché of a superlative landscape, a natural phenomenon, a striking local life form—even a pyramid of bikini-clad water-skiers emblazoned with a solicitous title like "Greetings from Orlando!"—all have clear meaning: *this* is what this place is about, come enjoy it!

Twentieth-century postcards romanticizing the state's white strands, spring-fed grottos, preternaturally pink flamingos, and leviathan alligators are read into our imagination as pop-cultural traditions. But the metaphorical postcards emanating from the Sunshine State these days are a paean to its dark side: highly modified landscapes, disappearing habitat, unnatural phenomena, and strikingly invasive life forms. So maybe it's fitting, after all, that the mascot of Florida's ecological abyss is a giant Asian snake.

# PART III

# The Python's Tale

## *Establishment, Spread, and Detection*

The invasion biologist is ecology's emergency-room physician, discovering how ecosystems work by watching how they fall apart.

*Alan Burdick,* Out of Eden

# Python Wars

"You know about fire ants, don't you?"

Skip Snow was staring at my sandaled foot, planted squarely on a sandy mound swarmed by thousands of red specks. "Of course," I answered, nonchalantly lifting my leg and surreptitiously shaking it just as searing pain erupted in a dozen locations.

We were scraping our way along a levee hemming the eastern edge of Everglades National Park, the dense overgrowth of an old banana plantation to our east, River of Grass to the west. The nearby canal had become a main conduit for the flood of invasive Burmese pythons flowing from the Everglades, a population that had swollen, by some estimates, to 150,000. It seemed like a good place to test some python traps.

As a wildlife biologist for Everglades National Park throughout the 1990s, Snow was among the first to recognize the magnitude of Florida's invasive python problem. Now, as director of the Everglades Python Project—a full-time responsibility despite numerous others—he investigated ways to capture and remove them. Control of the pythons would ultimately depend on what could be gleaned of their preferred habitats and distribution in the park (which was turning out to be everywhere) and how that population spread (which wasn't yet clear). In other words, *How do you hide 150,000 giant snakes?* was a question that Snow looked to answer.

"We haven't done well eliminating species we don't want unless they have a certain life-history trait we can exploit," he lamented as we walked. "And Burmese pythons don't really have that vulnerability. They survive happily here because they're generalists in diet, habitat use, and behavior. So dealing with them comes down to this: how do you effectively, *willfully,* achieve an extinction?"

At that, grim-reaper turkey vultures appropriately filled the air ahead. Maintaining the same distance from us with each step we took, they rose and alighted, rose and alighted, black tumbleweeds on the breeze.

"Those vultures were feeding on a big ol' dead python a few days ago," said Mike, a Florida University student working with Snow. "So we know for sure that the snakes are using this route."

"Still, trapping is a big experiment," shrugged Snow. "What kind of trap will work or will traps work at all? What type of entry design works best for a big snake—is it funnel, paddle, or door? What time of year is best—when the animals are mating, foraging, or seeking refuge? Do you set the traps out randomly or along fences that force the snakes toward them? Should traps be baited with live animals or not? We just don't know."

No one did. Of the assorted designs set along the levee, some were unbaited, others self-baiting—like the one in which a cotton rat huddled in a corner along with an ill-tempered Florida watersnake that Mike wrestled out, but not before the angry serpent sunk pushpin-length teeth deep into his hand, drawing beads of blood. There were, however, no pythons this day, and I found myself disappointed yet again.

On my own, I'd been searching diligently—and fruitlessly—for a week. In two days of canoeing from various public access points into the swampy gloom, I'd turned up nothing more exotic than a cottonmouth coiled around an empty Budweiser can. But squinting long and hard into the Everglades' watery umbra and chiaroscuro of vegetation had told me this: looking for a cryptically patterned snake of any size this way was like scanning a galaxy-sized haystack for a hay-colored needle. So I'd shipped the canoe and hit the road—literally.

With few natural enemies, roadkill remained the pythons' greatest source of mortality, and thus, the easiest source of census for park staff. Although pythons were, in 2008, becoming the most common snake on some Everglades road transects, I'd had no luck driving Tamiami Trail, the highway skirting the park's northern edge. Dead alligators, by contrast, abounded. Likewise a plethora of freshly flattened ratsnakes and watersnakes. Oddly, billboards still advertised "Alligator Wrestling," and I'd wondered if this anachronistic sideshow might someday be augmented with "Python Wrestling," a way for some of Snow's captures to earn their keep in the service of education. He'd certainly turned up some big ones.

In December 2005, four radio-tagged pythons were tracked through the park to get an idea of their movements and habits. Later studies would be more revealing on this front, providing an idea of home range and homing abilities (one displaced python would travel 78 km directly back to its point of capture), but remarkably, this first group of radio-tagged snakes had led Snow and his crew to twelve *untagged* pythons. Perhaps, they reasoned, this "Judas Snake" ruse could be used to round up larger numbers of animals during spring breeding season when snakes tended to aggregate. One of the betrayed was a five-meter female that, in addition to serious breeding potential (Burmese pythons lay up to a hundred eggs at a time and the 45- to 60-centimeter juveniles can reach two meters in a year) confirmed the worst of park officials' fears: in the snake's stomach was a juvenile wood stork, a federally endangered species. Snow had the snake professionally skinned, figuring the prop might come in handy. Later, as he unrolled the hide for me back at the lab—in appropriately slow, dramatic fashion to emphasize its truly impressive length—he recounted how he'd taken it to Washington and similarly stretched it out in the halls of Congress, an exercise in semiotics. The politicos were piqued but the implications lost on them until Snow pointed to cute little deer fawns on their office calendars, declaring: *Hey people—they eat those things!*

At this juncture, if Snow knew anything about Burmese pythons, it was what they ate. In a laboratory in the Dan Beard Center, an aging, nondescript bunker built during the Cold War's Cuban Missile Crisis and located far off the tourist path, the usually laconic ranger became positively animated as our discussion turned to eating habits. "Analyzing stomach contents is like opening Christmas presents!" he bubbled. "Most will be shirts and ties—you know, cotton rats and lots of mice—but once in a while there's a treasure like an alligator or a fawn deer." With that he'd brandished a Ziploc bag containing a young deer's tiny hooves and short hairs, now frizzled from the acid bath of the python's stomach.

"*Python molorus* is a trophic generalist that eats a broad spectrum of vertebrate prey," Snow continued. "That means wading birds, songbirds, bobcat, deer, any kind of rat, and a lot of rabbits. The last few years driving around here has been like, *Where the hell are all the rabbits?*"

A larger issue in the ecologically fragile Everglades was that pythons were also consuming species of special concern like wood stork and mangrove fox squirrel—even a sea-going frigate bird, which was found in a python far inland (they were still trying to figure that one out). "We're pretty much just waiting to find an endangered Florida puma inside a snake," added Snow. "One good thing about looking at stomach contents is that it makes some of our other work easier. Musk-rats, for instance, are hard as hell to come up with during mammal surveys, but easy enough to find inside pythons."

Not for long. A study in which Snow was involved at the time would eventually show just how serious a threat Burmese pythons posed to the ecological functioning of the Everglades. The resulting 2012 multi-authored paper documented catastrophic, decade-long declines in most of the park's mammal species. An experimental study with marsh rabbits published in 2015 further validated Snow's intuition: pythons accounted for 77 percent of marsh rabbit mortalities within eleven months of the rodents' being released into Everglades National Park, a finding that seemed to preclude the ability of rabbits to persist there. In contrast, no rabbits were killed by pythons at control sites located outside the park, where 71 percent of attributable mortalities were due to the usual mammalian suspects.

Not that ogling gut glop with Snow wasn't interesting and even fun, but as an invasives voyeur, by this point I wanted to see a live python. I mentioned this as casually as I could, not wanting to appear caught up in the lurid sensationalism surrounding a clearly serious issue. In response, Snow angled his head toward a pile of heavy plastic coolers stacked like kabuki Lego bricks at one end of the room, each with "Everglades Python Project" stenciled on the side in white. It was the mother lode. Cracking open the top cooler, Snow pulled out a mesh bag weighted with a lumpy, auburn stone. When he lay the bag on a table and gingerly unknotted it, however, the stone unraveled to violent life. Snow had a difficult time getting a grip behind the head of the lash-ing beast without being bitten. The snake looked to be just under two meters, hefty, hungry, and vicious as hell. "It's about a year old," said Snow between tail whips. "We find *dozens* like this."

In this telling tidbit lay the true size—literally and figuratively—of the Everglades python problem, as well as the size that any effective response should comprise. It was a clarion call that seemed to cast Florida's other invasive problems as small potatoes. Back in 2008, however, not everyone was ready to agree with that assessment, as I'd discovered a few days before I met up with Snow, at Meg Lowman's garden party.

## Lucy in the Sky with Children

You can see a picture of a big snake in a book. You can watch people wrestle one up out of the jungle on the Discovery channel. You can even look at gruesome web collateral of a snoozing Amazon oil-rig worker apparently swallowed by an Anaconda. But nothing can prepare you for the steel-cable feel of a 5.5-meter, 90 kg Burmese python in your arms. And if it hisses with every twenty-liter exhalation of its tubular lungs—a sure sign it's annoyed as hell and in the kind of unpredictable mood snakes are known for—all the more shock to your cerebral cortex.

"Yup, Lucy here bit and tried to constrict me three times already," drawled Justin, the man from Matthews Animal Rescue who'd fetched her up from a culvert she'd been living in beside a Walmart in Bradenton. "Otherwise, she's pretty chill."

With sleeveless muscled arms, a cropped beard, and a scraggly mullet sprung from a straw cowboy hat, Justin seemed a perfect cross between *Dog the Bounty Hunter* and *Crocodile Dundee*. Lucy had a name but not so another three-meter Burmese in Justin's rolling menagerie, a cold-blooded cabal that included a massive African spur-footed tortoise now happily mowing Lowman's lawn, flower petals stuck to its face like cupcake sprinkles at a birthday party, and a 1.5-meter green iguana that Justin clutched like a baby he was trying desperately to burp—or keep from tearing him to shreds (driving here, sitting in the passenger seat beside him as a dog might, the iguana had suddenly attacked Justin, biting, clawing, drawing blood, and, as he sheepishly admitted, dumping a slurpee into his lap).

Meanwhile, a handful of young folk hefted a now somnolent Lucy for photos in Lowman's driveway, one cradling the serpent's head while the others supported its massive body. An ecstatic five-year-old boy held up a middle section with hands raised overhead. As I snapped their picture, it was hard *not* to dwell on how easily he might fit inside the snake and barely create a bulge. That was no abstract thought: ten months later, in July 2009, a two-year-old Tampa child was killed by an escaped pet Burmese that had found its way to her crib.

The uproar over the toddler's death was immediate, prompting officials to announce intentions to stringently enforce an existing ban on the unlicensed purchase or possession of several large "reptiles of concern" (as of January 2008, seven snakes and one lizard had formally—though rarely in practice—required registration, an annual hundred-dollar fee, and the meeting of inventory and caging requirements). Now, Florida's multitudinous hobbyist snake-keepers heard, many for the first time, that it was also unlawful to allow the escape or release of these animals into the wild. Blowback came from both pet owners and importers unused to reactionary politics in a state that (1) was full of snakes anyway, and (2) boasted a Barnum and Bailey heritage that had long tolerated dangerous pets aplenty. But with the Everglades' pythons making daily news around the globe and the extent of the problem fast becoming apparent, the possibility of another child being killed—this time by a wild snake—forced the hand of reticent legislators. A licensing program complete with digital tags implanted under the animals' skin would, so the thinking went, both aid in keeping track of large, dangerous snakes that might be stolen or released, or that might escape into the wild, and help in dissuading more irresponsible types from keeping them at all. The zebra, of course, would be well out of the barn by that point. Not that people hadn't tried to get a handle on the problem earlier. After all, think-tanky meetings like the one where I met Justin had been going on for years.

## It's Too Late, Baby

The vista out back of Meg Lowman's Sarasota home was like many in Florida's labyrinthine subcoastal suburbia: viewed through a screened-in porch constellated with scurrying, splay-footed anole lizards, a freshwater canal meandered lazily between a strip of jungle-dark forest on one side, and the sweeping, manicured lawns of palatial homes on the other. Sandhill crane, American white ibis, and various heron species stood at attention in the shallows. Otters fished surreptitiously near the forest. Alligators cruised with regularity, often lumbering out of the black water into reeds on the forest side, and, very occasionally, pulling up on the edge of a lawn to sun. But unless you fed them, gators' inherent fear of humans kept them at bay; the slightest approach usually caused them to slip quickly back into their submarine world.

Absorbing these swampy bucolics, you got the idea that no matter how much Florida's abundant land-gobblers scraped clear the landscape (which they did with abandon, thoroughness, and insane regularity), benign nature would barge right back in. Which is why the neighborhood was still abuzz over the four-meter Burmese python a woman found in her garage a couple streets over. It was nature, sure, but not Florida nature. And certainly not benign. For the python, a garage was simply a warm place to shelter, but if the woman had a cat or a dog, either could have found its way into the snake's stomach. As Skip Snow had shown me, the otters and birds around the slough were also fair game; even alligators were a legitimate meal for the extra-large and suddenly abundant serpent. Now that the Burmese python was "officially" out of control in South Florida, specimens turning up in suburban garages were no longer ascribed to anomalous pet-trade escapees or foolish owners turning them loose. The species was established, breeding, and spreading rapidly, a bona fide nuisance in that it demonstrably ate almost anything. By the time the Tampa incident had suggested a potential danger to humans from this feral swarm, everyone from university biologists to wildlife managers to politicians had been called on for opinions, each suggesting that radical measures be taken. What they wouldn't know was that it was almost certainly too late. Coincidentally, many of those same people were attending the gathering at Lowman's stylish domicile when I'd visited.

Lowman was at the time director of environmental initiatives and professor of biology and environmental studies at New College of Florida and Sarasota County's regional science outreach partnership. Her research interests in insect-plant ecology often found her hovering high above the ground in tropical forest canopies, hence the home decor of Amazon chic—walls adorned with tribal masks, blowguns, and other handmade weaponry. In September 2008, ever-gracious "Canopy Meg" was hosting this synod of students, scientists, trappers, politicians, and policymakers as a prelude to a formal conference she'd organized to address Florida's broader invasive "herp" problem. She was chairing both

gatherings because—with practiced leadership and get-things-done skills
garnered through several university posts and a stint as director of internationally renowned Selby Botanical Gardens—she'd also taken on the
job of heading Sarasota County's newly minted Invasive Herp Task Force,
the latest group to declare war on pythons and a legion of other aliens.

For my part, I was still mesmerized by Lucy, all 90 kilos of which
were now bobbing in Lowman's swimming pool along with a gaggle of
her smitten students while Justin tugged nervously at a beer in the background. As I downed my own brew a thought tumbled into reach: it
was surely a personal indictment that nothing about this scene seemed
unusual to me.

As a kid with a nose for lizards and snakes who often vacationed in
Florida, I can testify that, back then, had there been even the *remotest*
chance of finding a python, my parents would have seen little of me in
the span between debarking a plane in Tampa and reboarding one for
Toronto. I would have been out, dawn to dusk, trashing through the
palmetto in hopes of finding a massive constrictor, fancying myself a
Marlin Perkins (of *Mutual of Omaha's Wild Kingdom* fame), unconcerned that my quest involved a nonindigenous species that was devastating native wildlife and capable of eating a curious child. Buried in the
underbrush, pythons would have been yet another exotic treasure to
unearth. Indeed, widespread inculcation of just such an ethos is part of
Florida's dirty secret. It's more than likely that many of the reptile dealers and amateur herp aficionados combing the state's lush back lots
for exotic animals—and in some cases seeding the area with them—
thought this: it's cool to have pythons here.

Of course that didn't explain why people and agencies that should
have been concerned largely hadn't been. But maybe this did: in a state
where gators swallow golfers, hurricanes swallow towns, and hanging
chads on presidential ballots swallow democracy, what was another
beckoning maw?

Gathered around Lowman's large table after dinner, politicos and biologists argued long into the night, both sides surprisingly informed. The
university types led respected research programs on various aspects of herpetology; several had authored papers on invasive species. A representative
from the Florida Wildlife Commission once worked in the increasingly

suspect pet trade. Even the Sarasota County commissioner was an amateur herpetologist who'd spent his formative years crawling through scrub after the *ne plus ultra* of Florida serpents—the preternaturally beautiful and now vanishingly rare indigo snake. As the commentary circled, each decried the accelerating disappearance of native species under a rising tide of aliens. Likewise, each had no idea what to do about it. Which was cheaper and more efficient in stopping new introductions—public education or law enforcement? How to deal with the animals already present—kill them outright or simply try to stop them from breeding? Much discussion centered on erecting an actionable definition of "invasive": in order to fund a fight against a nonnative species, should it pose a threat to human health, the economy, or an ecosystem?

Then it got interesting.

"I don't think the python is our *biggest* problem," stated one, missing his own pun.

"Right," said someone else, "my vote goes to the Cuban treefrog." (Picture an arboreal bullfrog gobbling up unsuspecting native frogs, lizards, and birds.)

"But," countered another, "Cuban treefrog and brown anole [a finger-length invasive lizard] have both been here a century, so they must've achieved *some* level of co-existence with native species by now. What about the Nile monitor?"

One of the world's largest lizards, this two-meter eating machine was afraid of nothing and known to be breeding in at least two places in South Florida. Everyone nodded. Silence.

"Well . . ." one biologist finally offered, a bit too quietly, "I think *Ctenosaura* could be the biggest disaster of all; it might just be on a few islands now, but if it spreads inland, it's game over. For everything."

I had no idea what *Ctenosaura* was, but after what I'd see the following afternoon, I was prone to agree.

## Save Florida, Eat an Iguana

We were hiking a trail in Blind Pass State Park on Manasota Key just north of Fort Meyers, the ground a crunchy matrix of dried palmetto, sand, broken shells, and beer-bottle glass. Shuffling behind professional trapper George Cera—adorned in a floppy-brimmed hat and carrying a fishing rod dangling a rubber-frog lure—we resembled a press corps shepherded by Huckleberry Finn. Suddenly, there was movement to the left. I turned to see a large gopher tortoise *running* (I kid you not) across the sand toward its burrow. No one else paid the slightest attention; they were too busy listening to Cera describe how aggressive black spiny-tailed iguanas are.

To start, they heard, *Ctenosaura similis,* a Central American native, had never been as common in the pet trade as the familiar—and coincidentally also invasive—green iguana (*Iguana iguana*) because it was considered mean, toothy, and untamable (relative terms when it comes to reptiles). Much like cats, black spiny-tailed were also pugnacious, predatory hunters that enjoyed chasing things down. The species' typically insectivorous hatchlings were also happy to subsist on invasive brown anoles (a small mercy), and though they switched to more vegetative pursuits in adulthood, they remained functional omnivores that consumed anything that fit in their mouths (which explained their local notoriety for dumpster diving).

With his audience now rapt, Cera further dished on how he tracked *Ctenosaura* by their guano (poo), footprints, tail marks in the sand, and trails of chewed-up sea grape and hibiscus (their favorite food—which everyone had yards full of). We were twenty meters off a thick palmetto grove when, eyes to the ground, Cera held up his hand to halt the group. Absurdly, he cast his rubber frog toward the edge of the scrub then furiously wound on the reel. The bait bounced awkwardly across sandy duff. Because it was both crude and cartoonish, we were all stunned and amazed when a perfectly camouflaged, dachshund-sized lizard materialized from the gloom in a ground-hugging blur to attack the frog mid-bounce, snapping its jaws around the lure before spitting it out and dashing back to cover.

"*That's* how aggressive spiny-tailed iguanas are," Cera nodded smugly. A pudgy, wide-eyed reporter from a local paper wheezed out a long, low "*Jeeeezuz H. . . .*"

It wasn't clear whether the lizard spat the lure because it didn't like the taste, or because it already knew what a rubber frog in the mouth—or a floppy hat at the end of a fishing rod—meant. Whether casting rubber frogs across the ground or fluffy imitation birdlings into a tree, Cera dispatched any spiny-taileds foolish enough to follow with the gun he carried. Believing the lizards to be smart enough to recognize him, Cera often changed clothes, or sometimes just a hat, hoping to get close enough to blast what was listed—with a clocked speed of 34.6 kph—as the world's fastest lizard by the *Guinness Book of World Records.* The odd fact that spiny-taileds were spooked by people but unafraid of vehicles led to wildlife officers cruising neighborhoods with fishing rods, casting into people's yards from their windows, a sight so bizarre you could only wonder how such a circus got started.

Like most invasive species forensics, it took investigation by both biologists and local historians to work out the precise emigration-immigration scenario for the black spiny-tailed's Florida sojourn. In the end, the story turned out to be pretty typical.

From a mere three *Ctenosaura* loosed by a boat captain near the Boca Grande lighthouse at the south end of Gasparilla Island in 1979, the lizard spread north through densely vegetated areas, multiplying obnoxiously and island-hopping along the Boca Grande causeway to become the chain's dominant predator. By 2006, the area was almost devoid of other lizards; most birds that should have been present weren't; and the only snakes left were old—that is, the largest and most wary. It was, by all measures, an ecosystem in collapse, yet what really goaded the mostly affluent residents was the extensive damage wrought on their precious landscaping and gardens. In Lee County, which covered the southern 75 percent of Gasparilla, including the community of Boca Grande, a levy was raised to hire someone to address the problem; the resulting tender urged bidders to take "a business perspective." Cera proposed a per animal tariff, won the contract, and began work in 2007.

"A contract is a great motivator," he explained of the 13,000 feral *Ctenosaura* that he claims to have removed so far at twenty dollars a head. Indeed. More so when you consider the estimated 20,000 or so remaining.

Still, with a certain noblesse oblige, Cera was willing to share the burden. The work inspired him to pen *Save Florida Eat an Iguana: The Iguana Cookbook,* containing recipes for curried iguana, smoked iguana pizza, and iguana tacos. Though professing an overriding respect and affinity for nature ("there are no *bad* animals," he writes, which could apply equally to taste or behavior), Cera also rationalizes why ecosystem-destroying invasives like *Ctenosaura* should be whacked wherever and whenever possible. His arguments sound convincing—and delicious.

At this point, with what I felt was obvious tongue-in-cheek, I wondered aloud to the group why someone didn't just release a few Burmese pythons on Gasparilla to take care of the iguana problem. No one laughed.

Although his diligence had had an impact, and small animals seemed to be returning to the area, Cera remained a worried, self-appointed guardian. An example: after keeping an eye on a woodpecker nest high up a hollow palm, determined not to let any spiny-taileds near, he'd one day discovered the back end of an enormous male hanging from the lofty cavity, merrily scarfing woodpecker hatchlings. Cera has found everything you can imagine in *Ctenosaura* stomachs—including traces of juvenile gopher tortoise, a highly endangered (and, apparently, fleet-footed) species crucial to Florida's ecology. Like Lowman's dinner guest, Cera asserted his belief that *Ctenosaura*'s voracity posed an enormous threat. And, like another guest, one of our group retorted that the equally aggressive but far larger Nile monitor (*Varanus niloticus*) possessed even greater potential for destruction. Now found regularly on nearby Cape Coral, a first specimen some 1.5-meters long recently turned up near Boca Grande. Unlike black spiny-tailed iguanas, Nile monitors are relentlessly carnivorous throughout their long, large lives. And they positively *love* eggs, making hasty omelets of native bird, snake, lizard, and turtle nests—thus demonstrating their potential to impact fragile and recovering sea-turtle populations.

During a Lowman-organized invasive herps workshop the next day, Todd Campbell of the University of Tampa described how he once believed people would get serious about Cape Coral's Nile monitor problem when the first burrowing owl nest was wiped out; that had now happened, he opined, with no spike in concern. Thus Campbell felt strongly that native reptiles and ground-nesting birds—anything "stupid enough" to burrow or lay eggs where there were Nile monitors—were doomed. He then chastised those gathered that he was frankly tired of giving this talk, but would keep doing so "until you all listen." Five years later, he'd still be giving it.

That day on Manasota Key, however, Cera offered a measured response to the reporter's query, noting that *Ctenosaura*'s shockingly large populations and arboreal talent helped it to spread more easily than monitors, which were really too large to hide above ground.

While the discussion continued, I found myself tracking an aggressive creature of a different sort. Crawling up my shirt, stinging angrily at the fabric, was a massive, nasty, red, yellow, and black wasp. Fascinated by its size and rabid actions, I reacted with a swipe of my hand only as it neared the exposed skin of my neck. It hit the ground twitching, furiously still trying to sting. It was the length and thickness of a cocktail weenie.

"Whatever *that* is," said a student beside me, "it's not native."

Right. But along with hundreds of other insects; a litany of diseases; dozens of invasive fish, birds, and mammals; almost 2,000 plant species; *Ctenosaura*; and some fifty other reptiles and amphibians including the Burmese python, it was here. In all likelihood to stay.

Invasion biologists often reference a much-ballyhooed "tens rule" that goes like this: for every ten species transported outside their range, approximately one will be introduced to a new region; for every ten introduced, one will establish itself; and for every ten that are established, one goes on to be a problem (you could just cut to the chase and say that for every thousand nonnative species imports, approximately one becomes invasive). Applied to the myriad nonnatives successfully established in Florida, this much could be deduced: if the current catalogue was indeed the tip of the spear, then its base—the actual number of nonnative species continually being introduced to the state—was colossal. Which made something else certain: more were on the way. As one Fish and Wildlife manager at Canopy Meg's invasive herps workshop put it: "If it has *ever* been in the pet trade—and what hasn't—it has been introduced to Florida."

## The Blame Game

Still, the big question lingered. In Florida parlance: *Where the heyall did all them goddamn pie-thons come from?* Though theories abounded, the primary explanation was at odds with the numbers: could this plague of behemoths really have been clumsily introduced solely by uninformed hobbyists whose pets outgrew their owners' capacity to deal with them or, apparently, to think clearly?

Recall that from the mid-1990s through 2003, Everglades officials removed around fifty pythons, but in 2004 alone, they turned up sixty-one. Until then, biologists had logically dismissed them as discarded pets; after all, almost 100,000 Burmese pythons were legally imported to the United States in the period 1996–2006, and further bred by the thousands for the domestic pet trade. But when hatchlings were found deep within the park that same year, authorities had to admit that the adaptable species appeared to be firmly established. In January 2003, aghast tourists witnessed the first python-alligator battle from a raised platform at the Royal Palm Visitor Center (the gator won). Epic tilts between the two apex predators became more common. The most infamous involved a four-meter python that burst open after swallowing a two-meter alligator; officials came across the grisly scene during a helicopter survey and a graphic picture of the mess went viral on the Internet. Suddenly, eradicating pythons from the Everglades seemed as intractable as the park was vast. Authorities doubled their efforts and even trained a snake-sniffing beagle—Python Pete—to help out. But python removal in and around the park continued to accelerate: 248 in 2007; 343 in 2008; 367 in 2009. Numbers decreased to 322 in 2010 after a severe January freeze and over the following two annum dropped to 169 and 152, respectively, bouncing back up over 200 by 2014. Though the rebound may have been due to increased removal effort in particular areas, it was clear that the python population was large and well disseminated in South Florida. What the hell *was* going on?

Global trade and travel have spread the problem of introduced species everywhere, but Florida—with its subtropical climate, mix of terrestrial and aquatic habitats, busy ports, and high human immigration—is particularly vulnerable. Escapees from individual pet owners, ship cargo containers, parks and attractions (*Visit Reptile World! See mighty pythons and cobras! Buy pecan pies and fireworks!*), as well as exotic plant and animal farms, have all contributed. It's not so much the abandonment of unmanageable pets (like the large snakes left behind in foreclosed homes—true story), as much as it is escapees (snakes are notorious

Houdinis) and the purposely loosed animals of overburdened dealers. Take the example of Strictly Reptiles in Hollywood, Florida, the country's highest-volume reptile importer. It ran afoul of the law so brazenly, and on so many occasions, as to inspire two books whose titles pretty much say it all—Jennie Erin Smith's *Stolen World: A Tale of Reptiles, Smuggling, and Skullduggery* and Bryan Christy's *The Lizard King: The True Crimes and Passions of the World's Greatest Reptile Smugglers.* South Florida's herp cognoscenti all knew that the parking area and surrounding woods at Strictly Reptiles *brimmed* with exotic species that had escaped from it. The Fish and Wildlife guy at Lowman's conference even insisted that the Hollywood facility accounted for at least nine of Florida's world-leading fifty-six established nonnative herps. "I have nothing to do with law enforcement but the sloppiness of this place is astounding," he'd said. "Go around at night with a flashlight and you'll find *everything* coming out of it . . . kids do it to get free pets."

What was to be done? Florida Statute 372.265 makes it illegal to release an exotic animal, but it is unenforceable unless authorities actually *see* it happen. Burden of proof. Thus, at the time, no one had ever been prosecuted under the statute. And were they to be, punishment was only a maximum $1,000 fine and/or up to a year in jail if convicted—the legal equivalent of littering when what was happening was tantamount to ecological sabotage. Nevertheless, importers and members of the huge herp-husbandry industry were wary of any government interference and considered being questioned over their practices to be harassment; they'd even sued over it.

But if it's so easy for captive critters to get away, that's all the more reason for regulations and containment laws—like those enacted after Hurricane Andrew in 1992, when thousands of species of plants and animals housed and bred on the edge of the Everglades were literally blown all over South Florida. To this day government types point to Andrew as *the* cause of the Everglades python problem, saying the timing seems about right to account for the discovery and rapidly rising curve of captures, and that a known python-breeding facility in nearby Homestead was the likely source. Given the human penchant to embrace mythical catastrophes like the Great Flood as certain explanation, the Hurricane Andrew story has gained apocryphal traction, disseminated everywhere from the state legislature to reality TV shows like *Python Hunters.* Skip Snow's Popperian training, however, confined him to the skeptics camp.

"Andrew is an unlikely cause. Pythons found near Nine-Mile Pond in 1995 included hatchlings, mid-sized, and large snakes that don't fit a release profile. The state people know everyone is fingering their lax regulations and lack of enforcement on the pet trade, and so they basically like to point to Andrew in a throw-your-hands-up sort of way that says: *It's not our fault!* But here's the deal: whether an actual crime was committed or there was simply negligence, it's still on their hands."

## Due South

"You know about Poisonwood, don't you?"

Wildlife biologist Karen Garrod and I were taking our first tentative steps into the dense brush of Crocodile Lake National Wildlife Refuge on Key Largo to check on a python trap.

"No, actually." I'd learned my lesson with the fire ants.

"It burns and itches pretty bad, especially if you're sensitive to stuff like that. It's *this* but not *that*," she said, pointing at what looked like identical leaves. "It's shinier, see?"

Not really. Which led to a morning filled with complex ducking, pirouetting, and occasional pas de deux with Garrod while attempting to avoid anything with similar looking leaves—which, to my eyes, was almost everything.

Following the lengthening Burmese python trail out of the Everglades, I'd visited the Florida Keys in October of 2008. In a now-legendary discovery, Key Largo's first python was found in April 2007 by a visiting Scottish biologist working on the highly endangered Key Largo woodrat. Tracking a radio-collared rat, she instead turned up a python—whose stomach held the aforementioned rodent and its transmitter. As they'd done for Snow, alarm bells rang loudly for area ecologists and wildlife officers. After several more pythons turned up, potential effects on the woodrat as well as the endangered cotton mouse and American crocodile helped fully depress the panic button. As we plowed through the jungle-like forest, Garrod explained how the scenario was playing out.

Under the auspices of the Endangered Species Act and in partnership with the U.S. Geological Survey, the U.S. Fish and Wildlife Service had struck up a rapid response team at the end of January 2008. Garrod's experience made her a natural choice for inclusion. For two years she'd worked on a similar squad in Guam, where, with eradication deemed impossible, mouse-baited traps were used to control the troublesome invasive brown treesnake. Now, each day on Key Largo, Garrod checked thirty-three different traps along a stretch of County Road 905 between the two main north-south arteries into the Florida Keys—U.S. Route 1 and Card Sound Road. That effectively covered the area between the main overland gateways, but Burmese were also great swimmers, and the Everglades was less than twenty island-hopping kilometers away.

Much like Snow's traps, Garrod's had various configurations; unlike Snow's, all were baited with a large, live rat. Raccoons, possums, and even invasive green iguanas messed with the traps and sometimes got into them. Large black racers and yellow ratsnakes had been found coiled on or near traps, proof that the bait, at least, was effective in drawing rodent-loving serpents. So far, however, no pythons.

By the time trapping began, a total of eight pythons had been found on Key Largo—all roadkills save one beaten to death by an indignant woman who'd found it eating a dove in her backyard. Did these isolated cases represent individuals working their way down from the Everglades, with perhaps a smattering of released animals in the mix? Likely not: in July 2008, several dead python hatchlings were found on the road. Alarm bells became sirens.

Crashing through brush in what seemed every compass direction, Garrod and I finally reached our goal. Trap 12 of the USGS Burmese Python Control Project sat lodged on the brown forest floor beneath sun-dappled greenery, meters from where the brackish waters of Barnes Sound lapped a tangle of mangrove roots that once hemmed every piece of emergent land in this archipelago. The trap's importance was its even closer proximity to an open, sunny area where several American crocodiles nested annually. At the right time, crocodile hatchlings were easy to find paddling around the adjacent lagoon. At the right time, even a small python could consume a good number of the highly endangered animals.

The vulnerability of this small ecosystem seemed even more dire than that of the Everglades; this time, I wasn't disappointed that we didn't find a python.

Given the conservation value of Key Largo's indigenous animal assets, the plan was to extend the trapping project over all seasons and life phases of the pythons, and with a full array of capture methods. Perhaps employing unbaited, drift-fence-style traps like those in the Everglades, or placing some along the north-south strands of U.S. 1 and Card Sound Road, enthused Garrod as we retraced our steps back to the road. Emblematic of every invasive-species worker I'd met, she was nothing if not committed, perhaps more so in having opportunity to hold the EDRR front on python spread after years spent in the flummoxing, quasi-efficacious control of brown treesnakes. Like George Cera and Skip Snow, Garrod also believes that there are no bad animals. Ecological restoration is her only concern.

"If I actually trapped a python I'd definitely be torn," she memorably remarked. "I'd be excited the trap worked, but I won't lie to you—I'd feel remorse for the animal."

And herein lay both the problem and—as many increasingly saw it—solution to Florida's invasive herp issues. "I don't like to kill *anything*," Garrod sighed, with world-weary finality, "but sometimes you gotta do what you gotta do."

A year later, in September 2009, a first python appeared in a Key Largo trap. When you're holding the line on an invasive species, it's usually only a matter of time.

## Breach

The north end of the city of St. Catharines, hugging the southern shore of Lake Ontario, features all the hallmarks of the explosive suburban growth seen throughout North America in the 1950s and 1960s: lavish Victorian lakeside manors once far removed from the urban center now stand hemmed by bungalows, dense social housing, strip malls, innumerable crosswalks, and more churches than make any sense. When the city's population doubled over those two decades to its present 130,000 souls, this swath of once-agricultural land annexed from a neighboring township had been a logical area into which to expand. On its eastern margin, the North End merges with Port Weller, a hardscrabble community pocked by vacant lots and bisected by the Welland Canal, the waterway that circumvented thundering, 100-meter-high Niagara Falls to connect Lake Ontario to Lake Erie in 1830, eventually opening the Upper Great Lakes to shipping from the Atlantic Ocean in concert with the St. Lawrence Seaway.

I'd driven the two hours south from Toronto just to see the Welland, a passage I'd known of my entire life but never visited. First impressions were that the presence of such an auspicious piece of North American history seemed impressively understated: in most places you didn't ease up to its edge with any architectural ceremony or anticipatory signage—the canal just appeared, rather suddenly, at the end of any number of nondescript streets. If you lived in this low-slung neighborhood where you could see over the houses in any direction, you might, in any given moment, have your eyes on the world around you—a road-hockey game, the yard you were raking, the shoe you were tying—when out of nowhere a ship ten stories tall and as long as three football fields could slide freakishly through your vision, slipping stealthily along at six knots, about twice walking speed.

From a water treatment plant near the shipping entrance to Lake Ontario, you could follow the Welland Canals Parkway or parallel multi-use trails south toward Lake Erie along the canal's 43.4-km route. At the Twin Flight locks in Thorold, where the behemoths called "lakers" are hoisted or lowered over the bulk of the Niagara Escarpment—a 450-million-year-old limestone formation from which the eponymous falls tumble some 10 km to the east—tight passages and crumbling concrete speak volumes about aging infrastructure and restrictions on the ship traffic still using it. History is everywhere, from a tourist information center to murals on antiquated buildings depicting scenes from the canal's earlier incarnations. Given the Welland's celebration as a feat of engineering when it first opened, I could well imagine the swollen pride of both workers and community. What I couldn't imagine was how those same long-gone folks would feel if they knew their marvelous accomplishment had become a dark, silent juggernaut of environmental change, facilitating the exchange of everything from the pollution dumped in the lakes for generations, to the invasive plants lining its banks, to the myriad ballast demons inadvertently imported from other continents, to the nonnative fishes that would swim in from both directions as the lock doors closed behind them. Economists would tell you it had all been worth it for the nascent financial systems of the United States and Canada. Ecologists might argue otherwise—and they would need to cite only one example.

## Irony of the Ancient Kind

Back in the day—and I mean *way* back, at the end of the Devonian period some 360 million years before present (mybp)—the good times were winding down for the Agnathans. Having ruled and roamed the seas for some 150 million years, these jawless fishes had put in a decent shift as the first major radiation of early vertebrates. For the past 50 million years or so, however, evolution, as it tends to, had been tinkering with a new model—fishes with jaws. Would these things work any better? So many moving parts, so many new accessories, so much energy required . . . oh, but look—way more food options!

Indeed jawed fishes proved up to the task and built to last, giving rise to elasmobranchs (cartilaginous fishes like sharks and rays), teleosts (bony fishes; the largest, most speciose group of vertebrates on the planet), and the lobe-finned fishes that led onward to amphibians, reptiles, birds, and mammals. The fossil record suggests that as jawed fishes diversified through the Devonian, they thoroughly out-competed and likely consumed their mostly dead-end jawless predecessors. I say "mostly" because a few clearly slipped through the cracks of time. Though Agnathans disappear from the fossil record after the Devonian, descendants of one or more lineages can be seen in two living jawless forms: hagfishes and lampreys.

Hagfishes are the ultimate bottom-dwellers: eel-shaped, slime-producing mud-burrowers that feed on marine worms and dead and dying fish. Because mud and janitorial scavenging never get old in the ocean, hagfish have remained employed, empowered, and unchanged through 300 million years and four major extinction events. Ditto their also-eelish cousin, the lamprey.

Larger, pelagic, and more diverse, with thirty-five of thirty-nine species in a single Northern Hemisphere family, all lamprey undergo a radical metamorphosis from larvae to adult. The larvae—or *ammocoetes*—filter-feed in freshwater habitats, while the adults vary considerably in their mode of life: some are parasitic and anadromous (ocean-dwelling adults that migrate up rivers to spawn), while others are parasitic or nonparasitic (that is, nonfeeding, living off their own body fat and muscle for their remaining lifespan) but restricted to freshwater. General awareness of lampreys, however, centers on the tabloid-esque nature of those species that are ecto-parasitic on large, predatory fishes: latching on with a sucker-like mouth, such lamprey employ a circular, toothed tongue to rasp away scales, exposing flesh into which they secrete an anticoagulant to loosen the bodily fluids they feed on. Not pretty, but a decent strategy for circumventing the challenges of jawlessness whilst turning your tree-of-life nemesis into a steady food supply. You might call today's lamprey sweet revenge for the Devonian smackdown of Agnathans. Paleontologists call them living fossils whose bauplan (body plan), identical to that of a 360-million-year-old South African imprint, confirms that these highly specialized creatures are primitive holdovers of ancient marine ecosystems that have been making jawed fishes miserable since Pangaea was just a baby. Thus, loathsome as their habits appear, you have to respect lampreys' staying power. You also have to respect their adaptability and opportunism, qualities that helped one of their rank accomplish the ultimate environmental irony: the sea lamprey singlehandedly took down the greatest freshwater ecosystem in the world.

Sea lamprey (*Petromyzon marinus*) is native to the North American Atlantic seaboard. Like most anadromous fishes, however, were it to find itself in a lake big enough to stand in for an ocean, it had no problem shifting allegiance. The Great Lakes not only fit that bill, but came fortuitously stocked with naïve forage in the form of lake trout and lake whitefish, with the additional gravy of Atlantic salmon in Lake Ontario. Although the first sea lamprey was positively identified from Lake Ontario as early as 1835, its status at that time remains a subject of debate. Was it native to the lake, arriving early via the post-Pleistocene Champlain Sea, a marine inundation of eastern North America that once occupied the current St. Lawrence River valley? Or did it gain access only when the Erie Canal connected the Atlantic to Lake Ontario through the Hudson-Mohawk river systems? Regardless, the sea lamprey's ingress to Lake Ontario mattered little: the insurmountable barrier of Niagara Falls had ensured no farther wanderings. The Welland Canal would change all that.

Two early iterations of the canal used diverted stream water from a source higher than Lake Erie to fill the locks, but a third iteration altered the system to include Erie water. When a fourth and final version opened in 1919, cutting off diversion contributions entirely so that water flowed downstream directly from Erie to Ontario, it was party time for upstream-seeking anadromous fishes. Within two years sea lamprey were all over Lake Erie, perhaps unsurprising given that the species was already out of control in Lake Ontario, where the ecological dominoes had been falling for years.

Despite ideal conditions provided by Lake Ontario, the sea lamprey's presence there had been balanced during the mid-1800s by the fact that the clear streams it sought for spawning needed to be warmer than most of those available. In the latter part of the century, however, a shift occurred that proved fortuitous: land-clearing and settlement around Lake Ontario caused temperatures to rise in many of the streams flowing into it. Spawning opportunities for lamprey increased by orders of magnitude and their numbers followed; those same rising temperatures, however, reduced stream-spawning habitat for fish that required colder water—Atlantic salmon, lake trout, and lake whitefish. Already severely overfished, all three endured significant depredation by lamprey, which caused their populations to plummet. The rest is schoolbook history, but William Ashworth's 1986 telling in *The Late, Great Lakes,* sums it well: "By 1900 the [Lake] Ontario population of the Atlantic salmon was extinct, and trout and whitefish numbers in the lake were declining rapidly . . . The lampreys now had a growing population and a declining food source—the classic conditions for the colonization of new areas. All they needed was a place to colonize, and the . . . Welland Canal gave them that. Suddenly a lamprey-path had been provided into the fertile fields of the upper Lakes . . . as though a gate had been opened between a pasture containing sheep and another containing wolves.

It would take time, but the upper Lakes' fisheries were, by the opening of that gate, effectively doomed."

By 1936 sea lamprey had reached Lakes Huron and Michigan. Here, after journeying so long through time and space, our Devonian voyageur found its promised land. Populations exploded, and by 1948 lamprey were so abundant that they roiled inshore waters, raiding fishing nets and clinging to boats. The lakes' main fisheries experienced 90 to 99 percent declines over a decade beginning in the 1940s: for example, lake trout landings in Lake Michigan dropped from 7 million pounds annually to less than four thousand pounds; and lake whitefish slumped from 12 million pounds annually across the five basins to under a million pounds.

Granted, the sea lamprey's Great Lakes conquest was in no small part aided by a now completely unstable aquatic ecosystem. Ham-fisted attempts to stock Japanese salmon in Lake Michigan and support them with smelt, an anadromous baitfish, resulted not in the desired sport-fishery, but in a smelt population whose economic and cultural impact peaked with landings of 10 million pounds a year and festive "smelt carnivals" in towns all over Michigan and Ontario that celebrated its spring spawning runs. Meanwhile, common carp, intentionally seeded around the lakes, were now numerous enough to impact shoreline spawning habits for many indigenous species, including whitefish. But an even bigger shock awaited.

Like sea lamprey, alewife (*Alosa pseudoharengus*), a small oceanic member of the herring family, may or may not have already been native to Lake Ontario (recently, bones thought to be from alewife were found in a five-hundred-year-old native midden). Notwithstanding, it was quick to follow the former into the upper lakes through the Welland Canal. With native top predators now devastated by a combination of overfishing and sea lamprey, alewife quickly reached absurd abundance, with peaks in the 1950s and 1980s. These boom-and-bust cycles also had an annual component, because the species tends to exhibit seasonal die-offs. My childhood memories include walks along Toronto's Kew Beach boardwalk with my mother and aunt, the putrid stench from windrows of washed-up alewife sharp in my tender nostrils. Through the 1960s this nuisance and health hazard was ubiquitous around Lakes Ontario,

Erie, Huron, and Michigan. In 1967, a blue-green algae bloom off Gary, Indiana, killed billions of Lake Michigan alewife and caused a near state of emergency in Chicago: fly-infested piles of dead alewife two meters tall along the waterfront were shuttled by truckload to dumps dedicated to the rotting biomass; the entire city was under a boil-water advisory; and the north end had no drinking water due to alewife carcasses clogging intake pipes. With a public consensus around the lakes that something needed to be done, fisheries types again looked to Pacific salmonids to save the day, in short order introducing coho, then later Chinook, to take care of business. What to this day represents the world's largest bio-engineering/biocontrol project did a decent job of bringing alewife down to manageable numbers while establishing a salmon/alewife fish community that sent millions of much-needed dollars flowing into municipal, state, and provincial coffers from recreational sportfishing—a situation sustained only by expensive and concurrent lamprey control then occurring around the Great Lakes.

Alewife, however, also raised a different kind of stink: as a predator on native and nonnative zooplankton and larval fishes, it was soon implicated in the food-web restructuring behind declines in numerous native species. Furthermore, the "alewife resource" itself would eventually be impacted by invasive zebra and quagga mussels that concentrated nutrients in the lake bottom. The mussels' filtering activities not only removed plankton and particulates, but also made the water clearer. This, in turn, encouraged growth of the bottom-carpeting algae *Cladophora*, a dead-end for nutrients because it wasn't consumed by anything; it led invasive round goby to eat the now-more-visible fish eggs; and it coincided with the inexplicable disappearance of the shrimp-like *Diporeia*— which cycled nutrients from the bottom to the mid-water pelagic food chain that fed both whitefish and alewife.

With this black synergy, it's no surprise that alewife is virtually gone from Lake Huron and currently in sharp decline elsewhere, along with its salmon predators (Huron's Chinook crashed in 2010), victims not only of these same forces, but also of predation pressure exerted on the alewife from their own successful breeding in the wild.

Lamprey, alewife, smelt, common carp, Pacific salmonids, zebra mussel, round goby ... the state of the Great Lakes' aquatic ecosystem could hardly be described as unstable: rather, it was pure ecological mayhem.

## Finger in the Dike

In our talks at CCIW, Mandrak spoke often of the Welland Canal's crucial role in the dispersal of sea lamprey and alewife into the Upper Great Lakes. During one, on a frigid December day, Mandrak says he has a project going under the aegis of the Asian Carp Program that he wants me to see. He grabs his parka off the back of the door and down we go into the concrete labyrinth, through a cavernous open space where experimental flumes once gushed, past a wet lab where live animals are maintained for various purposes, past the regional CITES enforcement office (tellingly, the lights are out), through a door, and past the Canadian Hydrographic Service where an upbeat staff is celebrating its recent find of HMS *Erebus,* long-lost flagship of Sir John Franklin's ill-fated 1845 expedition to discover the Northwest Passage. Eventually we step outside onto an open dock area next to a large boat slip, across whose mouth is strung a tight, dense net. The dark waters transmit nothing of its purpose, and a stiff wind off Hamilton Harbour stings my eyes.

"There's a hundred-plus common carp in there, all tagged for 3D real-time positioning," motions Mandrak. Meaning that someone, somewhere, could sit down and enjoy a monitor screen constellated with carp blips.

This mesocosm experiment was part of a larger project to track and monitor the movement of tagged fishes in the Welland Canal. In that effort, currently under way, they'd surgically implanted transmitters in fish, and arrayed numerous receivers along the canal's length and ingresses. Because the canal was drained for maintenance January through March, they'd used tags they could turn off remotely for those months. The question was this: how many and what type of fish used the canal as a means of dispersal between the Ontario and Erie basins? Despite abundant global evidence of canals' role in facilitating the dispersal of aquatic invasive species, and obvious examples like sea lamprey and alewife, few studies had looked at direct movements through them.

In 2012, seventy-nine individuals of seven fish species were captured, tagged, and released at three separate sites in the canal. In 2013 an
additional one hundred individuals (ten species; two release sites) and
many more receivers were added to the study. The main species utilized
was common carp, considered the best proxy for Asian carps. Overall,
almost a million hits from these original 179 individuals were detected
along the length of the canal over the two years. Of these, only a handful
were found exiting from one end or another, but with preliminary data
from 2014 that included fish *reentering* the canal from both ends, they
had an answer: fish were indeed moving through the canal. The next
stage in that research lies at our feet.

"The second question and the focus of this mesocosm experiment
is this: if we wanted to, could we *prevent* movement through canals?"
says Mandrak.

The study slip was cut off from the harbor and the tagged carp
released into it. A barrier was installed in the middle through which
carp must pass in order to access feeding sites at either end. This barrier
zone could be used to test deterrents to movement that included electricity, pressure devices, low-frequency sonar (sonic boom plates), bub-
bles (air curtains), carbon dioxide, underwater speakers, and alarm
pheromones. The mesocosm was fitted with sensors that tracked fish
movements and responses to barrier deterrents in three dimensions.
After exploring each in turn, the plan was to next test combinations
of several measures at once—a practice known as "integrated pest
management"—to see if any of these were more robust than a lone
measure.

"What we hope to find out is whether we can prevent movement
through a canal in both directions, and, if we *can* and spend the money
on it, then, is there any way a fish could somehow circumvent those
preventions," Mandrak summarizes.

"You mean, as in hitching a ride with a ship or dispersal by fishermen?"

"No, as in could they go over Niagara Falls. It would be a tough
drop to survive but even a few people have done it. So right now we're
testing for directional gene flow between river-resident populations of
fish above and below the falls to see if any of this is happening."

Of course. Though the idea of gene flow over Niagara Falls left me with the absurd thought of a few barrel daredevils of yore surviving the plunge only to hook up in the swirling mists to found a colony of suicidal thrill-seekers at the bottom.

## Canalization

It would be remiss to discuss canals without mentioning granddaddies like the Panama and Suez. And not because these represent landmark engineering accomplishments or have significantly impacted history by changing the global flow of commerce. While not wholly unrelated, the importance here of these two in particular is their contribution to the movement of nonnative organisms from one ocean to another. Both were longstanding transit routes for hitchhikers carried in ballast or as fouling on ship hulls, and both were about to become virtual superhighways for invasive species.

The Panama system, which currently operates two parallel sets of locks to connect the Pacific Ocean with the Atlantic through the Gulf of Mexico, gained a third set with much wider and deeper channels in June 2016. Thus, the world's largest ships, many coming from Asia and currently forced to dock along the west coast of North America to unload, can now just steam on through the Isthmus of Panama directly to Gulf and east coast ports. Up to 25 percent of west coast ship traffic could be diverted this way according to a recent study, and that means many things. Some goods might cost less as a result (though I wouldn't trust importers to pass on any savings), but one thing was almost certain to cost more: surveillance for, and response to, a manifold increase in alien species. Estimates from the same study showed that not only would ballast discharge eventually increase by almost 80 percent at some ports, but the near tripling of "wetted surface areas" delivered by higher traffic in behemoth ships would also deliver a concomitant tripling of hull-fouling organisms and opportunities for introduction. Consider the U.S. Department of Commerce calculation of forty thousand gallons of foreign ballast currently dumped into U.S. waters *every minute,* and this increase is obviously highly significant.

It's a simple equation: the more exposure a port has to alien species, the higher the risk of invasion. Just ask California, whose centuries-long Pacific trade activity has landed it more marine invaders than any other region in North America: San Francisco Harbor, possibly the most invaded marine ecosystem in the world (surpassing even the Great Lakes; Figure 4), is, as a consequence, also the most studied. The Caribbean and U.S. Atlantic seaboards likewise feature no shortage of invasive species that found shortcuts from the Pacific through the Panama Canal. Risk will vary from port to port depending on the cargo and type of ship, but identifying hotspots for certain types of invasions could help focus prevention and EDRR efforts.

None of these problems, of course, is unique to the century-old Panama Canal—which gives us a logical segue to the elder Suez, completed in 1869 to cut the sailing distance from London to India by some 5,000 km. Linking the Mediterranean Sea (an offshoot of the Atlantic Ocean) to the Red Sea (an offshoot of the Indian Ocean), and already hosting 10 percent of the world's shipping, the Suez was also undergoing a major expansion that would double its size. Unfortunately, this was proceeding without a hint of environmental review, garnering the attention of invasion biologists. In a 2015 letter to the editor in the journal *Biological Invasions,* eighteen scientists decried the lack of environmental oversight, noting that the venerable canal was already one of the world's "most potent mechanisms and corridors for invasions by marine species." Fully half of the Mediterranean's 700 nonnative species were of Red Sea origin, and with the Mediterranean Sea warming due to climate change it had become ever more hospitable to tropical invaders. In fact, the waters the Romans called *mare nostrum*—"our sea"—would likely be all but unrecognizable to their first real student, Aristotle. "We are playing Russian roulette, not with a bay or a river, but with the entire Mediterranean," said lead author Bella S. Galil of Israel's National Institute of Oceanography in a news report.

To make points both practical and political, the authors offered examples of Red Sea natives that have become problematic in the eastern Mediterranean: the nomadic sea jelly (*Rhopilema nomadica*) gathers in stinging swarms as long as 100 km each summer, halting fishing, impacting tourism, and blocking intake pipes at desalinization and power

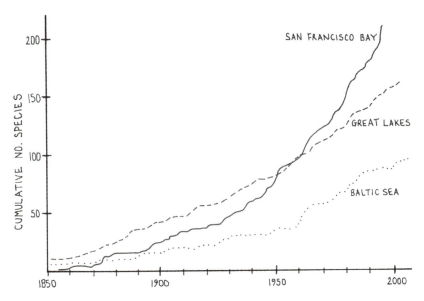

Figure 4. Cumulative number of invasive species discovered over time in select aquatic ecosystems. (Adapted with permission from Lodge et al. 2009.)

plants; the aggressive, highly toxic, silver-stripe puffer fish (*Lagocephalus sceleratus*), first spotted in 2003, was, but a decade later, among the region's most abundant fish by weight, encouraging consumption by people unfamiliar with its toxin and leading to widespread deaths in Egypt, Greece, Turkey, Israel, and Lebanon; off Turkey, two species of rabbit fish (genus *Siganus*) have virtually collapsed the Mediterranean's complex food web by clear-cutting seaweed forests, the native habitat for a multitude of species including the invertebrate prey of carnivorous fishes.

As the letter intones, the usual first step in a construction project of this size is to perform an environmental impact assessment, followed by a risk analysis and evaluation of measures to mitigate any harm. Furthermore, the United Nations oversaw three treaties—the Convention on Biological Diversity, the Mediterranean Action Plan, and the Division for Ocean Affairs and the Law of the Sea—with aegis over activities that affected the health of Mediterranean ecosystems, including invasive species. Though party to all three, Egypt appeared to ignore that fact—as did the United Nations, which seemed wholly uninterested in exerting pressure on Egypt to comply. The authors summarized marine scientists' consensus frustration over this absurdity: "Yet, despite these well-meaning international conventions, and a century worth of scientific publications documenting the spread and impact of [nuisance invasive species] introduced through the Suez Canal to the Mediterranean Sea—not to mention a vast literature that speaks to [their] staggering economic, cultural, and environmental impacts … we are faced with a seeming *fait accompli*. While global trade and shipping are vital to society, the existing international agreements also recognize the urgent need for sustainable practices that minimize unwanted impacts and long term consequences. It is not too late … to honor their obligations and urge a regionally-supervised, far-reaching 'environmental impact assessment' (including innovative risk management options) that would curtail, if not prevent, an entirely new twenty-first century wave of invasions through a next-generation Suez Canal."

It wasn't that solutions weren't within reach, as Mandrak's mesocosm showed: electricity, bubble walls, and low-frequency sound emissions were all known to deter animals, as were salinity barriers, something the Suez once featured in the form of highly saline central lakes, and that might now be reestablished by inserting a single high-salinity lock. Though projected revenues and the billions of bucks behind the Suez expansion suggest that the cost of an environmental review, at least, would barely register on the ledger, it remained to be seen if this would happen.

While other corridors—a new Pacific-Atlantic canal through Nicaragua, for instance, or the now ice-free Northeast Passage along the Russian Arctic coast—will undoubtedly accelerate homogenization of the world's flora and fauna if they receive no oversight, the sheer volume of ship traffic in the Panama and Suez Canals, and the opportunity presented by their expansions, make them prime locations for new technologies to be researched, developed, and instituted to minimize or eliminate the movement of alien species.

Of course, that doesn't address the problem of ballast water schlepped through them from one port to another. Growing awareness and regulations should help, but again, taking only the U.S. example, the vast volumes of ballast involved preclude any comprehensive prevention.

At least the International Maritime Organization—a specialized agency of the United Nations responsible for regulating global shipping—maintained a Ten Most Unwanted list of ballast-water villains: cholera bacteria, the spiny water flea, the Chinese mitten crab, toxic red/brown/green algae, the round goby, the North American comb jelly, the North Pacific sea star, the zebra mussel, Asian kelp, and the European green crab have all been widely transplanted via ballast water, causing measurable ongoing health, ecological, and economic problems. One of these whose larvae are readily transported with ballast and whose meteoric spread is a true symptom of globalization is the Chinese mitten crab (*Eriocheir sinensis*), which has taken over estuaries in Europe, the United Kingdom, and more recently, North America. While this large, fecund, migrating crab with furry-looking claws remains a food delicacy in Asia, it severely modifies habitats outside its native range, contributing to local extirpation of native invertebrates. Its intensive burrowing causes serious erosion, and by stealing bait and feeding on netted or impounded fish (it has no problem moving over land), the crab costs fisheries and aquaculture industries millions each year. Control strategies like electrical screens have failed to prevent its movements. More pernicious is the danger it poses to health: it's a host for Oriental lung fluke, which uses mammals in the egg-laying stage of its complex life cycle. Humans are infected by eating raw or poorly cooked crab.

The globalization of a long list of organisms via ballast may increasingly be on our radar, but this has been under way longer than most of us can imagine. Mitten crabs may be everywhere now, but tropical fire ants (*Solenopsis geminata*) were likewise globalized centuries ago by hitching rides on Spanish galleons.

The tropical fire ant is a pioneer species, a generalist keystone predator that readily occupies urban ecosystems, and a well-known agricultural pest both within and outside its native range. In an elegant 2015 paper, a team set out to determine just how this insect became so widely distributed. Analyzing tropical fire ant genomes from 192 different locations, a distinct pattern of global infiltration emerged that was highly consistent with historical trade routes, suggesting that the Spanish first introduced the ant to Asia in the sixteenth century. The study identified southwestern Mexico as the most likely source for invasive populations, consistent with Spain's use of Acapulco as its major Pacific port. From 1565 to 1815, one or two Spanish galleons set sail annually from Acapulco to trade New World silver for Chinese silk, porcelain, and spices in Manila. Not only was this the first regular trade route across the Pacific, but by linking with other routes through the Orient, India, and Africa, a first effective complete globalization of commerce. The data documented the corresponding spread of *S. geminata* from Mexico via Manila to Taiwan, and from there, throughout the Old World. And it made perfect sense: "ships, particularly if they were going somewhere to pick up commerce, would fill their ballast with soil and then they would dump the soil out in a new port and replace it with cargo," said Andrew Suarez, one of the study's authors, in a press release. "They were unknowingly moving huge numbers of organisms in the ballast soil."

As Spanish influence widened, so did that of tropical-fire-ant stow-aways transported to port regions in Africa, India, China, Southeast Asia, Australia, and the Americas. Doubtless it wasn't the only ant to travel by sea, but as the first documented evidence of human-mediated introduction of a costly, highly invasive pest through a truly worldwide trade network, the ants' movement nevertheless demonstrates one of globalization's earliest ecological impacts. There would, of course, be more.

## Ants Go Marching

It was bucketing rain in Vancouver. Not unusual for late October, but a miserable time to be looking for ants as biologist Sean McCann and I zigzagged backstreets and wheeled through roundabouts en route to the Arbutus Corridor Community Gardens. Parking on a side street near the mothballed railroad right-of-way in which these urban breadbaskets squat, we crossed to its jigsaw of plots, still filled with yet-to-be-harvested squash and pumpkin, grapes and berries, and, of course, a hundred varieties of kale. Amid them stood vine and tomato trestles, all manner of staking and fencing, and amateur greenhouses fabricated from clear plastic sheeting stapled to, among other things, broken hockey sticks. The inventive bricolage belied the gardens' venerable age, which was more easily ascertained by fig trees that had passed through several human generations. Given such a laudable use of greenspace, it was hard to see the railroad company's much-publicized recent decision to oust the gardeners as anything more than punitive posturing aimed at securing a better land settlement from the city. We weren't, however, here to consider politics.

A capable researcher who'd worked on everything from mosquitoes to tropical raptors to black widow spiders, McCann now flipped on the switch of a well-honed search image. Making our way to a small patch of grass between gardens and tracks where a stack of cheap plastic patio chairs abutted an elevated fire pit, McCann slowed to make tight circles, eyes scanning the ground. Nothing. Probably raining too hard. Finally, he motioned me over to a large rock at the base of the fire pit, squatting to roll it toward him. Beneath the rock, moist soil was riven by a labyrinth of tiny trails swarmed by orange pin-pricks of life. I felt a red-hot needle on my bare leg before McCann even had time to say "*These* are European fire ants."

I just couldn't seem to learn.

I was surprised how easy it had been to find *Myrmica rubra,* the Lower Mainland's latest invasive-species horror story and a critter that had stolen, however briefly, local limelight from the powerboat-transported zebra mussel and mystery-origin snakehead fish. McCann, however, who'd spent the summer mapping the spread of this problematic pest (generally referred to as EFA), expressed surprise that we'd had to work so *hard*.

"Usually, they find you first. Basically, if you stand on a lawn where EFA are present they're eventually going to come on over and sting you," said McCann. "Of course, this isn't the same behavior as invasive RIFA [South American red imported fire ants] in the United States. While EFA will crawl on you and sting at random, RIFA communicates with colony mates so they can all sting simultaneously; they'll swarm up your leg and then, at a given signal, *Zap!*"

McCann grinned at what I assumed was evolution's genius in momentarily converting a bunch of tiny toxic sparks into a giant flaming torch. I was all-too-familiar with the Borg-like RIFA from my time with Skip Snow in Florida, but appreciated his explanation of what was behind that painful experience—as well as what I could expect from this latest indignity. "EFA stings are about as painful as those of a narrow-waisted wasp, not as bad as yellow jackets or bees, and don't form pustules like RIFA stings . . . still, they can itch for a week."

That proclivity, more than anything, made these interlopers a serious and growing concern wherever they'd established themselves. Which turned out to be a shockingly large area according to McCann's boss, Robert Higgins of Thompson Rivers University in Kamloops, the province's resident ant expert. I'd called him up to see what had kicked off the project.

"Let's see . . . where to start? Probably with residential property issues: if you can't enjoy your property any more, it's a problem," he'd told me. "So when people say 'can't'—as in *can't* use the backyard, *can't* garden, *can't* get rid of—I know they're probably talking about EFA. It simply drives people out."

With few in the country working on ants, from time to time Agriculture Canada would send Higgins specimens to identify. Knowing EFA to be well established in eastern parts of the country, and with no interprovincial restrictions on the transport of soils or plants, Higgins figured it was only a matter of time before he saw it on the west coast. That day arrived in 2010 when he received some suspect ants from North Vancouver. After his EFA identification received the usual quasi-alarmist media attention, folks both interested and concerned sent Higgins samples from all over southwestern British Columbia. Some weren't EFA, but many more were; in fact, the ant appeared present in over half of Vancouver's twenty-four municipalities, as well as communities in the adjacent Fraser Valley and across the Strait of Georgia on Vancouver Island.

Higgins experienced what it was like to "suddenly" discover an invasive species that was already established over an enormous geographic area. He also found himself standing smack dab in the middle of the Invasion Curve.

## The Strange Case of European Fire Ants

First discovered in the Americas in 1900 at the Harvard University arboretum, and known from as far away as Quebec by 1910, EFA seemed to pose no real threat in either place. A century on, however, it had become a serious problem in populous areas of Ontario and New England—in large measure because, with no visible external signs of a nest, people didn't know that EFA had moved in until someone was actually stung. By way of example, in the Toronto suburb of Richmond Hill, ants had rapidly and surreptitiously distributed themselves along a trail bordering a blackberry patch where people had long taken kids to pick berries in the fall. Responding to vibrations from the foot-traffic during harvest season, the ants emerged to investigate, running up children's legs to sting them.

There'd been virtually no sampling of urban ants in British Columbia prior to 2010, but the North Vancouver hit opened things up. Given its current distribution, Higgins reasoned that *M. rubra* had been there for over twenty years but was only now making itself known. Smaller and easier to hide than a python, yes, but how did something so pernicious fly under the radar for so long? "The species isn't that aggressive when it first occupies an area; the behavioral shift occurs when densities get too high and colonies are competing," said Higgins, who also believed that recent invasion by a more aggressive subpopulation of EFA could be responsible for Vancouver's "sudden" problems.

In any given area of establishment, the time required to achieve dense populations was generally five years; a long while in ant time but fast from a human horizon. In any event, a widespread behavioral shift had, apparently, occurred. With EFA able to nest in any soil with proximity to heavy vegetation where it could farm root aphids, the lush and forested environs of North Vancouver were naturally heavily infested. Elsewhere in the region a litany of ant-specific problems had popped up: the world-famous Van Dusen Botanical Gardens had been forced to close both a quiet meditation garden and a meadow where people usually put down blankets; in the pastoral reaches south of the Fraser River, unwary horses were being stung mercilessly; dog parks everywhere were infested and canines were being ravaged (causing some veterinarians unfamiliar with EFA-sting symptoms to wrongly prescribe antibiotic instead of a simple anti-inflammatory treatments, with potential long-term ramifications for promoting antibiotic-resistant bacteria); and finally, a growing number of commercial plant nurseries had become Petri dishes from which new EFA infestations were being exported daily.

Higgins hired McCann to run a mapping program in several municipalities. "One reason was to get a sense of distribution and see if they were close to high-value areas for management reasons," said Higgins. "We were also trying to identify natural features that might act as barriers to dispersal. For instance, we marked ants with paint to see how far they foraged."

The work also included conducting basic counts to get a solid sense of the invader's natural history in an exotic habitat. "We used apple-baiting," McCann related. "We put out chunks at every house along a transect through a residential area, then came back ninety minutes later to see what was on them."

While the study discovered only three previously unknown infestations, it found that each could cover several blocks. Though it was clear that EFA didn't play nice with other ants, easily displacing most indigenous species, further ecosystem effects weren't well understood. McCann found colony densities as high as four per square meter, with each grapefruit-sized nest averaging 2,000 workers and up to forty-five queens capable of establishing new colonies. But without a plexiglass ant-farm ordered from the back of a comic book to dump a nest into, how were such counts made?

"You find colonies by stamping on the ground or lifting debris to expose the chambers," McCann told me. "With a shovel you dig around and under them for a fair distance—the nest itself might be small, but as soon as a colony is disturbed the queens escape down tunnels and workers start moving brood. You plunk everything you've dug up into a bucket while shaking it constantly to keep the workers from climbing up the sides, then seal it, take it to the lab, and freeze it. Afterward, you simply sift soil to count ants and brood. A queen pupae, male pupae, or newly metamorphosed versions of either are considered 'breeding brood.' A colony produces hundreds of queens a year. Interestingly, I noticed that some colonies produce only queens, some only males, some both. Because queens control sex through unfertilized haploid [male] or fertilized diploid [female] eggs, what are the implications?"

For me, the implication was that it was too complex to think about, perhaps one reason that McCann, a journeyman biologist, had long avoided ants. Since finding his way to EFA, however, he'd warmed to the taxon considerably. "The social insect world is infinitely fascinating," he said. "Look at the phenomenon of adult-larval and adult-adult trophallaxsis."

I think he meant food-sharing.

Ecologically, EFA were similar to other members of the genus *Myrmica* with their shallow-soil nesting, multiple queens, and an affinity for aphids and scale insects—utilizing the honeydew these produced and protecting them on the plants on which they dwelt. Some particulars, however, were quite different. An "invasion syndrome" of sorts, possibly caused by genetic bottlenecking according to McCann, was often seen in ants that had to shift activities in exotic habitats. EFA, for instance, were much more tolerant of local conspecifics in invaded areas than in native habitat. Like the Argentine ant (*Linepithema humile*—another invader and favorite subject of Charles Elton; a colony in Victoria, British Columbia, might be the highest-latitude infestation known for the species), EFA also formed supercolonies or displayed supercolony-like behavior in which queens struck out with a group of workers to found new colonies near the main one, with workers shared or at least tolerated between them. Ants from a place called McDonald Beach were tolerated by North Vancouver ants, for instance, suggesting they might be related. "It's possible that remediation work done there involved ant-contaminated rootstock from a North Vancouver nursery," McCann posited.

This pointed to another utility of mapping. Knowing precisely how EFA was distributed across the region would greatly aid municipalities if they ever deigned to quarantine soil that had been moved significant distances. Mapping had also turned up an unexpectedly high abundance of a *second* European invasive ant, *Myrmica speciodes*. "It was originally described from Holland," said McCann. "The first Pacific Northwest colony was in Bellingham, Washington. We also found a third introduced species of unknown identity at Vancouver airport."

The strange case of EFA got curiouser and curiouser. Another apparent facet of invasion syndrome in EFA was a loss of the ability to fly. Both *M. speciodes* and the new airport ant (fittingly) were able to take wing, however, and in large numbers, which is likely the reason so many were found. Colony types are roughly divided into two: those requiring a queen and workers to become established are considered *dependent,* whereas those that can be founded by a solo queen are *independent.* It was easy to see how species with flying queens and independent colonies spread farther and more readily: it meant higher propagule pressure in surrounding areas as the core infestation spread—the same type of pressure exerted by seed rain from an ever-expanding plant invasion.

While EFA's increasing ability to spread was an issue, of more immediate concern were the implications for homeowners—chiefly, that EFA might be listed as a reportable pest affecting property values. This had already proved a source of conflict in North Vancouver, with property owners antsy over municipal EFA-mapping because they didn't want others to know what was going on. Neighbors, Higgins confided, were apparently suing each other.

"EFA has upscale real-estate preferences," he said. "One infested street might have an elderly couple who employ gardeners, never go in their yard, and don't worry about ants, living next to someone smuggling pesticides in from the United States, who in turn lives next to organic greenies horrified at that thought. Then there are those who think they know *who* introduced ants to the street, and others who blame whoever sent the ants in to be identified."

One resident tried to solve the problem by building a deck over the entire lawn. It wasn't a bad idea. More than simply rising above the fray, so to speak, shading the soil below would ensure that it didn't warm enough to house a colony, and a deck surface was no place for foraging (as you'll know if you've ever watched the pin-balling, slack-mandibled wanderings of ants on a deck; I usually end up feeling sorry for them and kick crumbs in their direction).

Naturally, I wonder what this all means. "It means we have no chance of eradicating them," McCann told me as we wandered the tracks beyond the Arbutus gardens after the rain ceased, spotting EFA right, left, and center on the rotting railroad ties. He stopped to finish the thought, pointing and pivoting as he spoke. "So *that* property, *that* lawn, *that* garden, and soon, *that* park, are essentially unusable."

No wonder municipalities were tight-lipped about infestations. It explained why I heard back from exactly none that I contacted: EFA might rival Japanese knotweed in its perceived ability to lower property values and taint development potential. Yet another reason to regulate and monitor the transport of soil and plants between jurisdictions.

Maybe we were here for politics after all. Or maybe wherever non-native species become established, politics is sure to follow.

## Morphospace, the Final Frontier

A few weeks after our final session at CCIW, I visit Mandrak in his academic seat, located at the terminus of a long, echo-inducing hallway in the science building on the University of Toronto's Scarborough campus. Lodged behind his graduate students' office space, the bare-concrete workplace is stark and high-ceilinged, a wall-length window the only feature obviating its tomb-like feel. The simplistic environs channel our previous discussion of brutalist architecture, which suddenly seems relevant for another reason: the utilitarian style that proliferated on university campuses during the 1960s reflects a time when we thought we'd categorized all sciences into neat little modules that could be pursued within their own undistracted realms. Nowadays, growing awareness of the true complexities and fundamental interactions of nature, the extensive crossover these engender among disciplines, and the range of anthropogenic impacts and emerging threats facing all, make such bleak rooms seem like chiseled bunkers whose occupants find themselves on an eternal, unfocused war footing.

Earlier in the day Mandrak took his spot in the Asian Carp webinar rotation, delivering a talk on research into early detection monitoring for pathways like the Welland Canal. He looks tired, or at least relieved. Uncharacteristically subdued in any event, he confesses to not having slept well, both out of concern over his presentation . . . and a hockey injury. The quintessential Canuck researcher, Mandrak's foot had been struck by a slap-shot puck the previous night, tenderizing it to the point that even the gentle brushing of a bedsheet was excruciating. Now he dismisses it with a laugh: where should we begin?

How about what he was working on these days?

Fine. In a few months Mandrak would attend an American Society of Limnology and Oceanography workshop about reciprocal invasions—North American fish in Europe, European fish in North

America—where he was scheduled to deliver one of several invited addresses. "Probably something about risk assessment and the relative impact of different vectors," he says.

Interesting for attendees, no doubt, but a pedestrian rehash for Mandrak, who was more excited about readying a second talk, this one concerning trait-based invasion models and lessons learned from the Great Lakes. With a recorded litany of both successful and failed introductions, the Lakes' system is perfect for modeling; elsewhere it is usually only known what has succeeded in invading, not what has failed (if an alien tree falls in a nonnative forest . . .). His talk would look at how trait-based models evolved from discriminant function analyses through to classification and regression trees, and the statistical reasons for doing so.

"My assumption is that the important variables and traits will change between lakes: other models have looked at the system as a whole—that is, one big species pool. Those each identified a *different* suite of traits important to invasions. That's not surprising because if only one set of traits were universal drivers in Great Lakes invasions, they'd come up again and again. Still, the fact that they don't might be an artifact of methodology. For instance, if you try to predict which native species *should* currently occupy each lake, you can only hit an 85 percent match with what's there; different traits are important in different lakes. So I want to use a single database and a single analytical, controlling for method and trait data to address the huge question of *context dependence;* in other words, how important the 'receiving environment' is in filtering what's successful and what isn't. We have a handle on climate-match data as a higher-level filter, but no one has successfully incorporated these into models of invasion prediction with more localized environments. I'm also working on models with David Lodge, Julian Olden, and Michael Marchetti for California and the Colorado River system, with the idea of using the same database to look at context dependence."

The group had, in fact, already completed a basin-by-basin look at the Great Lakes that turned out to be an unintended and successful test of the idea.

"Another thing I might touch on is the effect of *data availability.* Kolar and Lodge published a prediction model for invasion of Ponto-

Caspian species into the Great Lakes but only used English-language literature when most of the data is in Russian . . . I got the Russian data summarized, and so I'll run it through their model."

In the event the outcome differs, it would finger data availability as an effect on certain models. The bigger question, however, was this: given the vagaries of context dependence and data collection, availability, treatment, what have you, could you *ever* come up with a *general* prediction for what made a fish—or any species—invasive?

From what he'd seen over the years, through what he'd learned about environmental variability, and the results of current analyses, Mandrak wasn't sure such a thing would ever be reliably accomplished. "But that's the very thing ideas like 'morphospace' are trying to get at."

Morphospace theory holds that the shape of a fish is an indication of its niche. A landmark 2014 paper by Ernesto Azzurro and a list of researchers including Daniel Simberloff compared shapes of nonnative and native fish in the Mediterranean, finding that more unique shapes allowed invaders to enter morphospaces in which they were able to exploit resources unexploited by fish of other shapes: "Biological invasions have become major players in the current biodiversity crisis, but realistic tools to predict which species will establish successful populations are still unavailable. Here we present a novel approach that requires only a morphometric characterisation of the species. Using fish invasions of the Mediterranean, we show that the abundance of non-indigenous fishes correlates with the location and relative size of occupied morphological space within the receiving pool of species. Those invaders that established abundant populations tended to be added outside or at the margins of the receiving morphospace, whereas non-indigenous species morphologically similar to resident ones failed to develop large populations or even to establish themselves, probably because the available ecological niches were already occupied. Accepting that morphology is a proxy for a species' ecological position in a community, our findings are consistent with ideas advanced since Darwin's naturalisation hypothesis and provide a new warning signal to identify invaders and to recognise vulnerable communities."

As this summary suggests, morphospace turns on the fact that most theories of invasiveness share an idea: successful invaders differ somehow from native species. This concept, the authors note, has deep origins in early ecology. In 1958, Charles Elton was already talking about a limit to the ecological similarity of coexisting species as the basis of a classical principle in community ecology. In fact, it was almost a century old: in his 1859 *On the Origin of Species,* Darwin pegged large phylogenetic distances (degree of divergence from a common ancestor) as benefiting invasive flora because the struggle for existence was inevitably greater between closely related species, an intuition now known as Darwin's naturalization hypothesis. Mandrak saw the morphospace iteration as promising, but in its infancy. "It needs refining to distinguish between species that will merely establish, and those that will become invasive."

That's an important difference: odd-shaped fish without analogs in a new environment might be more likely to establish themselves, but they won't have much impact if they aren't competing with anything; fish that are close in shape to an indigenous species in the new environment, but exploit the same resources *better,* can take over the niche to become de facto invasive.

"I have a student doing a morphospace analysis for the Great Lakes to see if it can overcome some of the shortcomings of trait space," he levels. "But there's another interesting problem: how do you make predictions for a species that does things in its new habitat that it doesn't do in its native habitat—like silver carp jumping, or doing well on a diet of blue-green algae? Basically you don't, unless you have a handle on context dependence."

Context dependence seems a fruitful area of investigation, but also fraught—especially in large water bodies where local "contexts" might be hard to circumscribe. Because we know that disturbance and other environmental characteristics facilitate invasions on land, I wondered whether there was anything equivalent to a vacant lot in the aquatic world.

"Hard to say. We haven't been looking at it that way . . . maybe we could. One thing we *should* be looking at is whether some of the predictions and generalizations of early invasive guys like Elton and Harold Mooney hold up [Mandrak fondly remembers sitting in a library aisle as a student, mesmerized by one of Mooney's monographs on plant invasions]. And if there *seem* to be rules, are there also exceptions to the point that there really can't be universals at all?"

This seemed like a good place to put on the boots and wade into something I usually avoided: the muck of policy. On my way to the office I'd spun through the campus bookstore, noting that Chris Turner's 2013 *The War on Science: Muzzled Scientists and Wilful Blindness in Stephen Harper's Canada* was a required read in EES3000H, Mandrak's Applied Conservation Biology course. Therein, the respected Calgary writer outlines the history and erstwhile progressive traditions of Canadian government science, whose researchers and departmental organizations had once been both trusted and internationally renowned. Internally, their output had also been respected by government ministers who, once upon a time, expected to be advised properly when drafting natural resource, agriculture, and health policy based on issues that involved water, air, soil, climate, food, pharmaceuticals, plants, or animals. At this point, however, the best science in these areas was being routinely ignored, because it posed inconvenient truths for the business-at-any-cost conservative government. I was interested in how, at this critical juncture in burgeoning invasive species problems, that approach impacted the drafting and implementation of overdue policies and actions. In particular I wondered how lack of funding for basic research on an invasive species might facilitate its spread. Mandrak's answer catches me off guard. He cites the example of the most epic fail in all of fisheries science: the collapse of the 500-year-old northern cod fishery in the northwest Atlantic.

"Why did cod stocks collapse?" he begins on unmistakably didactic footing. "Basically, because there was great uncertainty in the science around the species' ecology and, as a result, in sustainable quotas. With cod stocks inexplicably dropping—and recall this was being blamed on everything from the Portuguese fleet to water temperature to too many seals—fisheries scientists said, 'Okay, here's a realistic quota, but uncertainty means it also comes with huge confidence limits.' The federal fisheries minister at the time decided he couldn't sell that quota, so he chose a number *beyond* the upper end of the confidence interval. Even at that, the province of Newfoundland complained and the fishermen's union said they couldn't make a living off that number and it was increased *again.* So there's a purely political magnification of fishing pressure that ignores what the best science is saying."

In other words, with downward trending cod stocks and much un-predictability in the equation, scientists, as they tend to, were operating within the "precautionary principle," that is, recommending the lowest possible quota that would balance perceived biological and economic de-mands. Instead, the exact opposite was enacted with predictable results.

The bottom fell out in the summer of 1992 when cod stocks suddenly plummeted to a mere *1 percent* of their previous level. A subse-quent emergency moratorium on the fishery wasn't intended to be permanent: hubris drove the optimism that cod would recover within a couple of years, and with it, the fishery. But the moratorium came far too late, and the unraveling of Newfoundland's continental-shelf eco-system by centuries of rapacious overfishing of this single species now proved irreversible. Twenty-five years later, cod has barely rebounded and the fishery remains closed, the biological and socioeconomic col-lapse forever changing the lives of thousands of Newfoundlanders.

The main scientists involved were fired, thrown under the bus of public outcry to cover for ministerial, provincial, and industry incom-petence. Ironically, these same individuals are now considered among the best fisheries stock-assessment scientists on the planet. An infamous after-the-fact 1997 paper by Jeffrey Hutchings and others, "Is Scientific Inquiry Incompatible with Government Information Control?", con-cluded that a political environment in which information can be con-trolled, for whatever reason, is one in which the scientific process cannot exercise its full potential.

The Canadian government eventually responded in 2000 with the SAGE (Scientific Advice for Government Effectiveness) principles and guidelines. Meant to be adopted by all departments, to date only DFO has put it into practice. Which leads us back to invasive species, and the point that Mandrak is hedging toward: invasion biology is similarly rife with uncertainty, and accounting for this in any regulation is key to avoiding the Tragedy of the Commons errors of what historians now call "the cod fiasco."

Because regulatory impact assessments (RIA) are based on consulta-tions with relevant conservation, socioeconomic, and public stakeholders, it's imperative that science be a cornerstone of those RIA aimed at invasive species, which are little understood in the public realm. RIA are meant to provide a detailed and systematic appraisal of potential impacts of any new regulation. Because every new regulation commonly has impacts that can be difficult to foresee without detailed study and consultation, the central purpose of RIA is to ensure that the regulation will be of net benefit to society. The "Inflation Impact Assessments" of 1978 by the Carter Administration in the United States are generally considered to be the first RIA. The European Commission introduced an impact assessment system in 2002, while in Canada, most new federal regulations now require a Regulatory Impact Analysis Statement.

This has sharp relevance, Mandrak explains, because the federal government—ironically, after gutting the Fisheries Act of habitat protections in 2012—would soon post for public comment proposed regulations to address aquatic invasive species. In typical fashion this would be over Christmas, virtually guaranteeing that no one would see it. While some AIS can be controlled through fisheries regulations, the new legislations will clear hurdles or barriers to rapid response activities such as the need to apply toxins. In joint jurisdictions or shared landscapes, the weakest link is always the problem; Ontario may adopt an Invasive Species Act, but what about its neighbors? You couldn't possess live Asian carp of any kind in Ontario, Manitoba, or Quebec, making cooperative responses easier, but what did it mean if Alberta still allowed grass carp to be used for weed control in phosphorus-loaded agricultural ponds? It meant trouble in an age of cross-continental overnight transport, in which ethnic communities in British Columbia and Alberta conducted rigorous commercial exchange with their counterparts in eastern provinces. Within Canada, as within the United States, a lack of inter-jurisdictional barriers to moving live Asian carps (barriers that have since been enacted through legislation) would have made it inevitable that these would enter Great Lakes watersheds one way or another. Should they establish and spread, their presence would ultimately become so commonplace that subsequent generations wouldn't see them as aberrant: the pre-invasion concerns would be blurred by Shifting Baseline Syndrome.

"Memory creep changes everything; people our age say 'Oh, when I was a kid there was no *Phragmites*,' but no one who has grown up in

Ontario in the past twenty years can say that. Their visual memories will be full of dense stands of phrag the way pond edges in our memories are populated by native bulrushes," says Mandrak. "Shifting Baseline Syndrome is discussed big-time in freshwater circles. Look at common carp: it's been here over a century, and we just deal with it. No one back in the 1800s was documenting the catastrophic destruction it effected on much more naïve ecosystems. Those effects probably peaked a hundred years ago and can't be teased out from other effects that have occurred since, so we'll never comprehend the full extent of damage; all our studies have been post-common-carp-impact, so now that's our baseline. It's hard to get anyone in North America to look at common carp now and say '*Holy shit*, we need to be controlling these things!' But Australia is still trying to kill every common carp because they're only fifty years into its spread."

(Indeed they were, and by 2016 the country would earmark some 15 million Australian dollars to the task of common carp eradication via biocontrol, infecting the fish with a carp-specific, skin-and-kidney-destroying herpes virus that usually killed within twenty-four hours once it took hold).

Some folks around the Great Lakes, it seemed, still thought you could roll back that line as well. I asked Mandrak about "the Fishway."

Opened in 1997 at the mouth of the Desjardins Canal between Hamilton Harbour and Cootes Paradise—an extensive lakeshore marshland under the aegis of Hamilton's renowned Royal Botanical Gardens (RBG)—the Fishway was the Great Lakes' first two-way fish passage and carp barrier. Small fish including immature carp could move in and out through grids, while large fish were caught in lift baskets and sorted: native fish and carp exiting Cootes for the harbor continued on their merry way; incoming native fish were ushered through with open arms, while carp attempting to enter Cootes were summarily returned to the harbor. Before its installation, an estimated 50,000 adult common carp regularly tore up Cootes's shallows each spring breeding season, creating excessively turbid water and damaging or destroying aquatic vegetation used for reproduction by other species of fish, amphibians, and invertebrates. Attempts to control carp were nothing new, dating to the 1920s. In partnership with other agencies, RBG had been at it since 1950, employing commercial harvesting, the planting of wetland vegetation inside of carp exclosures, and other methods that all met with limited success. In the Fishway's first year of operation, however, 95 percent of immigrating carp (some fifteen thousand fish) were excluded, and the number of carp in Cootes dropped steadily, so that these days, roughly a thousand adults can be found at any one time. The Fishway remains the most visible manifestation of a decades-long ecological restoration project aimed at revitalizing marsh habitat used by a range of native mollusks, amphibians, turtles, fish, mammals, and migrating and resident birds.

But while some consider the Sisyphean task of completely eliminating carp from Cootes a vital step in marsh restoration, Mandrak is less bullish. "We *can't* eliminate them," he says. "It's too late. We had to do it a hundred years ago. We can make a *healthier* ecosystem by adding more native components and preventing future invasions, but we can't reverse things to what they were a century back. Even if we can get their numbers way down we're going to have to live with common carp."

Tys Theysmeyer, head of natural lands at RBG, accepts that even this will be a long, slow slog. I'd walked around Cootes with him one summer day, looking in on restoration projects that ranged from returning natural flow to the creeks feeding Cootes, to Canada geese exclosures planted with native shoreline vegetation, to the removal of *Phragmites*, to the first bald eagle nest in half a century, whose presence is helping vanquish irruptive cormorants and gulls. When we got to talking about the Fishway, Theysmeyer flashed a wan smile. At least for now, he'd remarked, it was a great interpretive location for interested members of the public to see what species were going in and out, and what was and wasn't native. But the Fishway's objective of ultimately eliminating carp? That was "something of a thirty-year adventure" whose success couldn't be known.

Mandrak has the closer, of course, and it circles back to Asian carp.

"You think *common carp* destroy native aquatic vegetation? They do, of course, but it's only incidental because they're rooting around in the shallows for food or mating. Just wait until something like grass carp that actually *eats* vegetation gets in. We're modeling the potential impact on Great Lakes wetlands right now—bioenergetics linked to habitat availability—which is mostly measured by the distribution of aquatic vegetation."

What was it showing, I'd wondered?

"It doesn't look good. The main difference between the four Asian carps are their potential impacts. Black carp is a molluscivore, silver carp a phytoplanktivore, bighead carp a zooplanktivore, and grass carp a herbivore. These very specific dietary niches are the reason they were brought into the U.S. to begin with: did you have too many disease-transmitting snails in your commercial catfish pond, too many algae blooms? One of these critters was the answer. So we're expecting grass carp's effect on the aquatic vegetation of wetlands to be devastating."

Discussion of biocontrol species like Asian carps that escape aquaculture ponds to become invasive segues quickly to the invasive potential of aquaculture species themselves. A 2015 study of aquaculture as a vector for invasions in California found that of 126 nonnative species associated with commercial aquaculture reported from state waters, fully 106 had become established—a shocking rate of 84 percent. As a measure of consequence, the authors conducted a literature search for studies related to mollusks and algae introduced in California, finding, unsurprisingly, that most demonstrated negative impacts.

The vast majority of these were unintentional introductions linked to historical practices of the aquaculture industry. And when *that* topic comes up, one always turns to the invasive organism whose very name captures the imagination: walking catfish.

## Walking Small

A widespread freshwater native of Southeast Asia, walking catfish (*Clarias batrachus*) typically occupy ponds, swamps, flooded fields, temporary pools, and otherwise slow-moving waters that might occasionally dry up. In the event, a facultative ability to gulp air into specialized pockets near the gills allows the animal, propped on spiny pectoral fins, to wriggle overland in search of a better life. Like many opportunists, walking catfish are omnivores, feeding voraciously on smaller fish, invertebrates, detritus, and aquatic weeds; that it cares for its young also ensures higher survival. Although these traits combine to make it easy to raise as a food fish, its catholic tastes and smooth-skinned appearance also make it a desirable pet—the main reason it has now been a Florida resident for half a century.

Walking catfish were first imported from Thailand as aquarium-trade brood stock in the mid-1960s. Yes, they eventually escaped—though it remains unclear whether this was directly from Penagra Aquarium, a Broward County fish farm, or from a tanker truck transporting a large number to the Miami area. Either way, the prodigious little catfish quickly reached such high densities that many of its rank and file were forced to go walkabout in search of food—easily found in nearby aquaculture ponds in the form of abundant eggs and fry of other fish. Area fish farmers cried foul, sensing economic collapse in this echo of the 375-million-year-old fish-tetrapod transition of the late Devonian. Similar to today's Burmese python scare-mongering, these fish-out-of-water concerns propelled walking catfish into sensationalistic headlines near and far. I remember standing with my mother in a grocery checkout line, all of age ten, eyeing the cover of a tabloid that featured the head-on closeup of a walking catfish appearing to tower over the land (with no attendant scale, there was no way to know it was only as long as my forearm). Accompanying the photo was a large headline announcing "They Walk Among Us." It was the first invasive species I had become aware of, and these were the terms of reference. And it wasn't just tabloids: an exotic fish researcher labeled walking catfish a "monster" in 1968, while a 1970 *Smithsonian Contributions to Zoology* article described its introduction to Florida as the most harmful of any yet witnessed in North America (the authors obviously didn't live near the Great Lakes). Though undeserved, it was perennially listed by the Global Invasive Species Database (available at issg.org) as one of the World's 100 Worst Alien Invasive Species.

Hysteria causes knee-jerk reactions that, in turn, often create further problems: in 1967, when the state slapped a ban on the importation and possession of walking catfish, it led to additional releases into the wild by pet owners and fish farmers worried of being found in violation of the law. In another foreshadowing of the python problem, walking catfish spread both by utilizing South Florida's abundant interconnected canals, and by moving over land—typically on rainy nights, when drivers might encounter "catfish slicks" at points where great numbers crossed a road. It covered a lot of ground for a fish, and quickly. By the mid-1970s the species was firmly established across twenty counties in the state's southern tier, including Everglades National Park and Big Cypress National Preserve. As of 2005, however, early fears of its ecological impact had failed to pan out, and Paul Shafland, director of Florida Fish and Wildlife's Non-Native Fish Research Lab, felt comfortable enough to downgrade the threat: "It's certainly not had any of the catastrophic effects originally associated with its find in Florida," he told a newspaper. "However, we still consider it problematic. We wish it weren't here."

Although continuing raids by walking catfish on aquaculture farms have led to owners fencing in their stock, the initial population boom wasn't sustained; instead numbers declined through the 1980s and 1990s. Such trends aren't unusual with introduced fishes, something an Everglades-based USGS research ecologist attributed mostly to predators (and here you can also include new invasives) that eventually keyed in on the catfish. Invasives often don't undergo a widespread crash so much as drift into some kind of ecological balance where irruptive super-abundances are limited to local, context-dependent situations.

Still, the walking catfish tale remains both classic and universally cautionary: a fish collector, whose property bordered a main canal, had stocked his outdoor pond with them for ornamental purposes. "One day it got dry and they'all jus' up and walk out of that pond into the canal," he'd drawled to a television reporter investigating the local invasion it set off. "Who knew they'd dew thayat?"

## Unnatural Natural Disasters

Mandrak has a point he wants to make about walking catfish. What the received view among biologists and managers holds to be a single Southeast Asian species might, he mused, include an African air-breather, the sharp-toothed catfish (*Clarias gariepinus*). A larger species, but also invasive outside its endemic range, it was amenable to small-scale aquaculture, and so had been raised and sold as food across the African continent since the 1970s; these days it was rapidly moving around South Africa via irrigation canals. The last time he visited, Mandrak was proudly served a smoked chunk of the creature and informed that some were hoping to develop a market for it. "It tasted good, but it was hard to enjoy knowing how bad a thing that would be," he notes of what might be viewed as a win-win situation but which, in reality, is complex and risky. "When you create a market for an invasive species you end up managing *for* it instead of against it."

The sheer abundance of many invasive species typically makes their commercial worth bargain-basement low, which only adds to the danger if there's reasonable downstream use of the product. For instance, current use of Asian carp as a high-tonnage, low-cost fertilizer in the United States creates a condition under which some might conceivably want to manage its availability into the future. A market for a food item can be even more of a direct problem, Mandrak believes, because once a taste is acquired, say, for smoked sharp-toothed catfish, it can become a regional economic driver with multiple stakeholders who are invested in its continued presence and who may have an incentive to maintain a natural source of the food outside of commercial control, that is, in the wild.

This Mandrak delivers while fronting a whiteboard awash in multicolored scribblings. The diagrams and formulas are leftovers from an earlier discussion with a student. Her thesis topic, "Changes in functional diversity in fish species in the Great Lakes 1870–present," is a chip off the ol' academic block that will also provide a de facto test of the morphospace theory by looking at Great Lakes fish assemblages in two ways: functional space occupied, by basin; and functional space occupied across the five basins homogenized. Launching into the detail of yet another Brobdingnagian analysis, Mandrak's voice recedes into the ether as my mind wanders from pythons to carps to ants to Scotch broom. For a moment, years' worth of similar discussions with a broad cross-section of people and an even broader range of problems seem to press down on me; with it comes the weight of revelation.

"Are invasive species problems actually *natural disasters*?" I blurt.

"Absolutely," he avers. "A Canadian Aquatic Invasive Species Network bulletin had a theme section on rapid response that considered AIS catastrophes in the sense of whether Ontario or Canada was prepared for them."

But when I then note that my thought has come courtesy of suddenly seeing large-scale AIS impacts in the Great Lakes as akin to meteorites striking an ecosystem, Mandrak hedges.

"There's still an issue with applying the term 'natural'—AIS disasters are *functionally* similar and the language is the same in terms of preparedness and rapid response, but unlike other natural disasters, they're not entirely stochastic. There's little anyone could do to minimize the risk of a catastrophe that can't be controlled—like a typhoon or a real meteorite striking the Earth—but that's not true here. Still, some say we should treat AIS as unavoidable catastrophes similar to volcanic eruptions or floods, and focus on how to minimize effects when they do hit. I agree on the response-preparedness front, but we have to continue being proactive in identifying potential invaders *before* they're introduced, for a whole lot of reasons," says Mandrak, doubtless referencing the money, manpower, and motivation issues plaguing rapid-response capacities.

He returns to the example of *Phragmites australis australis*, European common reed, whose thick, impenetrable stands have claimed shorelines and wetlands along the lower Great Lakes and St. Lawrence in mere decades. By any measure, *Phragmites* is a full-fledged disaster, and much like Asian carps, by dint of sheer volume of biomass the now onerous post-hoc scenarios for dealing with it well demonstrate the utility of EDRR approaches.

"I have a student looking at removal in Long Point Bay [on the north shore of Lake Erie] and it opened my eyes; *Phragmites* turns aquatic habitat into semi-aquatic habitat where turtles, amphibians, and fish basically can't go. For endangered wetland fish like spotted gar and warmouth, phrag is now the number one threat. Also, diked wetlands that were a last refuge for native mussels after zebra and quagga mussels arrived are rapidly filling with phrag. Both climate change and lower water levels are helping it creep farther and farther into the lake. I think *Phragmites* is a battle we're going to lose, so the outlook for wetland species in the Great Lakes isn't good."

Becky Cudmore's incident-command-training model for Asian carps now came into focus: just like any natural disaster contingency, preparation is key, someone has to be in charge, and you have to be both organized and ready to respond.

"If you really want a good discussion on invasive species as natural disasters," Mandrak finishes, "talk to my colleague Tony Ricciardi."

## Biological Spills

On the phone, Anthony Ricciardi was fast-talking and voluble. With his myriad examples and digressions augmenting a deep knowledge of invasion biology and its empirical underpinnings, I had the impression of information ricocheting with such rapidity and neural cross-referencing that he couldn't keep still. (Later, watching a podcast lecture from McGill University, I'd observe that he indeed had a semaphoric delivery.) Despite Mandrak's referral, Ricciardi, an associate professor and invasive species biologist at the Redpath Museum and McGill School of the Environment, was wary at first, uncertain that I was someone he should even speak with. Fair enough. Having read several of his elegantly argued papers and the subsequent commentaries directed at him and his coauthors, I couldn't blame him for aiming to divine my motivation. For scientists working in heated areas of research—under de facto siege from excuse-seeking politicians and rival contrarians looking to pounce on any misspoken line—it made sense to be cautious with an unknown journalist who might misrepresent their work. Though the pointed quizzing delayed my own standard opening query of how he'd found his way into the discipline, Ricciardi answered that without prompting, simultaneously sweeping me into self-examination the way an undergrad adviser might press a student on a degree choice.

"What *exactly* are you writing? I mean, *why* are you interested in this subject and *what* do you want to get out of it?" he began, shifting quickly to his own ingress. "I'm in it because I grew up on the St. Lawrence River, which has changed radically because of invasive species. In 1991 I found zebra mussel on Île Perrot [west of Montreal]. Though it was first found in Lake St. Clair in 1988 and spread quickly throughout the lower Great Lakes, it hadn't yet been reported from the St. Lawrence River. So I kept the find to myself because I wanted to bank it for a Ph.D. project and didn't want someone else coming in beforehand. Eventually, I was dismayed to hear it had been found clogging up intake pipes in nearby Pointe-Claire."

Notwithstanding, zebra mussel remained a gateway organism for Ricciardi, whose subsequent work with it befits the mollusk's own prodigious abundance. And he still carried a torch for the little striped devil. "When I first saw it I was in shock—but excited as well. Most biologists are enamored of biodiversity and like seeing new organisms. I don't know a single one who looks at an invasive species with disgust; there's respect for their ecological abilities, and so it's like this guilty pleasure where you're not at all OK with the problems it might be causing, but secretly happy you can study it."

Ricciardi's tastes in such simple joys were expansive; though rooted in freshwater ecology and the macroinvertebrate communities of Great Lakes–St. Lawrence ecosystems, he'd covered coral to fish to zooplankton, taking on theoretical considerations like propagule pressure, predictors of invasiveness, and the enemy release hypothesis. His was one of the top names in the field because, like his frequent collaborator Daniel Simberloff, he saw not only the scientific opportunity before him, but also the urgency of bringing understanding to the problem. Ricciardi's take on the growing worth and utility of invasion science was clear in his penchant for rhetorical titles: "Are Modern Biological Invasions an Unprecedented Form of Global Change?"; "Are Nonnative Species More Likely to Become Pests?"; "Should Biological Invasions be Managed as Natural Disasters?"

As Mandrak intimated, the last, a 2011 study with Michelle Palmer and Norman Yan, summarizes why and how biological invasions and natural disasters reflect similar phenomena requiring similar approaches. On the comparison ledger, both have well-understood causes that are generally unpredictable and uncontrollable, and both can generate immensely damaging environmental events whose frequency is inversely proportional to their magnitude. And then the crux: if this is so obviously true, why is our response to each so diametrical? Despite enormous global investment in training, preparedness, and response plans for extreme but rare natural hazards like earthquakes and tsunamis, apart from frameworks enacted for dealing with infectious disease, no nation has any similarly comprehensive program in place for animal or plant invasions whose impacts are "less predictable and often irrevocable," and whose annual worldwide economic cost *exceeds* that of natural disasters.

Preventing invasions, they concluded, requires the same commitments by nations as natural disasters. Ideally these would include international coordination of early warning systems; infrastructure that can provide immediate access to critical information; specially trained individuals (here they erected a neat analogy of oil spill versus "biological spill"); and rapid-response strategies. Furthermore, with the required expertise and personnel existing in a growing list of researchers on every continent, as well as national and international bodies dedicated to documentation and frontline actions, there was little excuse not to take these necessary steps. In fact, the sheer volume of current research and concomitant information it was generating would surely boggle the quixotic Charles Elton's mind: invasion science was a robust and mature discipline whose theoretical scaffolding had been clad in enough observation and experimentation to produce a solid edifice of practical application, something Ricciardi expended much ink in demonstrating.

"What we're talking about here is a lot of people as well-trained as any medical doctor, but who have ecosystems as patients," he said, singing their collective praise. "Only these patients are far harder to handle: there's so much variation, so many things going on at once, that anything we *do* know about them is context dependent. We're searching for predictable patterns masked by a lot of noise, so it's difficult to tease out cause and effect. Ecology is probably the hardest science in that respect, but one clear responsibility for ecologists is to provide advice to managers about threats, and so to *ignore* those threats and not pass them on would be irresponsible—our version of ignoring the Hippocratic oath 'Do no harm.'"

Ricciardi and Simberloff stated as much in the august journal *Science,* responding to an article that failed to consider potential downsides to purposeful introductions employed in restoring endangered species: the conservation modalities known as "assisted colonization" and "ecological replacement" could both yield consequences as devastating as those of the worst accidental invasions. Unsurprisingly, Ricciardi also had zero patience for apologists lurking on the edge of invasion biology who dismissed its critical mass of research and frank reflection of ecological calamity as alarmist poppycock, calling them out where he could. In "Misleading Criticisms of Invasion Science: A Field Guide," David Richardson and coauthor Ricciardi whimsically categorized these facile attacks and soft-headed calls for a hands-off approach as "a cottage industry of criticisms" and "premature obituaries." In this regard, invasion literature often showed Ricciardi playing like a bulldog Huxley to Simberloff's quieter (though still forceful and devastatingly blunt) Darwin. It was possible that at least some criticism of invasion science was, in part, a response to the inevitable sensationalizing by popular media of every example it got its grubby little sound-bites on. *Walking catfish! Giant hogweed!* and *Fire ants!* clearly sold papers and created click bait, but if arguments were being tacitly advanced that such distortions also masked ultimately benign incursions, Ricciardi was having none of it.

"There's media exaggeration of everything, *everything*—cancer, chemicals, climate—and you'd be crazy either to dismiss that reality, or to take everything you read at face value. Look at it this way: media always hypes up storms, but you don't *ignore* the fact these are occurring. Exaggeration doesn't make the *original* information less important. Most people grasp this."

It would be at our peril, for example, to ignore the broader, increasingly strident warnings of 1,500 International Panel on Climate Change scientists simply because media tended to fixate on only the direst of climate predictions. We'd be similarly *non compos mentis,* Ricciardi reasoned, to ignore the large number of respected researchers pointing to potential invasive species problems. "Here's an example: the Asian tiger mosquito vectors all sorts of diseases, and it's all over the United States now. Maybe the diseases haven't arrived yet, but the mosquito's presence there represents a *huge* potential problem."

Those words would be more prescient than he could imagine.

## Nonendemic Epidemic

Only two weeks later, chikungunya (pronounced chicken-*goon*-ya) landed on my investigative plate. Despite its sonic allusion, this wasn't some gustatory gumbo cribbed from the latest exotic foodie trend; it was, however, a bona fide recipe for disaster penned by the combination of rapid globalization, climate change, and invasive species. And that day, the U.S. Centers for Disease Control (CDC) was warning sunseekers up and down North America's eastern seaboard about a burgeoning outbreak of this little-heard-of pestilence.

Endemic to southeast Asia and sub-Saharan Africa, chikungunya was yet another in a litany of mosquito-borne viruses to jump oceans in recent years. The name derives from the Makonde language of southern Tanzania (where the disease was discovered in 1955) and means "that which bends up," a graphic reference to the cringing, arthritis-like pain it brings to the joints, in addition to fever and headache. Though it generally lasts but a few days, in some cases joint pain persists for months. There is no cure, no vaccine, no treatment, but fatalities are thankfully rare. Once infected you have lifelong immunity to the disease, but the only way to actually prevent it is to avoid its natural vectors—mosquitoes of the genus *Aedes*. Back when the African yellow fever mosquito (*A. aegypti*) and Asian tiger mosquito (*A. albopictus*) were found only in the tropical parts of those continents, avoidance elsewhere was easy. These days, not so much.

In 2006 a chikungunya outbreak in India infected over a million people. As humans are wont to do, one of these hopped on a plane and delivered the virus to Italy where *A. albopictus* had previously been introduced; with a ready-and-waiting vector, two hundred more subsequently fell ill. Further outbreaks in southern France and on Pacific and Indian Ocean islands likewise relied on the prior establishment of nonnative *Aedes* invaders—which had variously been transported via international trade in used tires and ornamental plants. With *Aedes*-transmitted encephalitis and dengue fever on the rise as well, global health researchers had long predicted that chikungunya would eventually show up wherever *Aedes* mosquitoes had gained a foothold. Given broader tolerance of temperate climates than its African counterpart, this meant pretty much everywhere in the tropical Americas for the Asian tiger mosquito, which has well-earned its perennial membership among the World's 100 Worst Alien Invasive Species.

Only a year earlier, on December 6, 2013, the World Health Organization had logged the first case of local transmission of chikungunya on the Caribbean island of Saint Martin. That is, the island's resident

*Aedes* mosquitoes were now infected and passing it to people—the first time this had been reported in the Americas. How fast could chikungunya spread around a 2,750,000-square-kilometer archipelago? Well, how well distributed were its invasive mosquito vectors?

Exceptionally well distributed. Only ten months later, on October 21, 2014, the CDC was reporting local transmission on twenty-five Caribbean islands, French Guiana in South America, and of course Florida, North America's welcome mat for tropical interlopers.

None of this was news to an acquaintance of mine, Isaac I. Bogoch. With a master's degree in clinical epidemiology from the Harvard School of Public Health, and a diploma in tropical medicine and hygiene from the Gorgas Memorial Institute and the Instituto de Medicina Tropical in Lima, Peru, Bogoch was now assistant professor in the Department of Medicine at the University of Toronto and a tropical disease specialist at Toronto General Hospital. In June 2014, Bogoch was a co-author—with Kamran Kahn and others—on an eye-opening paper, "Assessing the Origin of and Potential for International Spread of Chikungunya Virus from the Caribbean." Using airline arrival and departure information between the Caribbean and other parts of the world, the researchers constructed a predictive model explaining the likely origin and potential spread of the current outbreak. Their results were spot on.

"We've been tracking and modeling different emerging and re-emerging infectious diseases," Bogoch explained from his office at Toronto General. "These are influenced by human migration patterns, socioeconomic patterns, agricultural patterns, and climate and ecological change. Globalization and accelerated rates of air travel are the root of the problems we're seeing—and anticipating."

In other words, international travel by infected persons and the "increasing geographic availability of competent vectors"—Bogoch's polite way of framing invasive species—set the stage for chikungunya's introduction and spread in the Americas. Couple that with an increase in appropriate environmental factors for both vector and virus (for example, an average temperature above 20°C), and the result is over 16,000 confirmed and 900,000 suspected cases of chikungunya as of January 2015.

"Lately, a lot has been written about the expansion of various animals' ranges being coincident with the spread of certain diseases beyond their traditional geographic areas," noted Bogoch. "For instance, why are *Ixodes* ticks that carry the *Borrelia* bacterium responsible for Lyme disease expanding their range? Because of the movements of birds, rodents, and deer in response to climate change."

The takeaway on chikungunya from a public health standpoint? "Basically there's a new kid on the block to join mosquito-borne travelers like West Nile, dengue, and encephalitis. West Nile virus has been endemic in North America for decades now. Every summer it flares up, some years worse than others, but it's here to stay. Chikungunya is the latest but certainly not the last to fall into that category."

According to Bogoch, a 30,000-foot view of this trend raised only one question: what's going on? "And the answers are pretty obvious— globalization and mobility and interconnectivity and changing weather patterns—and none of those are going away either."

Not only were problem-making phenomena and organisms not going away, but they were daily joined by others. For many organisms, the Anthropocene has become a de facto age of dispersal; for the humans perpetrating these, an age of petulant arrival.

## Master Plan

In the terraqueous Finnish town of Savonlinna, occupying a series of islands in labyrinthine, post-glacial Lake Saimaa, a stone's throw from the Russian border, large, dense swaths of what the Finns call *kurtturuusu* (Japanese rose, *Rosa rugosa*) swirled like attacking huns around the base of castles and other historic buildings, spreading across parks and along right-of-ways. When my partner—a Canadian of Finnish heritage—and I left Savonlinna after a morning of sightseeing, kurtturuusu followed us west through the town of Juva toward the small city of Mikkeli. What began as kilometers-long roadside eruptions soon gave way to sporadic dollops seemingly dispensed from a small spoon— new infestations moved down the highway by wind and traffic. Driving toward my in-laws' cabin on a small lake near Mikkeli, kurtturuusu was everywhere: what seemed poorly landscaped meridians were actually troughs of tortuous invasive monoculture. I would not enjoy any respite from alien species here.

Despite living in Canada for forty years, my father-in-law, Reijo, who grew up on the lake, still maintained a property here to which he retreated annually. Behind hand-built cabins on a small point, the trees stood ramrod straight: poplar, birch, spruce, fir, Russian red pine. If you ignored the geographic specifics, however, you might be in any lake district in North America—Ontario's Muskoka, or much of New England, Wisconsin, or Minnesota. Key differences, however, highlighted the reciprocal nature of transatlantic freshwater invasions.

On either side of the dock from which we launched ourselves into the lake between sauna sessions, native European *Phragmites* hugged the shore, a third the size and nowhere near as dense as it grew as a pernicious invasive in North America. The lake was also home to an abundance of American signal crayfish (*Pacifastacus leniusculus*). As had happened across Scandinavia when populations of native noble crayfish (*Astacus astacus*) plummeted as the result of introduced crayfish plague, someone decided that the larger, hardier import was more commercially viable—certainly it was immune to crayfish plague (though advocates had no idea it was also the vector). With immunologically naïve noble crayfish as rare as hen's teeth, the bold, brash Americans now dominated Finnish lakes like this one, where boats creep along the shore each morning checking crayfish traps marked by homemade Styrofoam buoys— knowledge honed over generations.

Reijo's brother, who still lived here, was affectionately known as the Crayfish Baron. He'd trapped and sold them for decades and now his daughter Kaisa had taken over, baiting traps with cakes about half the size of a hockey puck crafted from the Baron's secret recipe. A daily haul of one good crayfish from a trap was sufficient, as each could fetch €2–€3.5 depending on size. Fishing pressure wasn't huge, just a smattering of cottagers grubbing for their own food. As we lounged on the dock one afternoon, Kaisa and her husband pulled up in a small outboard, passing over their catch bin for show-and-tell. Grasping the broad carapace of a large signal crayfish behind its flailing claws, its legs oaring forward as if it were still shooting through the dark depths of the lake, I thought this: after millions of years of our respective relatives evolving on various continents, this arthropod and I were now both North Americans on vacation in Finland.

Next evening the couple returned, this time with a bowl of bright-red cooked crayfish, boiled with a heavy infusion of dill to obviate any muddy bottom-dweller flavor. The dish is the central theme of a rowdy pan-Scandinavian end-of-summer bash called *kräftskiva* (of Swedish origin, as is the practice), where paper is spread on a table and guests set to claw-cracking, flesh-digging, and leg-sucking between rounds of vodka or schnapps—which seems the real reason for all the trouble. We ate and toasted heartily, inadvertently celebrating a tradition that had survived only by being transferred to the species that did in its original subject. What would happen with signal crayfish here if no one fished it anymore? No invasive species came without questions.

Indeed I had wondered before leaving Canada how a country like Finland—with abundant forestry and agriculture, and numerous maritime and transportation pathways—was handling its doubtless numerous invasive species problems. The question was answered online, at the government-sponsored public-outreach site vieraslajit.fi. Here I found a veritable invasives encyclopedia that included links to NGO partners like the Finnish Association for Nature Conservation and Finnish branch of World Wildlife Fund, plus a rogues' gallery of culprit species, each with background information and risk assessments rivaling anything I'd seen in North America. It was also interactive, with buttons for plant sightings: you could say what you saw, when you saw it, and how much was there; you could post photos and click on a map to register the location; you could also do it all anonymously so that if you were reporting your neighbors' giant hogweed plantation, for instance, they wouldn't know you ratted them out. I was impressed, but also depressed—many of the same invasives being dealt with at home, from plant to pathogen, were causing similar problems here, demonstrating their pan-global nature. There were also a few surprises: introduced North American mink and white-tailed deer, for instance, had become problematic, as had Canadian goldenrod, for which I felt the briefest pang of guilt for Finnish allergy sufferers.

Though much of the information was in Finnish, and thus indecipherable to me, it was thought-provoking enough that I contacted the person behind this marvelous portal—Johanna Niemivuo-Lahti at the Finnish Advisory Board for Invasive Alien Species—and made an appointment for a sit-down at the Ministry of Agriculture and Forestry in Helsinki.

On a blustery late-summer day, searching along a row of stately government buildings (no brutalist architecture here), I'd found the ministry, announced not by proclamatory sign, but on an understated, eye-level bronze plaque the size of a napkin. Behind the large wooden door, a motion-sensitive camera followed me through a narrow, marble-floored foyer lined with gilded pillars. I signed in with a grim-looking security guard, received a badge, and within minutes, Niemivuo-Lahti was escorting me up stone stairs into the more modern environs of her department. Passing doors and cubicles, I was ushered into a small, central meeting area enclosed entirely by glass. With the machinations of a busy office pulsing around us, it felt a bit like being in an aquarium. It was also very formal. Between us sat a tray with a smart porcelain tea set, and what looked to be blueberry scones. Otherwise the table was dominated by the hefty document that had brought us together: Finland's National Strategy on Invasive Alien Species.

Back home I'd poured over an electronic version, stunned by a detail and thoroughness orders of magnitude greater than for any organization I'd yet seen, particularly in that each of 160 nuisance species identified to date in Finland had, more or less, been assigned a current status (equivalent to a place on the Invasion Curve) and a specific plan of action—from risk-assessment to eradication—with prescribed timetables for the implementing of all measures. The timetables included something extraordinary: follow-up monitoring schedules with horizons of ten to twenty years that took the country's overall strategy out to the year 2030.

"I have to tell you, this is *amazing*," I said, balancing my pointed finger over the stack of paper. "I can't believe how comprehensive and forward-thinking it is."

"Really?" She seemed surprised.

"Really. No country, no state, no province, no region in North America has anything close. Especially my own country," I said, sounding like an invasive-management sycophant.

"I had the impression Canada was on top of things like this, that they are doing a lot," said Niemivuo-Lahti.

Other than economically impactive species like sea lamprey and Asian carps, the reality was quite the opposite under Canada's conservative government, I explained, which at the time was actually doing less than nothing on any biodiversity portfolio unless instructed to by the supreme court. Things were only slightly better in provincial jurisdictions, depending on government bent, and just approaching lip-service in municipalities. Any real heavy lifting was carried out by NGOs like the SSISC at local levels.

Niemivuo-Lahti remained implacable, nodding thoughtfully as I detailed the choking off of funding for any such measures. I took her silence to be disappointment over what she'd believed to be the case—a little unplanned bubble-bursting on my part. Instead, it was her sad realization of being in the exact same boat and, as she now expressed, if a place as resource-rich and well-off as Canada wasn't stepping up, then it was unlikely Finland ever would.

What did *that* mean? It meant what it meant everywhere: good intentions, great optics, no money for follow through.

At International Convention on Biological Diversity meetings attended by Finnish government delegations, the minister of environment had become distressed at having nothing to say about what his country was doing on the invasive species portfolio. "My boss eventually suggested we do *something*," said Niemivuo-Lahti, picking up the thread. "So we created a national strategy group of 100 experts and I spent three years putting this document together and another year getting a government resolution on it. Our ministry printed 40,000 brochures about hogweeds and other things. The discussion, of course, was how to pay for any further action."

In the meantime, there were forthcoming EU regulations on invasives that all member countries would be obliged to adopt—not just *directives* allowing countries to choose how to reach a target, but genuine regulations (these went into force on January 1, 2015). Knowing the pending legislation's content—what the new rules would dictate and where the gaps were in their own country—made it easier to address in the strategy document Niemivuo-Lahti had developed. "So yes, we have a plan, but there is no money. If we'd put taxes up ten years ago with this in mind maybe we could have funded it, but now—who knows? The forestry and fisheries people are worried about diseases of economic consequence. The transportation people are of course worried about roadways, and the marine sector has concerns, too; they're all aware but no one has a budget."

I felt something enormous had nevertheless been accomplished even in putting in place the framework for such a far-reaching strategy, more so in that Niemivuo-Lahti had spearheaded it on her own despite a raft of other duties. But where I hoped for insight into how *Section IV: The Action Plan* might proceed in the advent of some Revenue Department lottery win, bureaucratic procedure meant that she could only enumerate tangible accomplishments to date and locate Finland within the bigger global invasives picture.

"We are somewhere in a corner, or on the edge being so far north compared to Central Europe which has *really* big problems," she said. "So our list of invasive species is not so long. Of course this might change because of climate, travelling, globalization, and Internet shopping."

Internet shopping? Here was something I hadn't thought of.

"You can buy all pets, terrarium and aquarium species, live food for fish, live baits for fishing, and pretty much any other kind of plant or animal you want on the Internet," Niemivuo-Lahti elaborated. "I brought this up at a Convention on Biological Diversity meeting in Montreal because I think we need more cooperation on this front. No single country can do anything on its own and there should be a symbol required on all Internet transactions between CBD signatories to warn against selling or buying species that pose an invasive hazard. Many countries were on board but some Latin American countries were opposed. At least the recommendations are in a document and will be addressed. Right now nobody is controlling the trade of invasives on the Internet—it's like the Wild West."

A new look appeared on her face, a sad smile—worry stirred by a clever analogy?

"Do you feel pressure to get something done, regardless of the challenges?" I asked.

She held up two fingers several centimeters apart. I assumed it was the international gesture for a bit. "Look how thick this is," she responded, staring at the telephone-book-like volume before us. "That's a *lot* to do."

It was, and hard to imagine where to even start when every*thing* in that book could soon be every*where* in the country. It put me to thinking about a few days back, when I had attended a luncheon at Tertin Kartano, a traditional restaurant in a historical country manor near Mikkeli. Outside, guests had wandered through a large organic garden, flower beds, and neatly clustered topiary. Upsetting the otherwise bucolic scene, the largest, densest, most vibrant shrubbery was a heavily trimmed stand of Japanese knotweed hugging a century-old apple tree, visually bursting at the edges it had been assigned. The apple tree looked none too happy, but landscapers were clearly unconcerned, and that struck me as tragic. Later, on the highway, I'd spotted a lone Bohemian knotweed bursting from the base of a tall, dense hedge, pointing roadward like some giant, green, hitchhiking thumb. Though already large, the infestation was really just under way, which meant it could be knocked back with a quick herbicide treatment were the town to take it seriously—and could afford it.

But as with myriad similar explosions occurring simultaneously around the world, most folks here were deaf to this one, and, were they even cognizant of the concussive waves of a knotweed infestation, might be more inclined only to run for cover than do anything definitive to halt the ecological carpet-bombing to come.

## Item 34

Although a national magazine called it "The Plant That's Eating B.C.," a nickname that Pacific Northwesterners from Seattle to Squamish would heartily endorse, the accomplishments of Japanese knotweed far exceed such humble kudos: it is actually eating the world. Although the SSISC had done a laudable job of keeping it out of Whistler, other places hadn't been as lucky.

Over four decades of essays and scientific papers, botanist John Bailey had tracked the story of knotweed's migration from the Orient to his native Britain, where its human shepherds lost control and the plant became the greatest biological disaster that country has ever seen. The tale is instructive on many levels, but one facet seems most responsible for the thoroughness with which knotweed blanketed its new home: no one saw it coming.

Until its forced relocation to Europe, *Fallopia japonica* var. *japonica* was a towering herbaceous perennial minding its own business along riverbanks and forest edges in Japan, China, and Korea. There, natural checks and balances included competing for space and light with other plants, struggling with a cadre of allelopathic neighbors, rapacious invertebrates, and pathogenic fungi. When freed of such constraints, however, this hollow-stalked, luxuriously leafed, pleasingly florescent "false bamboo" was known by its Oriental cultivators to grow with startling vigor under a range of conditions. It was thus presented as a handsome garden ornamental to the colorful German surgeon, explorer, and naturalist Philipp Franz von Siebold by his hosts while he was stationed in Japan from 1823 to 1829. Most of an extensive collection of live plants amassed during this period, including the knotweed, eventually made it back to Europe, but according to Bailey, a series of "fascinating peregrinations" that included shipwreck, imprisonment, and civil war kept von Siebold from reuniting with them until 1830.

Having fared better than he over the period, the knotweed was soon flourishing in a "Garden of Acclimatisation" at Leiden in the Netherlands, feeding sales of a new horticultural venture, Von Siebold & Company, and winning a gold medal from the Society of Agriculture & Horticulture at Utrecht for Most Interesting New Ornamental Plant. Subsequent catalogue descriptions built on this accolade, while ascribing an eyebrow-raising litany of medicinal and practical properties, the only one free of hyperbole being the claim that knotweed was "inextirpable." Thus propelled, knotweed began its cross-continent march, laying siege to castles and communes, gardens and ghettos. It would eventually sail onward to the foreign holdings of most European countries as nursery stock—or vegetative bits concealed in soil ballast.

Exotic plant cultivation was a hot ticket in Victorian times, and the insatiable demand for unique forms meant that giddy gardeners snapped up von Siebold's knotweed despite a premium price (things that grew easily tended to be costly because customers were unlikely to return). Overlooked in the rumpus, however, were the habits and habitus of a bonsai version of the plant, *Fallopia japonica* var. *compacta,* whose often sole occupancy of the lava fields on Japanese volcanoes spoke volumes about the species' general capabilities. Knotweed's über-proficiency as an invader, in fact, lies in one such evolutionary adaptation: the ability to survive coatings of hot volcanic ash by storing enormous amounts of energy in an extensive woody rhizome that can reach depths of three meters and extend perhaps five times that distance horizontally—a veritable underground tree. Thus engendered, the rhizome, or parts thereof as small as a fingernail, can remain dormant for a decade to bud at only a moment's exposure to air or water in a shifting substrate. Though this seemingly miraculous regeneration lies at the root of knotweed's spread, also worth noting are its hydraulic abilities: its thick, blunt shoots and rapid rhizome expansion, employed to split difficult mediums like lava, could also crack concrete, asphalt, or masonry, making it problematic around dwellings and other infrastructure.

Sure enough, Bailey and fellow botanist Ann Conolly would trace Britain's first critical knotweed introduction to a single male-sterile plant (female clone) donated to the Royal Botanical Gardens at Kew by "a certain Mr. Siebold of Leyden on 9 August 1850," where it appears as item 34 on a manifest of Asiatic plants that the collector doubtless hoped would curry favor for the exchange of other novelties. Passed from public gardens to commercial nurseries to private and estate gardens, the popular ornamental seems to have thrived under the sooty, sulfurous, mid-nineteenth-century city conditions in which other exotics flagged. By the turn of the century, however, knotweed had escaped all but the most rigorous of confines in numerous locales. Some of the earliest recorded non-garden sightings would come from the volcano-like spoil heaps of Welsh coal mines, though it remains unknown whether this was adventitious or deliberate (it may have been planted to help stabilize friable slopes). Though few corners of the British Isles are these days free of knotweed, Wales remains an epicenter of infestation.

By the time the plant's true nature was apparent, it was too late, and attempts to dispose of it only caused its further spread: cutting it back could result in as many new plants as there were pieces, as well as redoubled growth from the rhizomes; digging it up and moving the soil was equally counterproductive, resulting in vigorous regrowth not only at disposal sites, but also from overlooked fragments in the dig. It was like trying to put out a fire by flinging its smoldering embers far and wide with a shovel, then adding more fuel to the original conflagration. Watercourses proved particularly vulnerable; pulling a chair up to the water table with no suitable competitors in sight, knotweed took off, shading out native vegetation while its prying rhizomes fissured even the most resistant riparian areas, tumbling itself into floodwaters to effectively spread downstream (something I would later observe firsthand in Squamish with Clare Greenberg). In the late 1970s, knotweed's U.K. conquest had not only captured scientists' attention (Conolly's 1977 opus on it was groundbreaking—pun very much intended), but also inspired the gentry to begin clamoring for action. The government's first overture, the 1981 Wildlife and Countryside Act, actually criminalized cultivation of Japanese knotweed or otherwise nurturing its growth in the wild. But the plant never got the memo, continuing its spread unabated—and not just because of toothless, difficult-to-enforce legislation that also failed explicitly to prohibit transport of contaminated soil. As would soon be discovered, more insidious forces were also at work.

Until this point, botanists like Conolly and Bailey had assumed that the few aberrant male-fertile knotweed plants being found were far outnumbered by the ubiquitous male-sterile female clone that had been loosed 130 years earlier. When someone got the notion to do a chromosome analysis, however, a frightening picture emerged.

(Reader warning: Cytogenetics Ahead. Buckle Seatbelts and Slow Down.)

The original clone that arrived from Japan—identified by its typically small leaves and truncate leaf-base morphology—was actually an octoploid, carrying 8 full sets of the same 11 chromosomes. This is written by biologists as $2n = 88$, where "n" equals the chromosome contribution of one parent. And here we'll work backward a bit. With the normal condition of most organisms being diploid (that is, two full sets of chromosomes, so that in humans the diploid number is $2n = 46$, 23 chromosomes from mom, 23 from dad), and a basic chromosome number of 11, the more distant ancestor of von Siebold's knotweed clone would have been $2n = 22$, meaning that two doubling events (first to a tetraploid 44, then to an octoploid 88) had occurred over evolutionary history. In contrast, both male-fertile plants and a litany of larger-leafed female knotweeds now gathered from the countryside by researchers proved a mix of tetraploids, hexaploids, octoploids, and even decaploids (you can do the math) resulting from crosses and backcrosses between either of the two known Japanese knotweeds (*F. japonica* var. *japonica* and *F. japonica* var. *compacta*) with another introduced species, giant knotweed (*F. sachalinensis*). These hybrids were collectively labeled Bohemian knotweed (*F.* x *bohemica*).

While the emerging backstory may have been complex, its implications were clear: in addition to its prodigious spread by vegetative means, knotweed was also reproducing sexually and had been for a while, generating both viable seeds and new clonal lineages, a phenomenon that Bailey and Conolly eventually traced, through plant collections, back to the late 1800s.

Plant cytogenetics is an exhausting topic that can quickly lose even the most biologically savvy reader (or writer). So let's recap and distill: rather than simply comprising descendants of the Japanese knotweed clone plucked from his Leiden nursery by von Siebold and gifted to London's Kew Gardens, as first thought, the U.K.'s knotweed knot was actually a mix of that pioneer lineage plus a growing "swarm" of interspecies and intervariant hybrids. Naturally (or maybe not so much), this was also occurring simultaneously throughout Europe—Bohemian knotweed was so-named for its original 1983 description from the former Czechoslovakia—and pretty much everywhere else knotweed occurred outside its native range, which was most of the Northern Hemisphere.

On the one hand, it helped to know what lay behind knotweed's hitherto inexplicable success. On the other, it was further troubling given what was known about *heterosis,* or hybrid vigor—whereby a biological trait is improved due to the mixing of different parental lineages (think how much better your no-name mutt is at *everything* than that neurotic, vacuous, but lovely purebred Irish Setter next door). In other words, Japanese knotweed may have been a formidable colonist on its own, and giant knotweed somewhat less so though still quite capable, but their hybrid form put both to shame: Bohemian knotweed was an invader on both steroids and speed. Experimental studies have quantified this, repeatedly showing that invasive knotweed hybrids are indeed more competitive than their parents and that hybridization has supercharged the invasiveness of the exotic knotweed complex. Seattle and Vancouver residents won't be pleased to know that, at a minimum, 80 percent of the knotweed tearing up their communities is Bohemian.

As any gardener will tell you, when you throw a mix of flower seeds into the ground, some do better than others depending on how well-adapted at the genetic level each species is to the soil type, location, latitude, and climate in which the garden lies. Similarly, when differing plant forms meet to hybridize widely and generate an ongoing range of genotypes, each variant is challenged by the same classic tenets of natural selection: certain genotypes will do better than others in particular circumstances, and come to dominate those environments. With a range of environments available and a constant supply of new genotypes searching them out, each eventually finds its best match: it's a dating app for plants. Most weeds—and no shortage of crop species—find their ideal terroir this way. With the knotweed litany as example, horrified human eyes can now see this process as selecting for ever-better invasion success, though the plant is simply doing what it does anyway: passively sorting through colonization scenarios involving whatever mix of vegetative and sexual reproductive modes are available to it, in this case turbocharged by our addition of several easily crossed forms that otherwise wouldn't be found together. As Bailey summarizes, with knotweed we've created a massive, inadvertent breeding experiment on a worldwide scale.

Revelation of knotweed's numerous cryptic forms increased public demand for serious action. That arrived when a duty-of-care obligation was finally placed on rhizome disposal by Britain's 1990 Environmental Protection Act, legislation that birthed a nationwide industry dedicated to eradicating knotweed from development sites. Indeed the chemical and mechanical warfare that ensued has, as of 2015, totaled the equivalent of $3 billion USD; among other high-profile projects, the 2012 London Olympics famously suffered costs associated with knotweed removal from building sites to the tune of $130 million USD.

Help, however, may finally be at hand: expeditions to Japan in the new millennium were successful in pinpointing precise locales from which the original Kew clone and other knotweed variations were obtained, with the added bonus of identifying an indigenous insect biocontrol candidate. The psyllid sap-sucker *Aphlara itadori* was, after extensive testing, approved for release in carefully monitored U.K. sites in 2010—the first use of a biocontrol organism in Europe's history. Controlled releases were now also under way in British Columbia, promising knotweed relief—but not freedom—for beleaguered Lower Mainlanders.

And what of the fate of von Siebold's infamous Garden of Acclimatisation at Leiden? In the year of his death, 1866, it impressively boasted nearly a thousand different plant varieties. By the time the eminent English horticulturists F. W. Burbidge and P. Barr showed up for a tour in 1883, however, it was naught but a tangle of neglected jungle, long-since overrun by a single energetic inhabitant—Japanese knotweed.

## Hide-and-Go-Seek

We've now learned something of how thousands of nonnative animal and plant species are introduced around the globe. Despite the high frequency of introductions, however, both a saving grace and demonstration of the difficulty of colonization is that few of these ever "take." The organisms are either ill-adapted to their new environment and climate, need a few goes at it (propagule pressure) under different conditions to hit it right, or encounter stiff competition from a range of native species and, increasingly, other invasives (biotic resistance—explained in Part 4). Interrogating that small percentage of introductions that *do* become established (viz the tens rule), we've seen that successful invasions aren't always those you'd expect; the fortuitous and extraordinary often play a role.

How indeed do you hide a giant python, let alone 150,000? The question is easier to reconcile when you consider the broader role of large wilderness and protected areas like the Everglades—or vast, contiguous ecosystems like the Great Lakes—as de facto incubators for the establishment and spread of everything from snakes and lizards to a full spectrum of fish, amphibians, mollusks, insects, trees, and parasites both macro- and microscopic. Given a spacious hiding place, the rest comes down to individual biology, ecological luck, and a fair quotient of human stupidity—as demonstrated in Florida.

Barring early intervention, pythons were free to become established in the Everglades because they (1) had the requisite life-history traits for rapid reproduction and growth, (2) had climate on their side, and (3) inhabited an ecosystem only slightly less visible to the American public than the soil underfoot—through which, coincidentally, invasive Asian and European earthworms now wend their way, drawing down leaf litter to fundamentally change the energy flow in forests, much as pythons have altered the energy flow in the Everglades by vacuuming small mammals from the landscape.

These cases, as well as the knotweed juggernaut, demonstrate that the establishment and spread of a successful invasive species almost always takes place unnoticed. Once detected, prevention of a widespread invasion may still be feasible, but the time-lag until public awareness ratchets upward through concern, political will, funding, and finally action usually makes it impossible to do anything more than lapse into a regimen of expensive control in perpetuity.

This is the point on the Invasion Curve where Florida's python hunters and those battling myriad other invasions around the globe find themselves. And it's something that needs to change if we wish to effect any widespread ecological salvation. As one pundit presciently put it, "If you need to have a meeting about an invasive species, it's already too late."

# PART IV

# Invasional Meltdown

## *Impacts of Alien Invasive Species*

Not only is the sea such a foe to man who is an alien to it, but it is also a fiend
to its own offspring—worse than the Persian host who murdered his own
guests; sparing not the creatures which itself hath spawned.

*Herman Melville,* Moby Dick

# Dominoes, the Game

Fittingly, I first heard the term "invasional meltdown" from Nick Mandrak. We were having a few beers with a couple of former labmates at a conference in New Orleans in 2006. By this point I was a writer living on Canada's west coast, Mandrak a freshwater biodiversity specialist with DFO out of Burlington, another a high-powered molecular geneticist at a noted Texas university, and the last a morphological taxonomist who still studied nano-sized tropical marine reef fishes despite being based at a Winnipeg museum. Thus, as the evening wore on, topics of discussion peregrinated all over the scientific map. At some point, invasive species came up and Mandrak took over, setting aside his glass to describe, with great animation and detail, the phenomenon in which two or more invasives teamed up to have an effect far greater than either would on its own. It had made abundant sense and, intrigued, I co-opted the conversation for an earlier book. Because it well serves our purposes here, however, allow me to paraphrase a reprise.

In that musty Bourbon Street bar, Mandrak painted an eye-popping tableau that began with the zebra mussel's unseen introduction to the Great Lakes. Flushed from the ballast of transoceanic cargo ships in the mid-1980s, it wasn't long before *Dreissena polymorpha,* native to Eastern Europe and the Ponto-Caspian region, made itself at home on this continent. First reported in 1988 from Lake St. Clair (the small basin that briefly interrupts the St. Clair–Detroit River system connecting Lakes Huron and Erie), by 1990 it was known from all five Great Lakes. A reproductive dynamo whose females produced thirty to forty thousand eggs per breeding cycle and up to a million each year, the fingernail-sized mollusk rapidly multiplied out of control, blanketing surfaces and substrates to wreak havoc on ecosystems, infrastructure, and economies in the Great Lakes and adjacent freshwater areas to which it was secondarily transported, mostly via recreational boating. (Already by 1993 it had made its way down the length of the Mississippi to where we now sat discussing its unwelcome success.)

The prolific breeder was also a prolific filter-feeder, with every individual processing around a liter of water each day. This proclivity had an unexpectedly welcome effect: notoriously turbid water bodies like

Lake Erie became preternaturally aquamarine, an illusion of health recalling the appealing-but-spooky clarity of acid-rain-affected lakes on
the Canadian Shield in the 1970s. Meanwhile, the round goby (*Neogobius melanostomus*)—a small Ponto-Caspian fish tumbled from ballast
into the Lakes in the 1990s—happily discovered a favorite snack from
home practically paving its new environment: zebra mussel. With a preferred, fast-growing food source in place, the equally prolific goby (females laid some five thousand eggs several times each season) also took
off, outcompeting native benthic fishes for habitat and eating their eggs
for good measure. Where at first it seemed that the consumptive relationship between goby and zebra mussel might apply a fortuitous brake
to the latter, it now appeared the pair had lapsed into a classic predator-
prey cycle of mutually exclusive boom-and-bust. But even that wasn't
quite so simple, said Mandrak, as he wound up for the finale to his tale.

    As zebra mussel peaked and the water cleared, the increased light
penetration helped to facilitate an invasion by the exotic aquatic plant Eurasian watermilfoil (*Myriophyllum spicatum*), which sent the Lakes back in
the direction of the weed-choked 1960s. The Lakes' newfound limpidity,
combined with accelerated enrichments from agri-runoff, also led to massive blooms of cyanobacteria (a.k.a. blue-green algae) that produced toxic
microcystins, which, when ingested, were harmful to animals, including
humans. But an even greater health threat loomed: when the blooms died
off, their decomposition rapidly burned up dissolved oxygen, encouraging
the growth of anaerobic bacteria like *Clostridium botulinum,* the source of
deadly botulin toxin. Concentrated via bioaccumulation by filter-feeders
like zebra mussel that were subsequently consumed, botulism caused unprecedented die-offs among fish and birds around the Great Lakes. At the
same time, the tendency of zebra mussel to anchor on any hard substrate
was putting native mollusks out of business. With bottom densities of up
to 10,000 zebra mussel per square meter, the invaders stole not just space;
several hundred could coat the shell of a single large indigenous clam, in-
tercepting its food. Yet another Ponto-Caspian transplant, the tiny hydroid
*Cordylophora caspia,* just so happened to feed on free-swimming zebra
mussel larvae—ironically, by using mussel shells as anchorage. As imagined, this one-two facilitation meant that hydroid populations expanded
dramatically in Lake Michigan after extensive zebra mussel beds formed.

Perhaps most worrying, however, was that this giant, basin-wide filtering machine was starving off the Great Lakes' plankton-and-bacteria-feeding benthic macroinvertebrates, the food source of baitfish like alewife. Though an invader itself, alewife had become the major food source for top predators like lake trout and introduced Pacific salmonids, so its rapid disappearance was already having a negative impact on those species.

*That,* Mandrak had said with finality, catching his breath and reaching for his beer, was invasional meltdown. World-weary biologists all, well-versed on a myriad of familiar ecological tipping-points, we sat in stony, contemplative silence. At last, someone spoke. "So . . . another round?"

## Dominoes, the Theory

The concept of invasional meltdown was first proposed by Daniel Simberloff and Betsy Von Holle in a seminal 1999 paper in the first edition of the zeitgeist journal *Biological Invasions,* helping kick off what quickly became one of the most important primary sources in all of conservation and ecology. For this auspicious contribution, the authors combed the literature to determine the phenomenon's frequency and range of processes that might contribute to it. But first they had to set the terms—literally: "We suggest . . . 'invasional meltdown' for the process by which a group of nonindigenous species facilitate one another's invasion in various ways, increasing the likelihood of survival and/or of ecological impact, and possibly the magnitude of impact. Thus, there is an accelerating accumulation of introduced species and effects rather than a deceleration as envisioned in the biotic resistance model."

Biotic resistance is the idea that species richness determines invasibility, such that places with fewer species—for instance islands and disturbed areas—are easier to invade. A corollary holds that as invaders accumulate in these habitats it should also become increasingly difficult for new species to muscle their way in, so to speak, due largely to expected resource competition and interference, that is, negative interactions. What Simberloff and Van Holle proposed was that there was also much evidence to support an invasion model of *positive* interactions, where a more-the-merrier scenario might make it easier for new invasives not only to join the party, but also to crank it up a notch. Or two—they also thought it possible that these interactions could trigger an autocatalytic process that would turbo-charge the wholesale replacement of native communities. Their literature search identified basic facilitative interactions of a few direct and indirect types, but the authors centered their discussion on mutualisms involving (1) plants and the animals that dispersed and/or pollinated them, and (2) various modifications of habitat by both animals and plants. They found "little evidence that interference among introduced species at levels currently observed significantly impedes further invasions," concluding that "synergistic interactions among invaders may well lead to accelerated impacts on native ecosystems."

The paper was significant in illustrating that not only could non-native species threaten through direct competitive interactions with natives, but that the complex, self-reinforcing mutualisms of entire suites of nonnative species also could rival other anthropogenic drivers like climate change and habitat loss as a force in biodiversity decline. (After writing this sentence, I had to re-read it a few times to let the implications fully sink in; you should, too.)

Although the authors came up short on a stated goal of determining the *frequency* of invasional meltdown (largely because the kinds of facilitations involved weren't a common empirical pursuit among researchers at the time), they listed sufficient examples to make this much clear: there were likely more invasional meltdown scenarios out there than would ever be known or even imagined—especially cascades facilitated by introduced microbes and other unseen organisms.

The zebra mussel debacle is frequently upheld as a classic invasional meltdown scenario, but others are just as instructive of the phenomenon's breadth. One of the most widespread landscape-level examples is the invasion of 400,000 square kilometers of the western United States by exotic cheatgrass (*Bromus tectorum*), which arrived in Eurasian grain imports during the 1800s and whose path to dominance was paved by the extensive grazing and trampling of native grasses by introduced livestock. Cheatgrass outcompetes native vegetation by sprouting earlier to "cheat" the native plants of their water and nutrients. It thrives on any kind of disturbance, particularly fire—which it also creates fuel and conditions for, such that a cheatgrass-fire-feedback cycle has transformed the American West from shrub steppe to grassland. Another oft-cited example comes from

Christmas Island, an Australian territory in the tropical Indian Ocean. Here, the introduced yellow crazy ant (*Anoplolepis gracilipes*)—which, like the European fire ant protects scale insects in return for honeydew—was already problematic when two other events occurred: numbers of a native scale insect soared, and a nonnative scale insect was introduced. The resulting explosive increase in the abundance of honeydew led to an equally explosive surge in the ant population, which, in turn, devastated a hitherto large population of herbivorous native red crab (*Gecarcoidea natalis*). The sudden decline in crab led to a massive increase in forest undergrowth, facilitating more scale insects and more ants, as well as increasing the island's vulnerability to plant invasions.

The invasional meltdown hypothesis offers great post-hoc explanatory power for why some introduced species thrive in their new range and others simply die out. But could the "adding fuel to the fire" nature of mutual facilitations by nonnative species be shown to hold *predictive* power when it came to species introductions and invasiveness? That hope was swimming upstream in the strengthening currents of invasion science. Like many young disciplines, its earliest ideas remained its most entrenched, and invasional meltdown fought for a place in the matrix. But as the science grew, and more-refined studies populated its literature, the bulwarks began to give. In a telling 2012 study, literature was again examined to determine if six popular invasion hypotheses were supported or refuted by experiments conducted across different groups of invaders and habitats. These were:

**Biotic Resistance**: Ecosystems with high biodiversity are more resistant to invaders than those with low biodiversity.

**Island Susceptibility**: Invasive species are more likely to establish and have major ecological impacts on islands than on continents (due to factors related to biotic resistance).

**Invasional Meltdown**: The presence of invasive species in an ecosystem facilitates invasion by additional species, increasing their survival or ecological impact.

**Novel Weapons**: Invasive species have a competitive advantage against natives because they possess a trait that affects natives negatively (for example, allelopathy).

**Enemy Release**: The absence of enemies in the exotic environment facilitates invasions.

**Tens Rule**: Approximately 10 percent of species transported beyond their native range will be introduced to the wild; about 10 percent of those introductions will establish in their new locale; and around 10 percent of those established will become invasive, or pests (that is, approximately 1 out of 1,000 species introduced will become problematic).

The truly global review revealed that those hypotheses which considered interactions of alien invaders with their new environments were also best supported by empirical evidence: at 77 percent, invasional meltdown won the contest hands down, followed by novel weapons at 74 percent, and enemy release at 54 percent. None of the other hypotheses was particularly well supported: biotic resistance came in at 29 percent; tens rule, 28 percent; and island susceptibility, 11 percent. The study also showed that empirical support for all six hypotheses declined over time, with support differing among taxonomic groups and habitats.

Because the effectiveness of any research, policymaking, and management of invasive species is dependent on sound hypotheses, the results of this review had far-reaching implications. For the Great Lakes, those included not only prevention measures around accidental introductions—for example, the ballast-water initiative—but also 20/20 hindsight for many intentional introductions. And when it came to a combination of the two, no single concern was more central than sportfishing. With its (largely) politically approved sequential introductions and reintroductions of large predatory fishes—often veiled as attempts to suppress or manipulate the invasive prey-base supporting them—and legion of rabid, often rogue anglers over which it had no control, the hubristic sportfishing industry virtually defined mismanagement.

## Outdoorsy

"If the sportfishing industry *knows* it's the primary vector of both authorized and unauthorized nonnative fish introductions, what the hell is it doing to mitigate its own impacts?" This was the kind of indignation you could express as a journalist who sees much of sportfishing in the same light as trophy hunting and monster truck rallies. Mandrak, to his credit, had a more diplomatic take.

"The Ontario Federation of Anglers and Hunters [OFAH] actually does great stuff around invasive species," he answers.

The organization did, in fact, pursue excellent and directed outreach, but I also knew this to be selective, purposely excluding from its purview nonnative sportfish that now resided at the top of the Great Lakes food chain. OFAH was obviously aiming to protect hunting and angling opportunities: invasive species that subverted this mission were declared enemies, while those invasives prized by its membership were a different story. If this seemed confused, at least I received confirmation from Mandrak that it very much was.

"There's really two questions. First, in the Great Lakes there's a legacy and tradition of legal stocking—rainbow trout, coho, Chinook, etcetera. Although most have become naturalized and self-perpetuating, stocks continue to be enhanced by some organizations," he says, referencing hatcheries in Michigan and Ontario that essentially raise invasive species to be dumped into the Lakes. "And you have to wonder what kind of message that sends, and if it sort of gives people the green light to do it anywhere on their own. The second question is about the movement of invasive species *inland.* OFAH has a 'Don't Move Bait' campaign to address that, and yet, though it's *also* illegal to move introduced sportfish, where's the outreach on that?"

Mandrak digresses briefly to ongoing efforts to return Atlantic salmon to Lake Ontario, where it was native until the turn of the twentieth century. "It's going to be hard for a whole bunch of reasons. There's competition with introduced Pacific salmonids, both in the lake for feeding and in rivers for breeding. And on some rivers they'll have to remove lamprey barriers. So how do you reconcile reestablishing native Atlantic salmon while managing for introduced Pacific salmon that represent a huge, multimillion-dollar recreational sportfishery? Thousands of folks who fish Chinook and coho in Lake Ontario every year are unlikely to be happy about any policy or initiative that changes their availability; those individuals can't catch politicians' attention, but OFAH, which speaks for all sportfishing with one voice, will."

He was suggesting that OFAH would have something to say about Atlantic salmon restoration—they would insist on protection of Pacific salmonids—and that they had the clout to do so. As for the outreach that OFAH *did* engage in, given how sportfishing already far outpaced other pathways for the introduction of aquatic nuisance species, one had to be thankful for any mitigation; if the industry wasn't already self-regulating to some extent, it would have been, as they say, off the charts in dwarfing other pathways.

"How much is the bait industry in Ontario worth—a few million dollars a year maybe? It's nothing, but killing it is off the table, so the industry is trying to work with the province to minimize impacts. For instance, you can't sell or use crayfish as bait anymore because of rusty crayfish invasions [the Ohio River basin native has been spread to Ontario and twenty nonnative states by anglers]; likewise frogs because of ranavirus [we'll get to this]. But those two are at the margins of the bait industry. When we talk along a spectrum from banning bait sales outright, to allowing only a few species, to restricting bait to same-lake-only, pushback is strong and we don't get very far. When VHS [viral hemorrhagic septicemia, a deadly infectious fish disease] hit the Great Lakes, commercial harvest zones that you couldn't transport fish outside of were put in place, but fishermen could *still* buy live bait in any of these zones and drive it to anywhere else. It's political. It always comes back to the whole reason something like the OFAH exists."

What he meant was this: a large, influential, and environmentally proactive organization like Ducks Unlimited might raise tens of thousands of dollars for wetland restoration and protection, and work in concert with government to effect it, but the noblesse oblige only went so far. Because its prime motivation was to conserve duck populations at a level that would allow its members to continue to blast them from the sky, if you sat down with the organization to discuss problematic crucial elements of the practice like, say, duck decoys slathered in lead-based paint that's toxic to aquatic organisms (now rectified) or blind-building because hunters were transporting DIY blinds fashioned from invasive *Phragmites* (true story), they might, depending on how big an impact it held for members, use their political "stakeholder" leverage to block any such initiative. In the fish world, DFO's Asian carp EDRR program was an exception in being a virtually unpoliticized entity with universal buy-in, likely because, in the event that Asian carp became established, the effects on all stakeholders would be so significant and so widespread. It was, in fact, a most pluralistic act of desperation, hammered home in the second of Asiancarp.ca's fall 2014 webinar series.

In "Multi-Jurisdictional Approach to Asian Carp in the Upper Illinois River and Chicago Area Waterway System," Kevin Irons, aquaculture and aquatic nuisance species program manager for the Illinois Department of Natural Resources, described why Asian carps were an issue in the United States, and why the battlefront in Illinois was so important. In a wide-ranging talk about monitoring and response that touched on CAWS, electric barriers, population control, and the development of partnerships across geographic borders, what stood out were Irons's comments on impacts.

His starting place was common carp—"grandpa's carp" as he called it—which had been introduced by Europeans for food and angling in the 1880s and subsequently "in a Johnny Appleseed fashion sprinkled across the U.S. and Canada," he'd quipped. The common carp has "been around a while and probably has a bad name; I often refer to carp as being a four-letter word." His point? The public needed to understand that while common carp was a familiar and accepted nuisance (per Shifting Baseline Syndrome), Asian carps were something *very* different. After they began spawning in high numbers in the new millennium, he noted, it took only until mid-decade before silver and bighead carp had displaced most other fishes in the Illinois River to numerically represent almost 80 percent of all individuals. In one stretch—La Grange Reach near Peoria—monitored landings for silver carp peaked at over 100,000 for the year 2009. By 2010, however, long-term monitoring was showing the condition of Asian carps to be declining on the Illinois and other rivers due to internecine competition caused by these same large numbers. Similar impacts were apparent in some of the few indigenous fish that remained in appreciable proportions, like bigmouth buffalo (*Ictiobus cyprinellus*) and gizzard shad (*Dorosoma cepedianum*), whose condition had measurably declined from pre-Asian carp days. The expected drawdown of zoo- and phytoplankton if these carps reached the Great Lakes ecosystem was predicted to have similar effects on a wide range of species.

Furthermore, Irons noted, what we *think* we know about these fish may not necessarily be helpful to predict outcomes: while they're clearly good invaders, their reproductive plasticity was confounding—some years had three or more spawning peaks, other years one, or none. If you didn't know when fish would breed and in what numbers, it was hard to maintain a level of scrutiny or outreach that could keep them from accidentally being moved around as baitfish. He finished with another blunt homily on fighting the spread of aquatic invasive species. "Be a Hero, Transport Zero" he reminded the webinarians, before flashing the sticker distributed by the group who'd come up with the slogan.

## Pyramid Scheme

Since sportfishing involves mostly top predators, the status of preyfish is always an important part of the equation. And across the Great Lakes, preyfish biomass has been decreasing since 1988, precipitously so since the early 1990s. All evidence points to a combination of pressures, including the impact of predation by introduced salmonids and the compounding effects of sequential explosive expansions of first zebra, then quagga mussel, as well as other invasive species. The only bright spot is the Lake Superior preyfish community, where the proportion of native species comprising that assemblage was on the rise, increasing the prey base's ability to support recovery of wild native lake trout in that basin. Looking over a troubling chart of the swooning preyfish symphony, I notice a particularly sour note: populations of commercially important lake whitefish (*Coregonus clupeaformis*) had dropped in lockstep with the smaller and more traditional preyfish species like alewife, and rainbow smelt. Why would this be so? Mandrak wasn't sure it was. He thought the whitefish had probably just moved.

"Lake whitefish originally fed on *Diporeia* so these numbers are likely an artifact of history and tradition," he tells me. (Amphipods of the genus *Diporeia* were the basis for the deepwater food web in the vast proglacial lakes that pooled in front of North America's retreating Pleistocene ice sheets, and along with the opossum shrimp, *Mysis relicta*, continued this role in relict waterways like the Great Lakes.) "When *Diporeia* disappeared about a decade ago, whitefish switched to eating quagga mussel, a deepwater species, which likely shifted their populations away from areas where commercial fishermen and researchers originally found—and counted—them."

Thus, by adapting to shifting conditions and food sources faster than we were aware, whitefish could be sending false signals from the deep to complicate the received view topside and, ultimately, any short-term management directives. There was, of course, more.

Silver chub (*Macrhybopsis storeriana*) is a large native minnow whose core distribution in the lakes centers on the lower Huron and western Erie basins. Linked to the Mississippi refugium—one of several geographic areas in which certain plants and animals persisted during the height of the last glaciation—its spread into the Great Lakes was limited by factors other than strict geography. A large river fish across the remainder of its range, silver chub is a broadcast spawner requiring a particular length of river of certain temperature and speed (much like Asian carps—a fortuitous parallel when mapping potential spawning areas for these invaders). In Lake Erie, these requirements confined it to the immediate area around the Detroit River. In addition, as far as was known, silver chub fed exclusively on benthic mayfly larvae of genus *Hexagenia*. In the late 1950s, both *Hexagenia* and silver chub disappeared, suggesting that the latter might have been extirpated. "The present status of the species in Lake Erie is in doubt," wrote the revered team of Bev Scott and E. J. Crossman in their 1973 tome *The Freshwater Fishes of Canada*. When silver chub then reappeared in the early 1990s with a resurgence in *Hexagenia* documented shortly thereafter, many saw these events as not only connected, but also related to the diversion of plankton foods to bottom substrates by invasive zebra mussel. Mandrak knew better.

"There's a strange phenomenology associated with invasives that causes people to make hopeful claims about benefits. It leads to patterns being misinterpreted, like 'Hey, *Hexagenia* is back and oh—hasn't its major predator, silver chub, been coming back, too?'"

Because scientific discovery is an intellectual journey that moves from waypoint to waypoint, it creates a de facto story. No surprise, then, that when recounting such discoveries, scientists often adopt the manner of entertainers, meting out information in such a way as to effect maximum drama. It wasn't hard to identify Mandrak's setup for yet another cautionary tale. It was New Orleans all over again.

"But when you map the return of silver chub," he continues, voice elevating in the way of a sermoning scientist about to deliver the moral eureka, "the population increase actually *predates* the increase in *Hexagenia*."

Aha! . . . but what did it mean?

"We did gut-content checks on expanding silver chub populations and guess what? They were loaded with zebra mussel. Me and two guys from USGS are writing a monograph about this because . . . well, we're the only people who care. My take is this: all cyprinids [the family to which chub belong] have pharyngeal [throat] teeth; in molluscivores like drum these are robust because they're used to crush shells. Not so much in silver chub, which largely eat softer-bodied invertebrates, but broken mussel shells in the gut suggest they'd have to be using those teeth."

Alternatively, Mandrak notes with little enthusiasm, silver chub may hang out in the spill zone where true molluscivores like drum are feeding, and simply clean up after them. Regardless, neither case explained *Hexagenia*'s spike in western Lake Erie since the early 1990s—or its current state of decline. That part, according to Mandrak, was simple.

"*Hexagenia* disappeared because of hypoxic conditions, and it came back in the '90s when those improved. But now we have those 'dead zone' conditions all over again; basically, we're right back to the '50s," he finishes. "We don't know how the rest will play out."

## Peregrine

Not knowing how things will play out is a hallmark of biological invasions. Maybe *the* hallmark. There's just so much going on out there to *not* be aware of.

To start, there was clearly much we didn't know about fish, a notion that always put me in mind of something my daughter, Myles, once said. As a child, she had a habit of stumping me out of the blue with sizzling bio-questions. "If you dropped a frog in the middle of a lake, would it know which way to swim?" had caused me both the ignominy of admitting I didn't know, and to ponder what was going on in her head. Years ago, on a canoe trip in central Ontario's Algonquin Park, after hours of paddling in contemplative silence, she'd delivered a similar zinger: "If a fish has never *seen* an earthworm before, how does it know to eat it?"

That was a good question given that no fish could claim a diet of earthworms. There were all sorts of quasi-answers I might have offered—evolutionary memory, generalized invertebrate predation, familiarity with smaller oligochaetes populating muddy lake bottoms and with freshwater leeches. I refrained, in the end, because her basic premise stood tall: why *did* we use earthworms as bait if they were a novelty to a fish?

In reality the question ran deeper than the obvious terrestrial-aquatic schism to which she alluded. Not only would the fish where we paddled be generally unfamiliar with burrowing soil organisms, they'd never have encountered even a stray washed into the water for one simple reason: earthworms didn't exist here.

Scraped from the land by kilometer-high ice sheets that lasted until ten thousand years ago, the native earthworm line in east-central North America now lay 1,000 km south, somewhere beyond Chicago. The billions of seemingly familiar earthworms in all of Canada, New England, and the upper Midwest were, to a one, species seeded on this continent by European settlers and moved around by the rest of us. Across Algonquin Park's 7,653-square-kilometer expanse, incursions of invasive worms might be found along peripheral roadways, in and around campgrounds and boat launches, and anywhere that buildings had been erected, but they would almost certainly be absent from the wild central areas of the park, a height of land for numerous watersheds where the only human traffic was by canoe.

That meant that the earthworms I'd dug from our Toronto garden as a child to feed my pet gartersnake, or whose post-rain sidewalk strandings always drew my sympathy, were actually invasive—a relatively recent revelation that, for me as for most, has been startling. There were some, however, who might not have been so surprised.

In the concluding chapter of *The Formation of Vegetable Mould Through the Action of Worms: With Observations on Their Habits,* the

1881 volume in which he summarized his forty years of studying them, Charles Darwin states: "Worms have played a more important part in the history of the world than most persons would at first suppose." Detailing their import in the decomposition of rocks, denudation of land, and improvement of soil for plant growth, Darwin makes the case for earthworms as a planetary agent, an "unsung creature which, in its untold millions, transformed the land as the coral polyps did the tropical sea."

This wasn't a new thought for Darwin, who'd presented a paper on earthworms' role in soil formation to the august Geological Society of London back in 1837. Although most of his peers found the topic "uncommonly mundane," respected geologist William Buckland praised it as "a new & important theory to explain Phenomena of universal occurrence on the surface of the Earth—in fact a new Geological Power."

Quite right, and one imagines that with his keen, calculating mind and eye for meticulous observation, were Chuck aware that earthworms had been scrubbed from some lands by glaciers, he might well have predicted the titanic effects their sudden reintroduction would have on ecosystems that had become established, via the tangled web of natural selection, in their absence.

Introduced exotic earthworms now occupied every continent save Antarctica (though I wouldn't bet against their introduction there, either), inhabiting every biogeographic region save the driest deserts and permanently frozen habitats. Early naturalists noted the pan-global distribution of a handful of species, rightly ascribing their presence to human trade and colonization pathways and labeling these "peregrine" for their ability to range widely and jump oceans in the process, a trait "denied to the majority of worms, and probably due to the direct interference of man," according to the 2009 paper by Paul Hendrix and colleagues titled "Pandora's Box Contained Bait: The Global Problem of Introduced Earthworms." Around 120 species were now known to have been distributed across regions or even globally, with—unsurprisingly—those adapted to adventitious transport and colonization of disturbed habitats most widespread and likely to be invasive. Two decades of concerted research clearly show that nonnative earthworms significantly alter the structural properties of native soil, organic matter, and nutrient dynamics, as well as the dependent plant and animal communities both above and below ground. With global invasions ongoing, the ultimate effects of introduced earthworms on soils, biota, and ecosystem processes in the temperate and tropical regions they inhabit cannot be known. But if someone had the patience, he or she could learn a lot by following this slow-motion tableau to see, well . . . how it plays out.

## The Lowdown

"Invasive earthworms are like the Asian carp of terrestrial ecosystems— it's too bad we didn't see them coming."

This was one of several memorable comments that surfaced during the time I'd spent discussing these creatures with Lee E. Frelich, research associate and director of the University of Minnesota Center for Forest Ecology in Saint Paul. A virtual quote machine when it came to invasive earthworms, he was also a sharp and dedicated researcher with an important theory. It was short, sweet, and somewhat scary: exotic earthworms + deer = major transformation of the forest. To wit: the bare soils and lower nutrient status resulting from worm invasions should favor hemlock and oak over maple, but deer will browse those species to the ground; thus, in partnership, deer and earthworms would help push the mixed deciduous forests of central North America toward savanna, and climate change would finish the job. Compelling and question-begging as this exploration is, we'll circle back to it later.

A forest ecologist, Frelich came by his interest in worms innocently enough. In the late 1990s, the owner of a parcel of old-growth forest called him up. "What happened to all my trilliums?" he'd wanted to know. Frelich decided to have a look. The first thing he noticed was that the forest was heavily infested with earthworms, and there was no leaf litter or duff (detritus) on the forest floor; the worms had drawn it all down. It would still be a couple years before the first media coverage of invasive earthworms, but this being previously glaciated land, Frelich knew the worms sure as hell weren't native. Graduate school had also taught him that trillium bulbs were typically embedded in the thick duff of forests. Putting two and two together, he saw an area of ecology worth investigating.

When Frelich got started, a lone worm research group was looking at invasion boundaries and nutrient dynamics in upstate New York; their early findings showed that worms were bad for a rare fern. Otherwise, though a vast literature on introduced earthworms in agricultural systems extended back decades, only a handful of other forest-related papers existed. "We were first to have a systematic, widespread program looking at the entire scope of native plant systems," recalls Frelich. "Now hundreds of studies a year are published worldwide."

These studies demonstrate big changes to groups of arthropods, small mammals, and amphibians in all earthworm-invaded forest communities. For instance, along several invading fronts in New York, earthworms led to a paucity of salamanders; the salamanders could eat smaller worms, but they were losing the shelter of the duff. Few analyses had been done for small mammals, in which the main effect appeared to be a shifting abundance among species; that is, changes to the *quality* of habitat made it "more interesting" for one species and less for another. Insects and fungi showed huge changes in their physical structure and their characteristics of habitat and community, some positive and some negative, but all confounded by the presence of exotics of unknown impact ranging from beetles to slugs. It was into this fray that Frelich happily waded. "We knew there were impacts on native species across all taxonomic groups, so our questions were these: Which species are the winners and losers when worms invade? Do invading earthworms facilitate the invasion of other species from their same ecosystems? Do worms impact productivity and how will this interact with climate change?"

In central North American forests, an additional factor raised even more questions.

"In a lot of cases, we saw major effects occurring because of interaction between earthworms and a high deer population," he'd said of early observations. "But there was a lot of variability across a landscape, so there was more of a mosaic of impacts: trilliums might be wiped out in some places and only mildly perturbed in others; some sites remained above minimum threshold requirements for trilliums either because the soil was very rich or the climate really good—not marginal like Minnesota where the prairie-forest border sees impacts from the interaction of poorer soils, a changing climate . . . and a lot of deer."

This piqued Frelich's interest, and he'd become evangelical about the threat that existed in this particular combination—easily demonstrated by known observations and phenomena. Ongoing investigations of the facilitative interactions among deer, earthworms, and various plants along the prairie-forest ecotone led to his 2010 paper with Peter Reich outlining the "Savannification" theory (Figure 5). Within fifty to a hundred years, they argued, a warming climate could be expected to have major effects on boreal and northern hardwood forests along the prairie-forest biome boundary, which, according to pollen data, had shifted northeast during past episodes of global warming, and should be expected to do so again. They believed that the future climate would bring "greater frequency of droughts, fires, forest-leveling windstorms, and outbreaks of native and exotic insect pests and diseases," leading to higher mortality among mature trees. Then the kicker: irruptive populations of native whitetail deer coupled with earthworm invasions would inhibit the establishment of replacement tree seedlings. With adult trees lost faster than they could be replaced, the net impact would be savannification of the forest, with a concomitant greater magnitude—and faster northeastward shift—of the prairie-forest ecotone than would occur solely because of temperature change.

As Frelich's academic marquee, the Savannification theory elegantly folds together the global phenomenon of climate change, the local problem of deer irruptions, an invasive soil organism, and synergistic effects of these on plants. Still, a few fundamentals might have helped my understanding—such as where the worms were, where they weren't, and where they were going.

"The site that got me started here in Saint Paul was totally invaded, but there are sites with ongoing invasion fronts about [240 km] north of here," said Frelich, happy to fill in the blanks. "You can detect the effects of worm invasions in northern hardwood forests through tree rings. We demonstrated that when worms invade, the ring pattern suddenly becomes different. If you reanalyze that data given what we now know, you can see a 30 percent growth reduction for in situ trees—that is, older stands of maple that were invaded several decades ago. The part of a

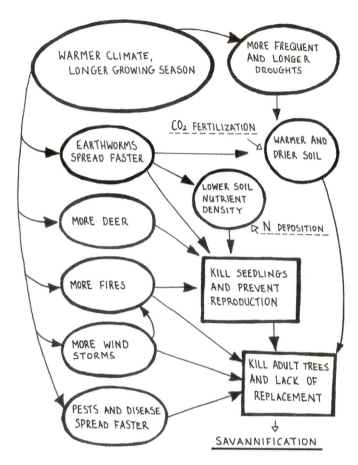

Figure 5. The Savannification theory of Lee E. Frelich and Peter B. Reich. (Adapted from Frelich and Reich 2010 with permission from the authors.)

stand where worms first invaded shows an almost instant change related to a decrease in growth, so you can track the front across a block of land with tree cores."

Frelich was also excited about a new, less labor-intensive way of tracking invasion fronts from a tablet screen. "We can follow worm invasions with photography using drones, measuring the Normalized Vegetation Index—which delivers information on how fast forests are growing by how much photosynthesis is taking place."

This was functionally similar to a satellite looking at the spectral reflectance of forests, but the greater proximity of the drone to the canopy provided much more detail. As with tree-ring measurement, reduction in tree growth signals that the entire ecosystem is simplifying, that native species are disappearing and making forests more vulnerable. This was big stuff, and not everyone was ready to hear it.

"A whole group of agricultural earthworm people don't like that we've found so many negative effects," Frelich admitted. "And of course there are the usual deniers who don't think *anything* is happening. We've got the attention of wildlife ecologists and conservationists now because of the links in the food chain and impacts on certain bird species, but we haven't gotten the attention of other key people—for instance those interested in water quality. Nitrogen and phosphorous leach out of forest soils when earthworms invade, so terrestrial-aquatic linkages and changes they could lead to need looking at."

In fact, he noted, what was occurring in Minnesota, Wisconsin, and eastern North American forests could be seen, in some cases, as wholesale ecological reversal: once-dominant plants now played a minor role, while once minor players now dominated. And there was, as always, more.

Recent papers showed that invasive European earthworms were facilitating the establishment of invasive European plants. "In many cases an invasive plant problem is actually an invasive *earthworm* problem. Propagule pressure isn't a big deal in this case because most of these plants wouldn't do well without worms: [native] plants evolved to germinate on duff soils, and these Euros evolved to germinate on bare mineral soils. That's why native sedge takes off—it was limited by leaf litter before. So it's one of the native winners. And Euro plants are winners because they can spread faster than natives."

One area Frelich studied was Minnesota's remote Boundary Waters, a cherished wilderness along the Canadian border currently under invasion with concomitant changes to the forest. His findings weren't good: 70 percent of the Boundary Waters had invasive earthworms and 30 percent of that was heavily infested. "This is highly related to campsite and portage areas. Worms go where people go, and eventually follow us everywhere."

So much for assuming that remote areas of Algonquin Park might be spared. Was anyone in government or parks departments on either side of the border listening? "They're thinking about it because nowadays there are all sorts of climate change and invasive problems for these places. In the U.S., some wonder if we need an amendment to the Wilderness Act to deal with it, but if you open *that* can of worms—sorry—it'll probably kill the whole thing given current politics."

What any amendment might achieve at this point was also debatable, according to Frelich. You certainly couldn't stop any invasion that was already under way on a large geographic scale. You *might* be able to do so locally, say on a private golf course that didn't want bumps caused by worm poo constellating the greens, but you wouldn't be allowed to use the kind of chemicals required to eradicate them elsewhere. Earthworms were on the march, and all we could do was watch—or report them to the authorities.

Erin Cameron, a forest ecologist who'd studied patterns and processes of worm invasions in Alberta for her master's, was one such authority. I'd tracked her down in Finland, where she now worked (on worms, of course) to ask about something she'd helped create—the Alberta Worm Invasion "Worm Tracker," a phone app that allowed the public to log photos and locations of worm sightings. Worm Tracker came about when an education outreach program encouraging anglers not to dump bait contacted Cameron's department about a citizen-science collaboration. She'd seen an app as a good way to get at this, plus it would be useful for her studies—worms, unlike invasive plants, are difficult to spot, so having people collect distributional data for you in the field was a huge bonus. Particularly when, like Cameron, you were detailing the dynamics of spread, where each additional datum counted, whether for mapping current species distributions or detecting new arrivals. Again, worms likely couldn't be removed once found, but any information gleaned was useful for management, adding to what Cameron and her supervisor had learned about the relatively recent phenomenon of worms in Alberta forests. Tracking invasions since 2006, they'd shown that without human help, the worm front moved about seventeen meters a year. Thus if you walked three hundred meters into a forest from the source of an invasion (say, a road) before worms disappeared, a reasonable estimate of how long they'd been there would be about eighteen years (three hundred divided by seventeen). Species large enough to use for bait were found mostly near boat launches, while the most common invader was a small litter-dweller whose eggs and cocoons were easily moved in tire treads. When I spoke with Cameron, data that would gauge the app's effectiveness was still being collated, but media attention alone had them planning to extend Worm Tracker another year.

Larger alien worm species were having the most impact: where they'd invaded, forest litter had been removed and soil horizons mixed: the Rocky Mountain foothills of southern Alberta showed effects on both soil structure and plant and invertebrate communities; lab studies showed substantial decreases in micro-arthropods; and a strong correlation was found between earthworm occurrence and American robins (wasn't there always?).

With earthworm invasions suddenly so hot they could form their own subdiscipline, Frelich had been prophetic in casting them as investigation-worthy. "One question we have is how big a role invasive worms are playing in the current Sixth Extinction given that they're so fundamental to ecosystems. What ecological cascades or meltdowns are taking place and how extensive are they? Just in terms of biomass, nutrient cycling, and other parameters, a 30 percent reduction in growth of the continent's sugar maple is *huge*," he'd said.

That, at least, was obvious, but there was much more that wasn't, including where these study areas sat along the Invasion Curve.

"In terms of resilience, we'd like to look at places in Sweden and Finland where worms may have invaded 3,000 to 4,000 years ago and see if any recovery or adaptation through natural selection has helped the forest back to levels of diversity and productivity. That would represent recovery on a millennial scale. In the meantime, if we lose species to this climate change/habitat-loss bottleneck—not unusual in the biological history of the world over broader time scales—that means recovery time will be extended to perhaps a few million years because it wouldn't just be *existing* species adapting, but whole new species as well. And that would represent recovery on an evolutionary scale."

This all seemed quite existential, but bald reality was simpler: invasive earthworms weren't going away anytime soon—and neither were deer.

## Case Study

I bagged my first deer on Anticosti Island from a plane. Not, like most who visited this unheralded reef in the Gulf of St. Lawrence, with a high-powered rifle, but with sadly under-powered irises. And I wasn't even trying.

Gazing down from the small plane in which I'd hop-scotched for hours along the ever-widening St. Lawrence River, I was beginning to cheer the appearance of azure reefs, intertidal seaweed rainbows, and jade-rimmed bogs when a lawn ornament appeared on the shore—a perfectly poised whitetail deer, head cocked, doe-eyed (of course), clearly unperturbed by the high-pitched whine of a descending turboprop.

Then another deer. And another. A dozen, at least.

Like most airfields, Anticosti's was ringed by a high fence. But not for security; it was specially designed to keep out deer. About 166,000 of them. Although *Odocoileus virginianus*—North America's most common large mammal—had become an over-populated nuisance in many continental areas that it was actually native to, its density on Anticosti approached twenty-one per square kilometer, the highest anywhere. There was a story there, and it was equal parts history, hubris, and biology, with a sprinkle of *The Secret Island of Dr. Moreau* thrown in for good measure. Yet it was only one of many tales imbuing an enigmatic outpost once labeled, perhaps too enthusiastically, "Strangest Island in the World."

As a Canadian schoolboy, certain shapes and patterns on classroom maps of the country had caught my attention: Hudson Bay and its swollen appendix, James; Isle Royale, the sinister eye in the wolf's head of Lake Superior; and, in the gaping mouth of the St. Lawrence, just above the unfurling tongue of the Gaspé Peninsula, a large pink pill about to be swallowed. That pill, I'd learned, was Quebec's L'Île d'Anticosti, but that was all I knew. Over the years, it remained a mystery. In graduate school, collecting amphibians in Atlantic Canada one April, I sought to make an exploratory voyage to Anticosti. But all inquiries were met with flat dis-couragement—the ferry doesn't run until May; there's nowhere to stay and no way to get around; it's only open during hunting season. The idea went on the backburner for a couple decades.

That made arriving in the summer of 2010—to explore Anticosti on assignment for an outdoor adventure magazine—a bucket-list op-portunity. Fascinated with how a limestone atoll 50 percent larger than the province of Prince Edward Island could be traded through time by colonial powers like a hockey card, yet remain virtually unknown to the rest of North America, I was also piqued by what Google had delivered where grade-school teachers had failed: a geology rich with fossils, stun-ning vistas, unexplored caves, massive waterfalls, and extensive beaches; as well as historical intrigue that included questionable commerce, de-lusions of grandeur, gruesome piracy, disease, starvation, cannibalism, lunacy, and sorcery. I hadn't even debarked my flight, however, before my attention fixed on Anticosti's plague of invasive deer. Thus my first observation of this uninhabited wilderness: it seemed a tad crowded.

Marie-Josée Légaré of Sépaq—the Quebec park service that ad-ministers the island—was helping me *découvrir* Anticosti. "M. J." had once worked in Parc National d'Anticosti for a summer, and her famil-iarity would minimize the time-wasting and lost-getting that often ac-companies travel in vast boreal landscapes. Maybe she could sand the edges off my certain incredulity with some ready knowledge. But even she was taken aback by the leviathan pickup trucks that every visitor was required to rent, with their giant tires, roll bars, and metal "deer-catchers" mounted to the grill like a locomotive. When a Sépaq rep walked us from the air terminal to our rig it was like a prelude to some reality-TV, monster-truck treasure hunt. Here's a truck. Here's a spare tire. Here's a map. Here's a radio to call for help. *Au revoir! Bonne chance!*

With the next plane seven days away, those who make the long haul out here have a week to check off as many experiences as possible. Often they're the only ones experiencing; in four hours of driving that day we saw only two other trucks. Mostly because, like us, everyone was headed *away* from the airport on the dusty, 220 km Trans-Anticosti road, and last week's visitors had already cleared off in the turnaround flight. My second obser-vation on Anticosti: this place would *never* get crowded—with people.

# Deer, Deer, Bambi, Deer

History has recorded how difficult it was for anyone to hold onto this resource-rich but cursed land. Over time, Anticosti was variously exploited, controlled, or owned by the Montagnais-Naskapi-Innu people, the Mi'kmaq tribe, France, Britain, the colony of Newfoundland (twice), various private interests, and the province of Quebec—which finally acquired the 8,000-square-kilometer island from lumber giant Consolidated-Bathurst in 1974. Eyeing it up as a strategic asset, even Nazi Germany tried to purchase what, at the time, was the largest privately owned island in the world (fortunately the Canadian government intervened). Despite this considerable transit, no more than a few hundred people had ever called Anticosti home. Indeed for long stretches the only people there were squatting fishermen, hermits, criminals, or the castaways tossed with regularity onto a shore that claimed four hundred *known* shipwrecks, making it a de facto cemetery of the St. Lawrence since the time of the first European explorers. (When Jacques Cartier sailed past in 1534, he famously cursed Anticosti and its pernicious reefs as "the land God gave to Cain.")

The island's strangest chapter, however, began in 1895, when French chocolate baron Henri Menier purchased Anticosti from a British lumber company, looking to turn it into a private hunting and fishing paradise. At that he succeeded, though biologists counter that what he actually did was turn Anticosti into the world's largest wildlife experiment—a lesson in the do's and don'ts of boreal ecology. Menier's network of local agents introduced elk, moose, caribou, buffalo, mink, red fox, snowshoe hare, beaver, and grouse. Most notably, in 1896 and 1897 they uncrated 220 carefully chosen, pre-quarantined, Virginia whitetail deer. Of his ungulate catalogue, only the artificially selected whitetail population would prosper, soaring to 50,000 by 1934, its only real check Anticosti's harrowing winter. Today you can't take five steps on Anticosti without seeing the influence of Menier's experiment. And you could hardly term what happened each fall in this cervid theme park "hunting." The undisguised cull was more of a carnival concession where you paid to shoot deer in a barrel.

Leaving the airport, we turned left at the deer. Literally. In Port-Menier, the island's sole town, we stocked up while seeing the sights:

general store, artisan shop, (non-ironic) eco-museum, and, of course, liquor depot. In the small plaza fronting what seemed like a massive church for a population of only two hundred, a totemic sculpture stood above a ground-mounted plaque in French paying homage to "Saint" Menier. It said (permit me both to translate and paraphrase), *Thanks Henri, for filling the island with nonnative animals that have turned this into an awesome place for wildlife viewing! Through a rifle scope!*

As I scanned, three grass-chewing deer watched impassively. How could they not know?

Lunching at Sépaq's Auberge Port-Menier, a utilitarian hotel that once housed Consolidated-Bathurst's lumbermen, waitresses in Sépaq ranger uniforms hovered, suggesting that, in some ways, it was still a company town. Outside, where a large buck was nibbling the lawn down to putting green, a fawn wandered into the scene; cavorting nervously, ready to bolt at a moment's notice. I finished my meal with an ode to Walt Disney as a backdrop.

Bouncing out of town on stiff springs designed to deal with gravel washboard roads, we registered ten deer grazing behind a large building, several in the field across from Sépaq headquarters, and a deer in every ditch. I was already unconsciously counting. Deer. Deer. Bambi. Deer. As the only wild things in sight besides trees, I now realized why the town appeared Nordic antiseptic: there were no food gardens, shrubs, or flowers (unless plastic)—the deer would eat them all. Later, hiking toward the grotto of Kalamazoo Falls, we viewed a dramatic example of the danger of being any kind of green on Anticosti.

Back in 1980, as biologists began to pay empirical attention to the deer's impact, someone decided to see what the island's plant life would be like *without* deer and cordoned off a large experimental acreage. Thirty years later the fence surrounding this exclosure bowed outward with the sheer force of vegetative industry within, a stunning juxtaposition to the bare ground on which we stood to view it. Where else did you have to cage a forest to protect it?

Even the birch trees—left behind after selective logging had razed their coniferous cousins—were decidedly un-birch-like. Instead, bunched crowns swayed a dozen meters off the ground; when these trees were younger, deer had relentlessly stripped their lower branches, leaving them to sprout into green lollipops. Elsewhere, deer had likewise sculpted spruce into pyramidal bonsais and so-called "ballerina trees" with noticeable waists. Balsam fir, the once-dominant tree in this boreal ecosystem, was scarfed so thoroughly none has grown naturally here since the 1930s.

Back on the road. Deer. Deer. Bambi. Deer. By the time we reached a seaside park lodge at Chicotte-la-Mer two hours later, the count was up to sixty healthy, preternaturally tawny, decidedly nonplussed deer— plus four skulking foxes and five panicky hares, all nonnative. Add four docile does munching quietly around the compound—including one with its wet nose in my pocket—and it was now sixty-four. We went directly from patting deer to the dining hall, where the venison steak on the menu suddenly lost its appeal. Until I was told it came from a deer farm on mainland Quebec; Anticosti's deer were considered wild game, which couldn't, by Canadian law, be sold in a restaurant. *Bon, moi je prendre l'entrecôte venison, s'il vous plait!*

I was more circumspect in menu choices when we dined later that week with Port-Menier's mayor. Not coincidentally, Denis Duteau was also a deer biologist, the vocation that had first brought him to the island as a graduate student. It made for interesting conversation about the dual commercial and biological role of the hunt/cull that took place each October to December. Not surprisingly, the island's threadbare economy hinged on those three months, and maintaining a deer density high enough to lure hunters who'd otherwise go somewhere cheaper (pretty much anywhere else) was key. It was becoming a challenge, Duteau noted, to balance this requirement with the need to lower the pressure on food resources so the deer would have enough to eat and plants would have a chance to regenerate. They had expended much effort to figure it all out, said Duteau, and in the interim, they'd learned a lot about the biology of *O. virginianus,* though most of what he detailed seemed a foregone conclusion: more food meant bigger deer; higher-quality food led to faster growth and reproduction rates; and too many deer resulted in too little preferred forage, which drove them to other, lower-quality nutritive sources. Which explained the deer I'd observed while flying in, reluctantly munching kelp on the beach: this was the very end of the Invasion Curve—maximum carrying capacity achieved, population self-regulation likely, but eradication impossible.

The 8,000 to 10,000 deer hunted off the island each fall made not the slightest dent in the continually rising population—certainly far less than the 20,000 to 30,000 that might die in a particularly savage winter. On that front, the mild, snowless winter of 2009–2010 left Anticostians alarmed: deer winterkill was way down, boding for another sharp rise in their numbers. And according to Duteau, the recently updated estimate of 166,000 deer was probably low—it might actually be more like 200,000. Some of the information came from air surveys, some from the larger exclosures used to regenerate forest patches with deer food: wildlife technicians would fence off a vast acreage, then open it to an intensive experimental hunt (a special license allowed you to bag three deer instead of the usual two, but two had to be female) to drive deer numbers below fifteen per square kilometer, the density at which forest could regenerate. Patches thus thinned of deer required twelve years of regrowth before the fence could be removed, a cycle that may represent the most directed long-term management *for* an invasive species anywhere in the world.

Prior to Menier, the only indigenous mammals eating the island's once-abundant vegetation were mice and black bears. By the 1970s, however, bears were almost nonexistent. The opportunistic bruins might have taken down the occasional newborn fawn to supplement their diets, but come autumn there wasn't enough food to fatten up for hibernation or gestation—deer had cleaned out the meager berry crops and further grazed the bushes to the ground. Bear birth rates dove and winterkill soared until *Ursus americanus* was officially extirpated: the last bear was seen sometime in the mid-1990s. Another dubious first for Anticosti: the only known case of an herbivore putting a large, well-established omnivore out of business.

## Listen Up

In his revealing essay "What the Deer Are Telling Us," writer Christopher Ketcham is quick to locate the important ecological role of predation among the observations of turn-of-the-century U.S. Forest Service officer, naturalist, and big-picture-thinker Aldo Leopold, author of the famed *Sand Country Almanac:* "In 'Thinking Like a Mountain,' [Leopold] outlined for the first time the basic theory of trophic cascades, which states that top-down predators determine the health of an ecosystem . . . the extirpation of wolves and cougars in Arizona, and elsewhere in the West, would result in a booming deer population that would browse unsustainably in the forests of the high country. 'I now suspect that just as a deer herd lives in mortal fear of its wolves,' Leopold wrote, 'so does a mountain live in mortal fear of its deer.'"

Ketcham went on to describe a startling litany of impacts from deer irruptions in various parts of North America. (Although its meaning is suggested by the context in which I've several times used it, "irruption" describes the sudden movement into, or occupation of, areas by an indigenous species within its native range but in abnormal numbers, a process usually driven by an ecological imbalance that allows a hitherto nominally distributed species to spread virtually unchecked— functionally equivalent to an invasion.)

Though not the largest problem, perhaps the most insidious associated with deer irruptions is the species' role as primary host for the adult blacklegged tick (*Ixodes scapularis*), a vector for such nasty (and nasty-sounding) human afflictions as Lyme disease, babesiosis, and granulocytic anaplasmosis. In the 1990s, as the incidence of Lyme disease rose across New England, tongues wagged about an association with concurrently spiking deer populations. After all, adult blacklegged ticks were being reported in densities of up to a thousand per deer. While more deer no doubt meant more adult ticks in an area, other studies suggested that any correlation with Lyme incidence was spurious, because Lyme was more likely tied to ecological conditions that were coincidentally favorable to more important links in the Lyme chain—shrews, chipmunks, and deermice. As preferred hosts of the tick's nymph stage, this tiny triumvirate acted as a collective reservoir for transmitting the Lyme-causing bacteria, *Borrelia burgdorferi*, to the 80 to 90 percent of ticks that carried it into adulthood. In this view, proliferation of deermice was encouraged both by the same fragmentation of woodlands that facilitated an overabundance of deer, and by the near-extermination of the rodent's primary predator, red fox—through hunting, poisoning, and trapping—and by infiltration of the peri-urban (interface between country and town) environment by coyotes. Coyotes, in turn, occupied these landscapes because *their* main predator (and, coincidentally, that of deer), the gray wolf, had long since been extirpated. This narrative briefly shifted focus away from the deer problem. In a 2015 paper, however, researchers returned to the association between deer density, tick abundance, and Lyme disease, evaluating one Connecticut community over the span 1995–2008. The number of resident-reported cases of Lyme per one hundred households was strongly correlated with deer density in the community. When hunts instituted during the period reduced deer density by 87 percent to 5.1 per square kilometer, the result was a 76 percent reduction in tick abundance, a 70 percent reduction in bite risk due to vector density, and an 80 percent reduction in resident-reported Lyme. This outcome demonstrated that while pushing back coyotes and restoring foxes would likely help, irruptive deer populations could—and should—be significantly lowered to reduce both human interactions and the environmental load of infected nymphal ticks, which, in turn, would significantly lower the local population's risk of contracting the disease.

## The Curse of Gamache

Back on Anticosti, M.J. and I clopped along a raised beach, our shaggy horses sidestepping mounds of seaweed. Though full-fledged ocean at this longitude, the Gulf of St. Lawrence was as still as a pond under clear skies; a giant reflector on which evening light swam like Quebec's world-renowned maple syrup—warm, rich, amber. When the horses stopped to reconnoiter a crossing of crystalline Rivière Chicotte, the bizarre depth of field made it feel like we were standing in a museum diorama. We actually were, for this island was a museum in every possible way. In one gallery stood artifacts of Anticosti's many human iterations and failed industries, as well as Menier's various enterprises. In another, nature's bounteous contribution: rare plants, 160 bird species, dolphins, whales, and untold lesser sea life; geological history and diverse fossils; even a hint of past and present life converging—atop a nearby bluff lay the bleached, fossil-in-the-making body of a thirty-meter blue whale that had washed up in 1991. Yet despite logging, rabid animal introductions, and a painful history of quasi-occupation, Anticosti also remained redoubtably pristine—a

Museum of Enduring Wilderness. Seals bobbed offshore, strung like whiskered buoys between mats of kelp, appearing to nod agreement. Beyond their curious crania, a ship materialized on the horizon to drift slowly by, a lingering thought in the sunset. It was remarkably quiet, as if the canyon's porous rock dampened even the sound of rushing water. Maybe clearer water made less noise: there was nary a speck of organic matter in the icy stream. Dismounting and stepping in up to my knees, I looked down to my feet as if there were no intervening liquid. Though I longed for a drink of this cool elixir, I didn't dare—the water table was contaminated with the prodigious daily shit of 166,000 deer.

We headed back on a height of land bordering the magnificent coastline. The still sea lay aquamarine over shallow reefs; it could have been the tropics save for umber kelp heaped upon the sharp, tilted ledges responsible for tearing boats asunder. Near the lodge, deep in cool woods, a pair of deer and two fawns moved not an inch as we brushed past. Well, *excuse us.* By the end of the day my count was 106—the hundredth being a trophy buck with a twelve-point rack. I had also acquired a second invasive-species hobby—frogs. Like the deer, frogs had been introduced only to run amok thanks to a lack of predators. Also like the deer, the plague was of biblical proportions—*tens of thousands.* No doubt precisely what Menier envisioned when he had his minions sow the island with northern mink frogs fetched from the Quebec mainland—a boreal amphibian swarm to control the usual boreal insect swarm. He succeeded at this, too; Anticosti is the *least* buggy northern wilderness I've ever visited. Islands in general are magnets for eccentrics and seekers, but Anticosti was on a different scale. Strangest Island in the World. Graveyard of the St. Lawrence. Mystery Island. Menier's Lost World. Or, as I saw it, Island of Enduring Invasions.

The original settlement, Baie-Sainte-Claire, lay 10 km northwest of Port-Menier on a pastoral half-moon reminiscent of the Irish coast. Long tidal flats of exposed rock had made for an impossible port, so when Menier arrived he simply moved the town. Now but two houses, a lime oven, and a few cemeteries remained—surrounded by waist-high meadows grazed by hundreds of deer. This part of the island was a protected area with no hunting, and, of course, no predators. Freed of the requirement to hide in the forest as small groups, deer here formed bona fide herds, sweeping over a Serengeti-like grassland constellated with Dali-esque ballerina trees.

We spent our last night in Port-Menier visiting the site of the baron's former house. Burned to the ground in the 1950s, the Scandinavian-style mansion with wood-carved wall murals had cost Menier more than the island on which he'd built it. Despite making only six visits to Anticosti before his death in 1913, his tenure was now referred to as the "Menier Epoch." Indeed the *Roi de Chocolat* poured considerable resources into making the island more habitable for all, but he'd also run it as a fiefdom, expropriating land in return for guaranteed employment in his invasive-species-hunting enterprises. He paid lighthouse and river keepers for their work and to run off poachers and squatters, but allowed no opposition to his grandiose plans. On the site stood a small tower emulating the turret of the old mansion. Atop, gazing toward town with salt air in my nose, I understood how Menier must have felt overlooking his unfolding creation. And I also noticed a cross, enshrined by a white picket fence tucked against the browsed-out forest. It was the final resting place of the mysterious Gamache.

After sailing for years on English frigates, Quebec-born Monsieur Louis-Olivier Gamache had settled into a strange existence on Anticosti. Heavily armed and garrisoned, his reputation as a sorcerer protected him from pirates and allowed him to live in peace until his death in 1854. Neighbors feared Gamache and the long, lonely path to his log castle. And there were rumors: he'd sold his soul to Satan for gold; he neither went to Mass nor confessed to the priest; he changed into a *loup garou* (werewolf); he caused cows to drop their calves, withhold milk, and bolt frantically into the forest; his evil eye summoned measles, smallpox, and other diseases. In short, many believed that he commanded the powers of darkness, with his enduring curse the island's de facto mantra: it was easy to founder upon these shores but difficult to leave.

According to others, however, Gamache was a softie, all ice-cream and balloons. Easily pacified by a small offering—that he'd strangely always refuse—he was but a simple wise man with insight into the ways of the world and the hearts of men: he predicted the weather, spoke to animals, understood medicinal plants, offered healing advice to the sick, turned up lost property, united estranged friends, helped the troubled find prosperity and peace. Maybe, some reasoned, Gamache wasn't *really* an evil sorcerer, but a practitioner of White Magic and a vessel of good.

Either way, neither the Catholic church nor the island's various owners liked the power he wielded, and each tried in their own way to vanquish him. Though what really happened remains a mystery, stories of Gamache far outlived the man. Even twenty-first-century residents recount his real or imaginary exploits. Indeed most odd occurrences on Anticosti lead quickly back to Gamache—even the Reign of Deer, which some see as Gamache's disseminate spirit taking over the island in one final, humorous curse.

It almost made sense.

## Sensibility

Sense, we all know, can be fraught with the hubris of the moment. It certainly *seemed* sensible, during the human Age of Exploration, to introduce both domestic and wild ungulates onto islands far and wide as a food source for colonists, outpost dwellers, and passing ships. Goats in the Galapagos, pigs in Hawaii, and, in poor ol' New Zealand, no less than seven species of deer—red, wapiti, sika, sambar, rusa, fallow, and white-tail. Unfortunately, each animal delivered the same suite of problems: free of predation, competition, and other natural checks, and occupying a wider range of habitats than would be expected, the deer flourished in an environment with abundant food. Such easy living for the deer drove up their birth rate until the rapidly expanding population, which foraged both selectively and ferociously, started damaging plant communities and nutrient cycling.

Though the primary motivation for introducing nonnative ungulates was food and/or sport hunting, these had in every case exceeded the capacity of regular hunters to maintain sustainable population levels, leading to the kind of costly and time-consuming control efforts suggested by the Invasion Curve. Lamentable as it is, however, ongoing control often provides unexpected economic opportunity; while traveling South America in the early 1980s, I met several intrepid Kiwis who'd funded their round-the-world adventures through government-sponsored bounty hunting for invasive European red deer (*Cervus elaphus*). Recently, a number of conservation trusts and governments had taken the stag by the horns and initiated full-out eradication efforts. Undertaken only in certain circumstances, these must be sustained and thorough—like the successful eradication, in only fifty-two months, of some 80,000 ecologically disruptive feral goats (*Capra hircus*) on 58,500-hectare Santiago Island in the Galapagos. Such efforts—which often involve helicopters and significant manpower—naturally become more costly the larger and more isolated the invaded area, as with introduced reindeer (*Rangifer tarandus*) on subantarctic South Georgia Island. From a few dozen individuals stocked by Norwegian whalers in the early 1900s as a source of fresh meat, the numbers of reindeer increased dramatically due to low hunting pressure and the reindeer's ability to evade hunters in the island's vast alpine spaces. Attention was paid to the growing population only when widespread, voracious, down-to-the-nub grazing began devastating the island's already minimal natural vegetation, which led to follow-on effects for burrowing seabird communities and the facilitation of numerous nonnative plant invasions. Though South Georgia also had issues with invasive rats (ubiquitous on any islands utilized by the whaling industry), reindeer removal was key to recovering the island's vegetation and to improving its habitats for breeding birds. Furthermore, the rapid retreat of glaciers due to climate change had created a high risk that the reindeer would expand into new areas. Carried out in three phases, the South Georgia Reindeer Eradication Project eventually resulted in the successful removal of 6,600 reindeer at a cost of some $2 million USD.

Both Santiago and South Georgia are remote islands with few humans where the issue was primarily ecological restoration. This isn't always the case. Consider the unique impact—literally and figuratively—of moose (*Alces alces*) in Newfoundland.

In the spring of 1904, two bull and two cow moose from New Brunswick's Miramachi region were released near the small community of Howley on the western side of the great island situated off Canada's east coast in the North Atlantic. It wasn't the first such experiment on the 111,390-square-kilometer landmass, which had been considered a peripatetic sportsman's paradise (that concept again) until native woodland caribou were hunted to near extinction in the late nineteenth

century. Unlike previous failed efforts, however, this one went swim-
mingly. By 1920 fair numbers of moose were recorded as far as 80 km
from Howley, and by 1935 they were occupying much of the island—
aided in no small part by extirpation of the wolf around 1930. Regulated
hunting began in 1935 and by 1957 almost 8,800 licenses were issued;
nevertheless, populations increased steadily. As hunting pressure increased, wildlife officials in the new province (Newfoundland joined
Canada's confederation in 1949) instituted a lottery system focused on
areas where moose needed thinning. With an abundant food supply and
negligible predation by the island's lynx, coyote, and bear, that was soon
everywhere.

Today, perhaps to avoid politically charged language, moose is
deemed "hyperabundant" by provincial wildlife ecologists, and the estimated 150,000 descendants of those original four represent over 10 percent of *all moose in North America*—stuffed into less than 2 percent of
the species' continental range. Aside from strategic culling in national
parks, hunting is the only control; like Anticosti, however, it has proved
inadequate. To begin, hunting achieves only a 73 percent success, with
an average 20,000 moose taken annually out of 33,000 licenses issued;
with over 40,000 calves born each year, it's not hard to plot the upward
trending growth curve. Regardless of efficacy, the annual fall hunt has
become a deeply entrenched cultural tradition celebrated in story and
song, with moose featured everywhere from ski-area cafeterias and
backyard barbeques, to community festivals and restaurants promoting
locally sourced foods. And while I've enjoyed many a delicious moose-
burger on the island, it's harder to swallow the rest of the story. The
newcomer that made an indelible mark on Newfoundland's identity has
also indelibly marked a landscape where it's now the driving ecological
force.

With an average 1.7 moose per square kilometer, a density three to
twenty times greater than other boreal ecosystems in which it's native,
Newfoundland's moose have exceeded their adopted environment's carrying capacity. Intensive browsing has reduced the abundance and distribution of native species like balsam fir and mountain ash, while
severely inhibiting forest regeneration in both timber-harvested and
insect-disturbed areas, all of which impacts both the ecology and econ-

omy of the area. The alteration of forest ecosystems by moose has affected the habitat of everything from birds to lichens and, in a nod to invasional meltdown, facilitated the establishment of exotic plants for which it coincidentally functions as the main vector for seed dispersal.

Newfoundland's imported cultural icon packs a serious ecological punch. But there is another problem, one also associated with irruptive deer. Only here, it is so much bigger: moose-vehicle collisions.

As the continent's largest ungulate, adult moose can weigh up to 450 kg, the bulk of it propped high on stilt-like legs used for wading and feeding on aquatic vegetation. That's a lot of mass coming through a windshield. The main cities of Corner Brook and St. John's are located on the island's west and east coasts, respectively, and with more drivers than ever plying the 900 km between them, and more moose than ever strolling the island's roadways, the risk of hitting one rose more than 50 percent between 2005 and 2009. Although a hazard that plagued the province for decades had now reached crisis proportions, the government seemed disinclined to spend money to fix it.

In response, the Save Our People Action Committee (SOPAC) was formed in July 2009 to address the uptick in moose-vehicle collisions through outreach and lobbying for government action. Though the provincial government refused to supply SOPAC with collision statistics, the Royal Canadian Mounted Police constabulary that mopped up the mess was happy to do so: its data showed an increase from 474 moose-vehicle accidents in 2008, to 726 in 2009, and 741 in 2010. The annual average from 2000 to 2008 was 455; from 2009 to 2010, 734, around which the average still hovers (this doesn't include accidents with damage under $1,000 CAD or near misses). All in, with police, insurance, health costs, and road-clearing (four hundred moose carcasses carted to garbage dumps each year), the academically estimated cost of each moose-vehicle accident is about $31,000, or, at the current collision rate, around $23 million a year, much of it from provincial coffers. Yet wildlife fencing being demanded by SOPAC and other groups costs only around $50,000 a kilometer, meaning the government could spend a similar amount each year to install some 460 km of fencing. Proponents argue that this option makes sense given the human costs: hundreds injured or maimed each year, and an average of two deaths.

Knowing that the only truly effective way to lobby government is in court, SOPAC launched a class-action lawsuit that opened in April 2014 in the capital of St. John's. Alleging negligence on the part of the province for failing to manage moose populations, the class was limited to injuries requiring hospital admission and involved 135 plaintiffs—including some fifteen estates. The court heard emotional testimony from people whose lives had been devastated; it also heard from a scientist who'd studied wildlife overpasses, underpasses, and fencing in Banff National Park. His affidavit alleged that road safety measures adopted by Newfoundland—signage and brush-clearing from highway margins—had little scientific validity, and that, by contrast, linked passage and fencing systems had effectively reduced moose collisions in British Columbia, Quebec, New Brunswick, Sweden, and Alaska. Notwithstanding, negligence on the part of uninformed bureaucrats could not be proven; the suit was dismissed.

Though the verdict was immediately appealed, SOPAC and the government put their enmity aside to partner on public awareness. Radio and TV ads abound, and there's a hotline for reporting moose on the highway or crossing hotspots. The SOPAC website includes a map of all known collision sites. Recently, a few pilot-study wildlife fences were erected in key areas. When some of the public complained, saying that fenced highways gave the impression of driving through a prison, Ben Bellows, a sixty-year-old who had been rendered quadriplegic after a 2003 collision, took umbrage. Addressing critics on the SOPAC website, he invited them to sit by his bed to experience the very *real* prison of immobility he lived in.

Though its math has been challenged, the government's stated position remains that wildlife fences will never be cost-efficient. Thus, when Newfoundland tabled a budget in May 2015, to no one's surprise exactly zero dollars had been set aside for moose barriers. Instead, authorities vowed to increase the moose hunt, particularly in hot zones, perhaps fix a failed electronic wildlife-warning system, and continue public awareness campaigns that now included a finger-wagging "Slow Down!" message. In all of this, an obvious control measure seems to have been overlooked, perhaps because few would be interested. When the first wolf seen since 1930 appeared on Newfoundland's Great Northern

Peninsula in March 2012, after crossing sea ice from the Labrador mainland, it received a typical welcome from hunt-happy islanders: someone shot it.

## Moose Crossing

While invasive moose made a mess of Newfoundland, the species was paradoxically experiencing steep declines elsewhere in North America. What was happening varied according to location, but even in areas of relative abundance or natural increase, signs pointed to imminent collapse. By 2013, alarming declines were seen in all provinces from Quebec westward, and several northern U.S. states. Although the answer to "Why?" wasn't entirely clear, like most other animals in decline it appeared to be a variety of factors—disease, predation, and climate change chief among them. And you didn't need to look far into that troika to find an odd connection: propelled by climate change and other factors, deer were overpopulating and expanding their range everywhere, spreading disease to moose and increasing the abundance of predators affecting them.

In Minnesota, for instance, 40 percent of adult moose were dying from brain worm (*Parelaphostrongylus tenuis*), a tiny parasite transmitted by deer. An additional 20 percent were dying from higher-than-normal winter tick loads, and a final 20 percent from a combination of stressors. In Canada's Jasper National Park, moose decline was attributed to the deadly giant liver fluke (*Fascioloides magna*), more wolves, and continued—albeit reduced—collisions on the railways and highways traversing the park. Another emerging problem was decreasing calf survival. In Minnesota's Grand Portage Trust Lands, for instance, 90 percent of moose calves died each year—75 percent lost to predation during the first weeks of life, the remaining 15 percent to subsequent health issues.

Though brain worm doesn't affect its deer carriers (it lives benignly in the connective tissue around a deer's brain and spinal cord), the parasite has a typically complex life cycle: the worms release eggs that hatch in the deer's lungs; after being coughed up, swallowed, and excreted, the larvae find homes in snails and slugs. By accidentally consuming these mollusks while grazing, 90 percent of all deer are infected in their first two years of life. In areas of deer abundance, moose pick up the worm, too. But brain worm isn't happy in moose: it just burrows around looking for a whitetail deer brain that it never finds. Not only is the moose a dead-end host, it also usually ends up dead. (On Lake Superior's Isle Royale—free of deer and thus brain worm—the trend is opposite: collapse of the wolf population from inbreeding sent moose numbers soaring. As on predator-free Newfoundland and Anticosti, the moose irruption now threatens to exceed carrying capacity by wiping out its main food, balsam fir, another example of why predators are critical to ecosystem health.)

Ditto for winter tick (*Dermacentor albipictus*), another devastating pest moved around by fellow cervids like deer and elk. Larval ticks climb onto moose in the fall and feed themselves into adulthood over the winter. Moose with particularly heavy loads—over 100,000 ticks were counted on one decidedly miserable animal—simply can't eat enough to replace the blood loss, and end up cannibalizing their own muscle tissue for protein before, inevitably, dying.

Despite proximal parasite problems, however, the underlying cause of moose declines is climate change: brain-worm-carrying deer populations expand with shorter winters and longer growing seasons, and tick populations explode with earlier snow melts. Historically, deep winter snow kept deer from moose country, with less opportunity for the animals to mix. But a series of warmer winters in parts of North America allowed deer to move north into the boreal forest and to higher elevations in the west (Leopold's mountain's mortal fear realized). Finally, in areas where they'd become prevalent, deer fed and subsequently boosted populations of apex predators that were also more than happy to prey on moose calves.

The recommendations from moose-management advisory committees? Hammer deer through any means possible. And yes, they used that word. The problem with hammering deer, however, is that they've proven adept at hammering back. Much like humans, deer modify the habitat they occupy and eliminate biodiversity, though not to their immediate detriment. And when habitat quality declines, they simply trek to a new area—or swim to one.

## Trouble at the Edge of the World

The tent door rustled in an unfamiliar way. Unfamiliar because after several days on a remote North Pacific beach I was pretty sure it wasn't the inevitable morning breeze trying to poke its head through the un-zipped mesh door—or snort its arrival.

With barely time to bolt upright, I prepared to scream bloody murder at what was almost certainly a black-furred, pig-eyed intruder. But before I released that tremulous yell, an auburn muzzle appeared, followed by dewy chestnut eyes and twitching ears. Improbably, like some hand-puppet theatre, another slipped in beside it. A pair of curi-ous deer found themselves staring at a naked and cowering man.

Satisfied I wasn't edible, the deer withdrew, content to nibble grass sprung around a massive, weathered log to which the tent was anchored. They paid little heed as I donned shorts and slipped out to photograph them against the driftwood-crossed beach. Behind me, gentle breakers foamed across a glistening mudflat while Alaskan peaks deliquesced on the northern horizon. It was then I'd noticed the dune-line grasses, uni-formly shorn in either direction as far as the eye could see, and the pel-lets constellating the ground anywhere you gazed. I could have acted on the limbic urge to shoo the deer away to assert that this was *my* spot, but it clearly wasn't. Besides, having animals treat you as part of their envi-ronment stirs a deep-rooted connection. While it wasn't exactly a vi-gnette from the indigenous Haida people's creation myth, in which raven the trickster pries open a giant clamshell to release the first hu-mans on these shores, the deer's entreaty to emerge from my nylon shell into a new dawn seemed wholly natural—even if they weren't.

A six-hour ferry ride from the port of Prince Rupert, the almost mythical archipelago of Haida Gwaii (née Queen Charlotte Islands) floats off the coast of British Columbia like a 250-km-long slice of pizza. Two main islands—Graham to the north and Moresby wedged below—taper southward to a point at a small, windswept scatter known as the Kerouards. Deer were but one of many changes to wash over these islands since the late 1700s, when Europeans had arrived: within a century, intro-duced disease had decimated the Haida, greed had vanquished the sea otter, and colonial zealotry was well on its way to claiming renowned forests of giant cedar, hemlock, and spruce. Despite millennia of Haida seafaring and trade with a mainland where it abounded, Sitka black-tailed deer (*Odocoileus hemionus sitkensis*) was introduced here only during the accelerated period of modern settlement and resource extraction of the early 1900s, where it prospered in the predator-free, increasingly clear-cut environment. Other purposely introduced critters—raccoon, red squirrel, beaver, muskrat, wild cattle, Pacific chorus frog—likewise experienced little competition from the few endemic land vertebrates like western toad, woodland caribou, river otter, marten, short-tailed weasel, a few species of bat, mouse, and shrew—plus the extra-large variety of black bear that I'd anticipated at my tent. Though this native cabal seems a pedestrian lot, many claimed membership among the forty-odd endemic plants and animals whose measurable divergence from continental forms have earned the archipelago its nickname "Galapagos of the North." By any measure, Haida Gwaii comprised a geographically isolated, evolutionarily unique, and sensitive ecosystem.

Certainly those species that were indigenous earned it, finding their way across after the height of the final Pleistocene glaciation, when the shallow Hecate Strait now separating Haida Gwaii from the rest of British Columbia was an undulating tundra riven with serpentine channels and wandered by mammoths. Everyone has a theory about what the islands were like during the Ice Age: the Haida sing and dance stories of ancient times in *Xhaaidlagha Gwaayaai*—"islands at the boundary of the world"—suggesting that they've borne witness here for some 10,000 to 13,000 years; scientists back that up with analyses of bones and lake-bottom pollen profiles, but ultimately know little more than the Haida. The dates each agree on, however, are those that changed life for all of the archipelago's original habitants.

Dispatched at the behest of the viceroy of New Spain, Mexico-based ensign Juan Perez sailed north in 1774 to investigate reports of encroachment by Russian and British fur-traders in the Pacific Northwest, an area claimed for Spain by Balboa 250 years prior. Charged with reaching 60° north latitude, Perez made it to 54°40′N, off Langara Island on the archipelago's northwest aspect. Intercepted by Haida paddling enormous canoes, legend holds that the natives sprinkled eagle down—a commodity equivalent to gold dust—on the waters in greeting. Perez traded with the Haida but infamously never went ashore. The Brits who followed weren't so shy: itinerant explorer Captain James Cook landed in 1778, and in 1787 surveyor Captain George Dixon christened the islands after his ship, HMS *Queen Charlotte*. Soon hunters of whale, seal, and otter were making frequent landings for trade and provisions, rats pouring from their ships the moment their mooring lines touched land. It was the beginning of the end.

By the time I'd pitched my tent on the beach at Tow Hill, adjacent to Naikoon Ecological Reserve, I'd been delivered this history in two short days of hitchhiking around Haida Gwaii, where everyone—from Haida mask-carvers in rough-and-tumble Old Masset to the fishermen and hostel operators in Queen Charlotte City—helped fill in the blanks. It would be my first of several visits, and each subsequent trip left me more curious about not only this mist-shrouded enigma and its indigenous dwellers, but also its nonnative inhabitants, whose influence was on display everywhere.

On assignment in the fall of 2011, I was fortunate to have the opportunity to accompany a Parks Canada–led squad of restoration ecologists to several remote islands for the wrap-up of a historic rat-eradication program aimed at saving burrowing seabirds (described in Part 5). But not until 2014 was I able to sit down with the person who had launched the program and who continued to monitor the ecosystem's recovery— Carita Bergman, Parks Canada terrestrial ecologist for Gwaii Haanas National Park Reserve and Haida Heritage Site. I caught up with her on a drizzly July day at park headquarters in Skidegate, a cluster of modern buildings marked by the broad lines of traditional Haida architecture, abundant totems, and the lapping waters of the North Pacific.

As Bergman told it, during her first trip here in 2001 she had been unprepared for the experience—at least biologically. Her group had sea-kayaked the length of Gwaii Haanas without knowing that the myriad deer peering back from beach edges and verdant rainforest were nonindigenous. "So here I am staring in complete wonder at these forests but having no idea how much they've been altered from their original form," she recalled.

Eventually, she would know *too* much. After working on the ecology of other terrestrial vertebrates, mostly in Alberta, Bergman took up a post with Parks Canada in Gwaii Haanas in 2007. It took little time to realize that terrestrial ecology here entailed monitoring not only the archipelago's longstanding native ecosystems, but also numerous non-native species now chopping at that trophic scaffolding the way an out-of-control logging industry had leveled most of Graham Island. Topping the list of concerns were rats that had devastated offshore bird colonies—and deer, which had destroyed everything else.

Like anywhere that deer are a problem, the forest understory had gone missing on most of the 150 or so islands large enough to support the animals and close enough to swim to or between. As Lee Frelich's earthworm studies showed, deer impacts weren't solely structural or concerned specifically with plants. In a 2008 paper coauthored with Sylvain Allombert, Haida Gwaii deer researcher Jean-Louis Martin of the Centre National de la Recherche Scientifique in Montpellier, France, compared invertebrate communities in understory vegetation and forest litter across six islands with varied browsing histories. Unsurprisingly, longer browsing histories correlated with a sharp decrease in both invertebrate abundance and species richness. Over-browsing also affected vegetation and litter differently: while understory habitat was quantitatively reduced and simplified in a linear manner over time, litter remained widespread, undergoing qualitative changes only because of impacts to the vegetation producing it. Overall, they concluded, "cervid engineering" had precipitated a cascade of effects across different components of Haida Gwaii's invertebrate community.

The impacts of deer weren't solely of ecological concern. On a previous trip, Gwaii Haanas's cultural liaison, Barb Wilson, had recited to me a litany of damage by alien species to medicinal plants, berry gathering, and various birds, mammals, and sea life important to the Haida. Her greatest worry was loss of the culturally significant cedar, from which the Haida derived art, clothing, housing, and transportation. Pointing to the cumulative impact of invasives, she feared that were nothing done, they'd soon be inhabiting rocky outcrops in the Pacific with few native plants or animals—a situation she was loathe to bequeath her grandchildren. As with balsam fir on Anticosti and Newfoundland, cedar on Haida Gwaii was too intensely browsed to regenerate on its own. The most dramatic examples of this were protected old-growth tracts in parks and other set-asides, where new cedar was essentially restricted to deer-inaccessible stumps and cliff faces; even where blowdowns had created forest openings in which cedar *should* regenerate, it wasn't doing so. Browsing had also effectively eliminated regeneration in logging cuts—that is, until the mid-1990s, when timber companies were required both to replant and protect cedar. As a result, second-growth stands younger than 120 years (after deer were introduced) contained little cedar, and virtually all of the cedar that had managed to survive was twenty years or younger. If the remaining inventory of harvest-ready cedar on timber-licensed land were removed, there'd be a lengthy gap before any more was available, with logging during that period dependent entirely on the economic viability of hemlock and spruce.

When the understory is eaten away by deer, something else happens on the smaller islands in Gwaii Haanas: seabirds become more vulnerable to raptors and other predators. In one sequence of camera-trap photos analyzed by Bergman's group, an ancient murrelet disappears into its burrow just as a predatory bald eagle makes a cameo. "That turned on the lightbulb about other potential threats to these birds," Bergman said. "Since then, we've documented nighttime predation on murrelets by otherwise diurnal species that's clearly exacerbated by lack of understory. Nest predation of songbirds by squirrels and martens also increases in the presence of deer."

Bergman sees much in her travels around the archipelago, and though not all of it makes sense, some points to potential avenues of understanding—like her as-yet-untested hypothesis about which islands deer swim to. Within spitting distance of the heavily impacted Swan Islands, for instance, she'd found a lush, unaltered, deer-free island ringed by a dense band of kelp, its closest point to the Swans draped in seals and sea lions. Was having a pile of large sea mammals in their face as they arrived turning back deer? Or did a thick, entangling kelp bed form a natural deer barrier? That is, could it be that elsewhere the loss of sea otter and subsequent decline in kelp forests—caused by the ecological release and irruption of the otter's prey, purple sea urchin—was *facilitating* deer dispersal?

In autumn 2009, Bergman attended an invasive species conference in New Zealand, a group of similarly isolated islands whose seventy-two introduced mammals had made a mockery of unique, ancient ecosystems harboring the world's most primitive frogs, ancient tree ferns, flightless birds, primeval insects, and a monotypic order of lizard-like reptiles whose only known relative was a 200-million-year-old fossil. Presenting a poster about where Parks might best accomplish full ecological restoration in Gwaii Haanas, Bergman had chosen Ramsay Island because it was far enough from invasive source populations and large enough to matter; and because although rat-free, the island was overrun with deer, which might fan out to surrounding smaller islands like Murchison and Faraday. After the conference, Bergman was invited to visit Secretary Island, where New Zealand conservation officials had eradicated deer. Secretary was larger than Ramsey, and the can-do Kiwis had cleared much bigger; they poked fun at Bergman's "bad attitude" in remaining unconvinced of the feasibility of doing so in Gwaii Haanas, in part because Jean-Louis Martin's group had tried on Reef Island and failed. "At that point I thought [deer] couldn't be eradicated, but now [I know] there are techniques to do it," she told me, her team having recently cleared deer off Murchison and Faraday ahead of a rat-eradication program that the cervids could interfere with. "The efforts are costly but worthwhile: there has to be a commitment to keep stepping-stone islands deer-free."

Back in 2010, however, when a pool of federal funding for ecological restoration became available under a program called Action on the Ground, Bergman found herself having to make a strategic choice. Parks Canada administrators across the country were directed to apply for money to deal with their number-one ecosystem threat; in Gwaii Haanas, quantitatively, this was deer. But, Bergman wondered, was that the *best* investment? Skeptical whether deer could be tackled in any meaningful way, she was also cognizant that the problems they caused were always local, while rats on offshore islands were impacting globally significant populations of ancient murrelets. She decided to go after rats.

At one point, conversation in full flight, me furiously taking notes, Bergman asked how I'd become interested in the issue of invasive species. It precipitated the same introspective pause which, months later, I'd experience with Tony Ricciardi. Fair question: why invasives? I explained how I kept encountering invasive issues on unrelated assignments, enough that they'd piqued my interest as an emerging global phenomenon. I believed the topic both appealed to my inherent biophilia, which had led me into science, and resonated with the subsequent training that had taught me to pursue investigations to a logical conclusion—a combination that, were I honest, also delivered me to interesting places, fascinating people, and remuneration for my efforts. Like so many others I'd met, I realized in that moment that I'd made invasive species part of my job: with problems so multitudinous and dynamic there would always be something to learn and/or experience, they offered a potent dialectic.

Bergman looked on stoically while I blathered self-analysis; she telegraphed neither perplexity nor empathy. "Has that experience *done* anything to you? I mean, I see everything through a completely different lens now, and a pretty depressing one; I can't take that lens off and it kind of upsets me," she'd offered, channeling the sensibility of many invasive managers likewise condemned to Armageddon goggles—a 3-D experience in which the writing on the wall hovers in front of every endeavor, background lit by a global homogenization bomb exploding in ever-faster motion. Recalling our model in Part 2, would future Earth look more like Jupiter, with distinct bands encircling different climatic zones between and beyond the Tropics of Cancer and Capricorn? The answer was unclear, but Bergman didn't feel particularly hopeful about being able to do anything about it . . . save on these small islands.

"The furthest perspective I can take is to view invasive species as a symptom of global overpopulation and people traveling. It's a simple equation: the lower the human population, the lower the number of introductions. But we've been doing it for a *long* while."

True. Even the Haida, who traded widely along the Pacific coast, had surely introduced a few critters and plants over their 13,000-year history. Notwithstanding, modern-day plant introductions to Haida Gwaii had visibly accelerated since my last visit. Comparing photos I'd taken in 2011 with the same spots in 2014, the most noticeable change was knotweed—Japanese, Bohemian, Himalayan—swarming in concert, forming thick clots everywhere, often obliterating what I was photographing, including, in one case, a four-meter totem. Leery of jumping on the herbicide train, citizens had experimented with cutting, seawater soaking, and other dubious modalities that, in a redux of Britain's pre-control days, had only spread the knotweed farther. Elsewhere, along even the remotest roads I'd tracked toxic foxglove and burdock, signposts of a growing tourism industry and the islands' logging juggernaut, both of which, along with the provincial government, had absolved itself of any duty of care. "Plants mostly come in on hay, with beetles and other invertebrates . . . I've seen bears with their faces coated in burdock looking *super* unhappy. I mean, that could kill a bear," noted Bergman, who, echoing Bob Brett's Whistler crusade, had a personal mission to eradicate foxglove from remote Hotspring Island, another potential stepping-stone to Ramsay. "It only takes a few hours of my time each summer. My dream with Ramsey is to restore its original ecology—and that includes plants."

When Bergman first began work here she'd become interested in plants and made an effort to learn about them. Out one day with an older biologist, they'd pulled their boat up at a rock swarmed by seabirds. At an oystercatcher nest, she photographed a plant she didn't recognize springing from a rocky crevice, later learning it wasn't native to Haida Gwaii or even Northern British Columbia, but to the Vancouver area, some 700 km south. "A month later I went back to collect it and there was now a gull nest on the spot; I moved the gull to get at the plant and it took a chunk out of my lip—that's another story—but the presence of a wide-ranging bird with a long-distance dispersed plant made me suspect the seeds arrived in mud on a bird's foot. And that's island biogeography. *That's* colonization. Which is another perspective [on invasives] if you want to include humans as natural agents of dispersal."

Anthropochory. And you could go one better: as natural agents of dispersal, humans and their global migrations starting in Africa have facilitated the invasion of one or many organisms that then went on to rewire ecosystems both local and continental, precipitating numerous documented extinctions. As we now know, in many cases anthropogenic invasions interact to create ecological cascades that facilitate even more introductions. In every case we are the linchpin; we *are* invasional meltdown.

## Factors Galore

Newsflash: Invasive plants are conquering the United States!

Oh . . . really? At least that was the tone struck by numerous stories to hit the Interweb after a first comprehensive assessment of plant distribution in the United States found that nonnative plants are more widely distributed than native plants. Media wasted no time in referencing the American West's plague of cheatgrass or the three-million-or-more hectares of the South now smothered in fast-growing Japanese kudzu (*Pueraria lobata*). Yet many made too-short shrift of the 2014 study's key tenet: in theory, nonnative species have had less time than co-occurring natives to expand their ranges, so it was often assumed that distribution models would predict disproportionately smaller potential ranges for nonnatives. To test this assumption, the authors compared geographic ranges of 13,575 plant species (9,402 native; 2,397 endemic; 1,201 alien; 755 invasive) in the continental United States using data from herbarium records. It showed distributions of invasive plants to be consistently broader, both climatically and geographically, than ecologically comparable native species, suggesting that distribution models for invasive plants at regional scales (the ones at which most management organizations operate) were *not* underpredicting their potential ranges relative to models for native species. In contrast, the comparatively limited longitudinal ranges of native species suggested a high degree of nonclimatic limitation—that is, context-dependent factors not related to climate—which, in turn, would likely cause distribution models to underpredict these ranges.

Broadly, the study showed not only that alien plants—whether rare or common on the landscape—have footholds in more diverse ecosystems than was previously thought, but also that the average invasive inhabited only about 50 percent of its expected potential range. Sadly, this meant only one thing: there was still plenty of space for invasive plants to spread.

Even if invasives enjoyed intrinsic advantages like broader climate tolerance or stronger competitive abilities, biological factors alone could not fully account for their success compared to natives. Instead, the key to their success appeared to be due more to propagule pressure exerted by repeated human facilitation—for example, widespread intentional planting of ornamentals with known invasive proclivities; the sowing of grass seed and, as Bergman mentioned, movement of contaminated hay. For any invasive plant story you cared to name, not only had humans introduced it, but once established, we'd done pretty much everything we could to increase its presence—as if unwittingly daring it to take off.

I spent the month of November 2014 in Southern Ontario, buried under serious snow while dashing around to visit various researchers and spend time with Mandrak at CCIW. While I was mostly focused on aquatic invasive species, I also began investigating invasive plant problems like suddenly ubiquitous *Phragmites* and garlic mustard (*Alliaria petiolata*)—a plant on every manager's lips and every community eradication program's list. With little on the ground to see at this seasonal juncture, I had to content myself with background reading. I came across an intriguing passage in Wayne Grady's 2007 compendium *Great Lakes: The Natural History of a Changing Region*: "Garlic mustard, a European wild brassica brought over by settlers as a potherb, has suddenly become an aggressive alien. Content as an edge plant for two centuries, growing abundantly but confined to the transition zone between forest and meadow, it has recently begun invading the deep woods in the Great Lakes region, where its tenaciousness and lack of natural enemies allow it to outcompete native flora."

I would find this meme of "sudden" aggressive invasiveness regarding garlic mustard repeated often, and wonder over this apparent phenomenon. Was it an actual not-so-sudden artifact of something else—perhaps another slow-burn potentiating effect of, for instance, invasive earthworms in combination with other single or multiple factors like deer or climate? Given what I'd learned, that seemed wholly possible, yet having grown up in the region it also seemed baffling; it's known that long-naturalized alien plants often become invasive after changes in land use, weather, climate, or genetics, but hadn't the earthworms and deer combination been a fixture in Southern Ontario for centuries? What, then, might have poured gasoline on the garlic mustard fire? I knew who to ask.

"Several things are likely going on," Lee Frelich responded to my email. "It's not quite as sudden as people think—garlic mustard has been invading woodlands for twenty to thirty years. It took time to build up a large seed bank and probably there was genetic adaptation as well."

At the same time, he noted, earthworms were spreading over large tracts of woodland. These worms had "many, many introduction sites" when European settlers arrived, and as they spread from each of these, the circumference of invaded areas increased exponentially (imagine a computer simulation showing a field of random dots that grow into circles, expanding at a logarithmic rate until they eventually all merge with each other). Given the simultaneous growth in the number of introductions in new areas due to second-home development in forests, a huge increase in earthworm-infested area over the past few decades was all but inevitable. Poised on every edge of the reticulate agri-field-woodlot-forest landscape, "garlic mustard was then in a position to invade those areas infested with earthworms. *And* deer don't like to eat it so that helped it spread. Like most phenomena in ecology it's explained by many factors."

## Gang Warfare

The idea that several synergistic causative factors could best explain the spread and impacts of certain invasives was as key to public understanding as it was to research, and is the reason Tony Ricciardi believes studying ecosystems is as important now as ever.

"Take the zebra mussel," said Ricciardi, reaching again for this bottom-dweller's bottomless well of exemplars. "It's obvious where it came from, and obvious how it got here. It had already been moved from the Ponto-Caspian area to ports like Rotterdam and Hamburg where a lot of trans-Atlantic and Great Lakes shipping was originating, and so its arrival in North America was very much predictable. In fact, someone did a risk assessment around 1980 that said as much—I'm pretty sure I have it somewhere."

His voice trailed off as I imagined him, like most biologists I knew, combing mental index cards for the physical location in a filing cabinet or a tottering stack atop a desk. "Uh, never mind . . . but the gist was that ballast water had already proven to be a big issue in introducing problematic invertebrates around the world, so these guys, a consulting firm I think, did a risk assessment of ballast water coming out of these ports. They found zebra mussel larvae in the ballast of many ships heading to North America, so their report stated it was likely to colonize. And a few years later, it did."

To the great detriment of North America's aquatic ecosystems but the pyrrhic benefit of researchers, who would, eventually, learn something of how to prevent future such disasters. Paradoxically, these same researchers were now also being delivered information that might help to "reverse engineer" a picture of the original Great Lakes–St. Lawrence benthic zone, where they might once have studied interactions between native, now-endangered mussels instead of watching the current cage match between pint-sized invaders.

"What's going on under the surface in the St. Lawrence is like watching a bunch of motorcycle gangs. There's massive warfare between the quagga mussel and the zebra mussel—and the quagga is winning. Experiments show zebras are still best at re-colonizing after a disturbance, but quaggas come in later and push them out," said Ricciardi, quickly donning his fascinated-biologist hat. "Both have now been moved west past the Rockies, and in some places the quagga invaded first—an interesting phenomenon that would be great to experiment around. A whole new living laboratory that might help us understand and prevent what happened here."

Ricciardi's emphasis on the heuristic value of ecological research into invasions addressed not only the misguidance of those who dismissed it as so much pissing in the wind, but also the fact that our need to understand what transpires when you put novel biological entities into an environment wasn't diminishing, but increasing.

"Nonnative species are only one example, but what happens when we start pouring engineered GMOs or even synthetic cells into the environment?" he said, alluding to the paramount need for risk-assessment of potential continent-level effects prior to enacting any such hubris. "Chestnut blight is a good example: purely because someone wanted an ornamental tree in some garden, a fungus gets loose that takes out 800 million trees in only a few decades. That left a huge hole in the ecosystem, but without a very precise 'before' picture, we *still* don't know what loss of that tree has meant to food web and habitat structure."

We did, however, know enough to guess.

## Nuts and Dolts

A century ago, the majestic American chestnut (*Castanea dentata*) represented a quarter of all trees in a forest swath that ran from Georgia to
Ontario. Measured by the biomass of its prodigious mast alone—a primary food source for a range of invertebrates, birds, and mammals—
the tree's importance to forest ecosystems was undoubtedly tremendous.
But a fungus introduced on Asian chestnut trees imported from China
proved catastrophic; between 1900 and 1950, the beetle-vectored blight
killed off virtually all American chestnut trees. Subsequent attempts to
breed a blight-resistant tree proved laborious and difficult until recently,
when a team led by William Powell of the SUNY College of Environmental Science and Forestry in Syracuse produced an American chestnut carrying a gene from wheat that confers resistance to the fungus.
Currently navigating USDA regulatory approval, it is being hailed as the
potential first use of a GMO in an environmental restoration project,
highlighting both of Ricciardi's points: the need to understand original
ecosystems, and to prevent, through vigorous testing and risk assessment, introduction of anything that impacts these negatively. In other
words, like zebra mussel, many invasive problems might have been anticipated, as encapsulated in the old chestnut (couldn't resist) appropriated by generations of comedians: "Poor Lou Gehrig. Died of Lou
Gehrig's disease—how'd he not see *that* coming?"

Like their American counterparts, under the auspices of the Canadian Chestnut Council, University of Guelph researchers were also
working on restoring the chestnut. As the blight raged through Ontario
in the early 1900s, many chestnuts were preemptively cut down in hopes
of both salvaging the wood and halting the disease's spread—much like
a firebreak. Not only did this fail, but wanton cutting likely knocked out
the few disease-resistant trees that might have endured. The whole sad,
calamitous scenario was repeated with the introduction and responses
to Dutch elm disease (DED).

Arriving in Ohio around 1930, DED reached Canada by the 1940s,
vectored by elm bark beetles both native and exotic. American (white)
elm hit by the disease could die in as little as a few weeks; within three
decades American elm had virtually disappeared east of the Mississippi.

In the epidemic's early years, disease spread was rapid because many elm (and their root systems) lived in close proximity as stately arbors lining streets in hundreds of towns. As with chestnut blight, when diseased trees were cut down they were neither disposed of correctly, nor the tools cleaned, missteps that also aided spread of the disease. American elm hangs on in central North America where towns now take rigorous sanitation measures; when disease manifests, the affected portion is cut out and burned, the tools sterilized, and fungicides injected into the base of the tree, a process repeated every few years. Though it has largely vanished elsewhere, elm isn't considered endangered because it can reproduce before succumbing to DED. Younger trees seem more tolerant, but they don't last; the species' lifespan now averages fifteen to twenty years. Because of rabid felling, however, researchers were again left in the dark as to how many trees might have been resistant and to what degree. With both chestnut blight and DED, no individual tree displays true resistance, only varying degrees of tolerance in which disease might be isolated and contained by the tree's immune system. With few North American trees having *any* response to foreign pathogens, however, the vast majority succumbed—billions gone in a veritable geologic flash.

One person sorting through the forensics of latent potential resistance was Sean Fox, who, as manager of the University of Guelph Arboretum, was involved in the longstanding Elm Recovery Project. On another cold November day with a blizzard on the horizon, I'd gone to see him, driving an hour west from Toronto to the arboretum, whose trails I often ran as a master's student there, where I was met by the sprightly, youthful Fox, looking natty in a newsboy cap. Leaving the Arboretum center, I followed his small blue sedan with its yellow "Brake for Turtles" bumper sticker down a dirt track. A couple hundred meters later he turned left, and we crossed frozen, flattened brown grass between rows of elm saplings decorated with various colors of surveyor tape, their trunks encircled with chicken wire to protect from—of course—an irruptive local deer population.

The Elm Recovery Project was a pet initiative of the late, much-respected horticulturist Henry Kock, who'd once described it to a reporter as "A dating service for lonely elms." In his work around Ontario's southern tier, Kock had discovered a few large elms that seemed to have weathered the disease. At first it was thought they'd survived due to isolation, but it turned out that other elms around them had succumbed. Farmers, for instance, often mentioned how their entire lane was once lined with elm, but they'd all died one year save the one remaining. Was there something special about these singular trees; did they hold the key to recovering the species?

A search was mounted for more, with a baseline criteria of three-meters circumference (which meant they were old enough to actually be survivors). Of some six hundred trees visited, Kock and associates—including a young Sean Fox—narrowed the candidates down to three hundred, and began testing them against disease. This laborious, time-consuming approach was accomplished via vegetative propagation—grafting an apical stem from each tree onto a sapling of seedling rootstock (the seed source being a large tree at the University of Guelph); everything from about three centimeters up on these new chimera saplings was a clone from one of the three hundred. When the trees were five to seven years old (about 2.5 meters in height), each was inoculated with Dutch elm fungus, the way a beetle would deliver it to a tree; then, using various experimental designs, trees were tested for different strains of the fungus.

"As expected, all trees become infected," said Fox. "So true resistance doesn't exist. But for trees that tolerate it, it's like us getting a common cold; they compartmentalize the disease and continue to grow—like trees generally do with wounds, insects, and other fungi. Our goal is to take these survivors and cross them to develop a healthy, genetically diverse population."

Much time and resources are being poured into selecting, growing, and breeding these trees with no idea how many will survive. Fox acknowledged that this longer-term approach—unusual in the world of ecological restoration—drew criticism for being too slow and old-school given current DNA technologies. But it also received a lot of praise for its intentions.

"It's a more natural way to develop resistance. Yes, it's slow, but still way faster than if nature was left to accomplish it," Fox noted. "Sure the immune gene pool would naturally strengthen over time as the

remaining elms bred, but it would take centuries under the best conditions. Now, with the survivors so spread out due to mass development
and a fragmented landscape, sharing pollen is tough, making recovery
even more difficult. What we're looking at with disease-tolerant trees is
putting the genes back out in their natural environment to bridge gaps
on the landscape. Maybe eventually we'll be able to sell seedlings that
people can plant. It'll be a matter of time before the trees can do this
themselves, so in the interim we can link some of these fragmented habitats back together to help that along."

What they want to avoid, however, is using a single clone that will
have low genetic diversity and be vulnerable to another disease or pest.
One of the great failings in rush-to-innovate agriculture has been to
breed monocultures newly resistant to some *current* threat; ten years
later, when something else hits, they crater. Wheat strains resistant to
rust, for instance, are only good until the rust mutates. "There's a good
chance that if we find a tree that's amazingly good at coping with Dutch
elm disease, then clone it and put it out there, it'll get taken out by something else," said Fox.

Convincing arguments for slow resistance didn't obviate the fact
that other, more high-tech, recovery initiatives aimed at both elm and
chestnut were bubbling away in labs at this very campus. I'd wanted Fox's
take on the entire molecular bailiwick. "To start, the difference between
the chestnut and elm recovery projects is that with elms we were able to
find a lot of mature, disease-surviving adults to work with. There's also
more wiggle-room with elm than chestnut because it's a more adaptive
species; elms are good on wet or dry ground, in fields or ravines, and
along all kinds of streets. They are actually great at exploiting disturbance.
This isn't the case with chestnut; even if you eliminate the blight issue, it's
a more particular species that requires an ecologically mature habitat for
reintroduction and, unlike elm, which is seeded by the wind, chestnuts
have to be moved around by animals to seed the tree. That being said,
people are working on introducing blight tolerance into American chestnut from Chinese chestnut by crossing them; it's easy enough to produce
a hybrid, and then you backcross to the American chestnut as much as
you can while always breeding for the resistant expression until you end
up with a 99 percent American chestnut with Chinese resistance. But

what kind of habitat would you have to introduce these to, and does that even exist? Henry Kock liked looking at things from the ecological scale and was a big advocate of maintaining genetic diversity in populations. Once we were chatting as we drove around and he said he didn't think artificially introducing Chinese chestnut genes into the American chestnut gene pool was the way to do it. He thought you should work with indigenous genetic diversity and the tools you have."

Similar hybridization ideas were under way for the American elm, but a complicating element was its tetraploidy, while resistant Asian and European elms tried in crosses were diploid. Two other Ontario species—slippery elm and rock elm—were likewise susceptible to DED and had been all but wiped out. "But no one really cares because they're not majestic," Fox allowed. "Most hybrids released on the landscape are Asian-European elms. There's also a research program here at the university trying to get around the ploidy issue by using protoplast fusion of Asian elm with American elm."

Finally, needing to get on the road before the blizzard hit, I left these two classical invasive beetle-vectored fungal disease examples for a minute to discuss emerging invasive issues that Fox might be worried about locally, for instance, on the arboretum lands themselves.

"European buckthorn is an invasive shrub in Southern Ontario that's eliminating a lot of native flora. Unfortunately we can clear buckthorn from the arboretum, but birds feeding on it at the edge of the city just reintroduce it. So at what point do we acquiesce? When do we say 'OK, it's naturalized?'" contemplated Fox, suggesting the kind of prioritizing and triage undertaken by many managers these days. "We're not going to get rid of it or be able to quarantine it in a small area, so how long until a balance is achieved? Will it really matter in the long term? At the ecosystem level there'll be changes, there'll be losses, but there will also be adaptation to some degree. But species *disappearances*, now those are hard to deal with, especially with the rate it's happening from us moving things around and hitting species from all fronts. And yet *we're* the ones that are most impacted."

"Then . . . why save *elms?*" I blurted, for no obvious reason.

But there was a reason. It was the devil's advocate on my shoulder jumping in, knowing this was precisely how some people would paint the issue. It wasn't easy to answer, but Fox wrestled with it for only a moment.

"It's about everything down the line that has been, or will be, impacted. About the passenger pigeon I'll never see and the chestnut that was once its main food. I traveled the province with Henry Kock and stood under these amazing elm canopies and looked up and said 'Wow.' The current generation is never going to have that opportunity. Ever. But *if* we succeed, there's a chance for future generations to experience something that's part of the heritage of this place. It's like that with all sorts of species. And because we're responsible for most of it, it seems we should be doing something about it."

"And should because we *can* do something?" I wondered. "Maybe that's the very definition of eco-responsibility."

"Yes."

The remaining issue, per Ricciardi, is how deeply we understand any particular problem at the ecosystem level, which in turn affects what we choose to do. By now it should be clear that in the widespread absence of such knowledge, foresight and proactivity in the way of prevention are our only reliable tools; everything else is naught but well-intentioned folly.

Driving home, I chewed on a final morsel of Fox's food for thought: the pernicious emerald ash borer arrived from Asia around the millennium to settle in the Detroit-Windsor area, where it began killing ash trees before being outed in 2003. As with chestnut and elm before, attempts were made to save a greater number of ash by cutting a 5-km-wide swath to stop the beetle's spread east; inevitably and spectacularly, it failed. Indeed, Nick Mandrak had two of the living dead in his backyard—a daily reminder of how little we've learned from our mistakes.

## Tantrum

Here, a savvy reader might think to finger humans as the most invasive species of all, a point admittedly never far from the discussion. As we've seen, the phrase "invasional meltdown" cannot be used without evoking the hapless naked ape—at least without also acknowledging that humanity comprises the only invasive capable of recognizing and consciously meliorating its own effects. But this brings up the million-dollar question that lurks beneath any consideration of invasives: can one species change an entire ecosystem?

It can, and we have. But we've seen that there are others as well.

Zebra mussel, salmonids, ungulates, cheatgrass, chestnut blight. The closer you look at almost any ecosystem, the longer the list of invasives grows. Could anyone have predicted the many innocent ways that deer mess things up on their own or in concert with diseases, earthworms, climate change? Maybe—but only if we'd taken seriously the important top-down role of predators in maintaining their balance. We did not, despite Leopold's entreaty. Though biological bookmakers would have offered very long odds (see how I did that?) on Burmese pythons pulling off a complete ecosystem reorganization in Florida's Everglades, as increasingly appears to be the case, these would have been considerably shortened had they heeded the rapidly multiplying effects of some of the state's other invasive predators—including walking catfish, cane toad, Cuban treefrog, black spiny-tailed iguana, Nile monitor, and nine-banded armadillo (this last might have caught you off-guard; I haven't mentioned invasive armadillos, but, well, here's a headline for you: *newsweek.com/spitting-armadillos-blamed-floridas-emerging-leprosy-problem-356823*).

And let's not forget Florida's extensive Little Shop of Horrors—plants like paperbark (a.k.a. Melaleuca), Brazilian pepper, Japanese kudzu, water hyacinth. Could the state have seen *any* of this coming? Perhaps. Could it have at least adopted a precautionary stance against that which they couldn't see? Well, shouldn't we all. In this regard, Florida is best regarded more as example than scapegoat, because most subtropical ecologies have been similarly affected by globalization. If we shift our scrutiny from generalist to specialist organisms, adjusting our lens from wide-angle to zoom, we might see pythons as occupying a layer of macroscopic invaders that conceal (or distract from) an equally potent invasive underworld that may be interacting in whole or in part to support them: earthworms, nematodes, bees, ants, beetles, moths, mites, spiders, mollusks, plants, fungi, and bacteria.

Some of these less noticeable invaders arrived in Florida as adventitious hitchhikers on agricultural enterprises, whereas others were introduced intentionally to those same enterprises as a form of "biocontrol" on pests both native and non. Biocontrol, as we'll hear in the next part, is a double-edged sword. On the one hand, historically it has been one of mankind's most ill-advised acts, with myriad of these "pest-control" organisms themselves becoming problematic: Indian mongoose and cane toad, for instance, were loosed to control rats and insects, respectively, in sugar cane the world over, only to extirpate dozens of native species instead (#epicfail). On the other hand, biocontrol has racked up many successes, particularly in the area of invasive plants, where it has been used to at least draw down some problems to levels where ravaged ecosystems are better able to restore themselves. If past lessons learned could be coupled with new technology and more stringent study, testing, and regulation, biocontrol might flourish in the future.

Direct loss of native species isn't always the only issue with invasive predators. By wiping out the island's native bird, mammal, and reptile faunas, for instance, Guam's brown treesnake infestation changed the composition of the island's invertebrate community. Suddenly-silent forests became web-strung galleries filled with irruptive spiders feasting on an equally abundant supply of quasi-carefree insects, now that traditional predators of both had been vanquished. The mildly venomous snakes, meanwhile, moved into urban areas in search of food, swallowing everything from hamburgers to used tampons and developing a penchant for frying themselves on utility poles in spectacular fashion to cause regular and annoying blackouts. Tens of millions were spent on studying the problem, developing traps, and fruitless eradication programs (in one flight of fancy, drug-filled mice were airdropped over Guam in an effort to kill snakes). Even snake-sniffing dogs were employed—a model briefly adapted to Florida's python problem.

The Guam example emphasizes how there is seldom a single, measurable impact from an invasive species; more commonly a range of effects interact in a cascade. When sea lamprey infiltrated the Great Lakes through the lock system of the Welland Canal, it took only a few decades for them to become abundant enough to destroy several multimillion-dollar fisheries and set off the first widespread invasive-species alarms in North America. An individual lamprey packs a big punch, destroying up to eighteen kilograms of fish over its parasitic life, with only one in seven fish it attacks surviving. Sea lamprey impacted or reduced stocks of not only apex salmonids, but also chubs, walleye, catfish, sturgeon, cisco, and burbot. Despite a measure of success in controlling lamprey, the lakes have been further compromised—in some instances through ham-fisted attempts to compensate for the effects precipitated by lamprey—by a spectrum of invasive organisms whose interconnected effects are the epitome of invasional meltdown.

"Of course, they're still stocking Pacific salmon," Mandrak lamented during one of our chats. "The original rational for introducing *Oncorhynchus* species was to control planktivorous fishes like alewife and smelt while they worked on restoring native lake trout to previous levels. They also thought the presence of salmon would reduce lamprey predation on lake trout—you know, another target—and that might also help that species recover. If that worked, they could stop stocking *Oncorhynchus,* lake trout would re-establish as top-predator in the upper lakes, and they could try restoring Atlantic salmon in Lake Ontario. It worked to some extent, but the recreational fishery for *Oncorhynchus* became so economically and culturally entrenched it's now impossible to treat these species as invasive. I think we're into an era of revisionist history where people are saying: 'Oh, it was always our plan to establish a recreational fishery.'"

As with young children, meltdowns are not uncommon in ecosystems rendered eternally youthful by repeated invasion. Ultimately, however, our look at how invasives are introduced, established, and spread, as well as the impacts they can have, has been a mere academic walk in the park compared to issues around the costly efforts of control, which have caused more than one frontline worker to throw a tantrum.

Native to eastern North America, the American bullfrog has been introduced to at least twenty countries, six western U.S. states, and British Columbia.

A group of kids heft Lucy, an abandoned five-meter Burmese python, during an invasive species gathering convened by Meg Lowman in Sarasota, Florida.

Everglades National Park Ranger and Python Project Director Skip
Snow unrolls the tanned skin of a 5.5-meter female invasive Burmese
python captured while guarding its eggs.

The Everglades' interconnected canals have become highways for water-loving Burmese pythons.

A forest "exclosure" on Anticosti Island, Quebec. Inside the fence, zero white-tailed deer; outside the fence, 160,000 of them.

Fearless white-tailed deer in the Baie-Sainte-Claire area of Anticosti Island, Quebec.

Lexi Forbes refills a bait box with the poison brodi-facoum during rat-eradication efforts on Arichika Island in British Columbia's Haida Gwaii archipelago.

The Italian wall lizard has found new homes in the United Kingdom, California, and Ohio, as well as southern Vancouver Island, British Columbia, where it is absurdly abundant.

*(Left):* George Cera, the professional trapper and bounty hunter charged with ridding Boca Grande, Florida, of invasive black spiny-tailed iguanas.
*(Right):* Wildlife rescue expert Justin with a green iguana during a gathering of invasive species experts convened by Meg Lowman in Sarasota, Florida.

During Meg Lowman's pool party in Sarasota, Florida, the author checks out the "steel-cable feel" of Lucy, a "chill" five-meter Burmese python.

Field biologist Karen Garrod checks on trap 12 of the USGS Burmese Python Control Project on Key Largo, Florida.

Laurie Wein, Parks Canada project manager for rat-eradication efforts on Haida Gwaii, goes over bait-box arrays on the Bischof Islands.

B.C. Ministry of the Environment herpetofaunal and small mammal specialist Purnima Govindarajulu eyes an invasive American bullfrog during a photo shoot on Vancouver Island.

Whistler's Angels—gunslingers Sam Cousins, Clare Greenberg, and Sharon Watson of the Sea to Sky Invasive Species Council.

*(Right):* Richard Beard quit his life as a dentist to start Green Admiral, an eco-restoration company. Here he's picketed by the dead stalks of knotweed that his crew sprayed a few months prior.

*(Below):* Itinerant biologist Sean McCann identifies a specimen of European fire ant near the community gardens of Vancouver's Arbutus rail corridor.

*(Left):* Invasive plant crusader Bob Brett on a roadside in Whistler, British Columbia, with a tenacious sprig of Scotch broom—the only one he was able to pull out that day.

*(Below):* Ichthyologist and aquatic invasive species doyen Nicholas Mandrak in his former DFO office at the Canada Centre for Inland Waters in Hamilton, Ontario.

*(Above):* Epidemiologist Isaac Bogoch of Toronto General Hospital monitors, maps, and models the spread of emergent diseases, including those driven by invasive species vectors. *(Below):* Clare Greenberg treats a Bohemian knotweed in full florescence in Squamish, British Columbia. Injection guns fill the thick stems with glyphosate, where it's then carried to the roots.

A highly toxic giant hogweed grows in a yard in Britannia Beach, British Columbia, while an SSISC crew member logs its location into a GIS database.

*(Left):* An example of how the seeds of invasive plants are easily moved around by humans—in this case, the author after a short hike in Pemberton, British Columbia. *(Right):* The invasive plant that moves itself: A Bohemian knotweed sprouts from a piece of driftwood along a river in Squamish, British Columbia.

*(Above):* A large Bohemian knotweed infestation in Burnaby, British Columbia. Rhizomes from the mother plant *(right)* have sprouted inside the electrical conduit on the telephone pole *(left)*. *(Below):* Three different kinds of invasive ivy engulf trees and pour out of a yard in White Rock, British Columbia.

*(Above):* A rare giant knotweed plant near Port Renfrew on Vancouver Island, British Columbia. This species hybridized with Japanese knotweed to produce a super-invader, Bohemian knotweed. *(Below):* Dog-strangling vine carpets a ravine hillside in the Rouge River Valley in Toronto, Ontario.

A dense stand of *Phrag-mites australis australis* towers over a visiting researcher near the Metro Zoo in Toronto.

Scotch broom releases chemicals that make it difficult for other plants to grow. This dense, impenetrable stand on Salt Spring Island, British Columbia, has shed its leaves for winter.

Florida has numerous invasive species problems, many of which originated from the release of unwanted pets.

Sign of the times: the borders of almost every state, province, and country feature billboards urging people to take steps to stop the spread of invasive species.

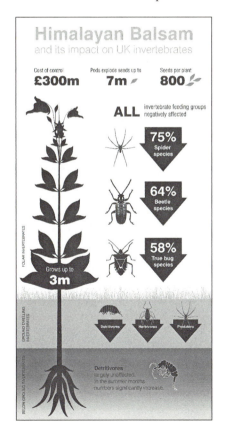

A poster from the European organization CABI explaining the ecological effects on the native U.K. invertebrate fauna of invasive Himalayan Balsam.

What you don't want to see if you're looking to buy a house: someone treating knotweed in the yard.

Fanciful outdoor art adorns a power company box in an industrial area of Vancouver. Unfortunately the plant depicted is invasive English ivy.

Serious crime: don't get caught with live Asian carps warns a billboard in Kingston, Ontario, along the Great Lakes St. Lawrence Seaway.

One of the easiest ways to move highly destructive forest pests around is with firewood and wood products like crating, pallets, and bark mulch.

It's all fun and games until someone steps on a carpet burweed. State, provincial, and national parks increasingly have signs warning of various invasive dangers within their boundaries.

Invasive plants can't go into regular landfill or be composted because seeds can survive the process. Many municipalities now provide special disposal facilities for invasive plants.

# PART V

# The Sisyphus Files

## *Eradication and Control*

For in all countries, the natives have been so far conquered by naturalised productions, that they have allowed foreigners to take firm possession of the land. And as foreigners have thus everywhere beaten some of the natives, we may safely conclude that the natives might have been modified with advantage, so as to have better resisted such intruders.

*Charles Darwin,* On the Origin of Species

# Horror Show

Hugging Toronto's western lakeshore, Marilyn Bell Park is a modest greenspace tribute to the plucky sixteen-year-old who was the first swimmer to conquer Lake Ontario in a twenty-one-hour marathon from Youngstown, New York, to this location in 1954. Lionized for her heroic accomplishment, news of the day highlighted that Bell had battled pollution, strong currents, even stronger winds, and large waves in completing her historic crossing. Universally, reports also noted another challenge: sea lamprey, which Bell had to repeatedly peel from her legs and stomach as she swam. It was a nightmare that might not have happened had she attempted the swim only a couple years later.

In 1955, the United States and Canada established the Great Lakes Fishery Commission (GLFC) with twin intentions: researching and controlling sea lamprey, and jointly coordinating fisheries management. Sixty years later, of 180-plus aquatic invasive species known to have become established in the Great Lakes, sea lamprey remains the only one controlled basin-wide (if Asian carp enter the lakes that will likely change), making it the world's only aquatic vertebrate pest-control program carried out with any kind of success at an ecosystem scale. Thanks for that go to the usual actors in the Great Lakes' serial enviro-drama: GLFC delivers sea lamprey control in partnership with the U.S. Fish and Wildlife Service, Canada's DFO, and the U.S. Army Corps of Engineers. Not to be left out, the U.S. Geological Survey also contributes critical field research to the binational effort that has led to a reduction in lamprey populations by 90 percent in most areas of the lakes.

Tremendously successful, the work nevertheless conjointly costs the United States and Canada about $30 million USD annually (for example, $26.9 million in 2009, and $34.6 million in 2010). Some might call this pricey, while others consider it cheap insurance given that the figure represents but 0.0075 percent of the annual $4 billion Great Lakes commercial fishery. Political pocket change or not, $30 million represents better-spent-elsewhere tax dollars that put the much lower cost of prevention in context: what if several lamprey-like problems required basin-wide control at similar price tags? Furthermore, necessary as it is, the sheer complexity of lamprey control creates a raft of other costly problems.

Adult sea lamprey spend only a year or so of their five- to eight-year life cycle feeding in the open "sweetwater seas" (*les mers douces,* as French explorers labeled the Great Lakes) before migrating up rivers to spawn in tributary streams. Females deposit, in gravel, up to 200,000 eggs, which hatch into filter-feeding ammocoetes that spend three to six years burrowed into soft-bottom substrates. The GLFC's integrated lamprey control program targets various parts of this breeding cycle. The first line of attack are lamprey barriers erected in spawning streams—hydroelectric and other dams serve this function, as do trap-and-sort fishways (recall the anti-carp Fishway of Cootes Paradise), and low-head barriers that most fish species can readily jump over but lampreys, with their limited fin power, cannot. The good: barriers work spectacularly well. The bad: barriers inhibit movement for numerous other organisms, and so ultimately are ecologically disruptive. Lampreys can also be trapped moving in or out of spawning tributaries; while the success rate here is not optimal, it's also a way to gauge populations and make decisions about where to employ other methods—like lampricides.

Application of 3-trifluoromethyl-4-nitrophenol (TFM) is the primary method employed to control sea lamprey. Some two hundred Great Lakes tributaries and high-value larval habitats are regularly treated with it to kill sea lamprey larvae before they have a chance to return to the lakes. As "lampricide" suggests, TFM is a lamprey-specific metabolic enzyme-inhibitor that doesn't affect most other organisms. A less-costly compound, Bayluscide, is often combined with TFM to reduce the amount of TFM required. Sand grains coated in time-release Bayluscide can also be dumped into sluggish or deeper waters, where they sink to the bottom before dispersing the chemical, thereby enabling larval control in areas where TFM cannot be used.

Despite the reasonably high efficacy of sea-lamprey management in the Great Lakes, outcomes had plateaued long-ago, as ascertained through metrics of abundance and wounding rates on lake trout. This brings to mind a common invasion-science aphorism: knocking down the last 10 percent of your problem will cost you 90 percent of your resources. A new weapon was required, and GLFC—thinking outside the control box with everything from mobile inflatable barriers to ways to sterilize lampreys—was now, through various research partners, seeking a magic bullet in the potential exploitation of sea lampreys' incredible olfactory powers.

I heard about these efforts firsthand in July 2014, at the annual American Society of Ichthyologists and Herpetologists meeting in Chattanooga, Tennessee. One morning I sat through a session on general freshwater issues that Mandrak kicked off with a talk about how each of the five Great Lakes had undergone extensive but separate changes to their fish faunas without overall homogenization—his potent argument for applying ballast-water rules to inter-basin travel. A subsequent lamprey paper, by Thomas Luhring and Michael Wagner of Michigan State University in East Lansing, sounded heavy but intriguing. In reality, "Chemical Risk Information Guides Migratory Movements of Semelparous Sea Lamprey: Implications for Control and Conservation," was promising in the simplicity of its argument.

As a primarily nocturnal organism (the cover of darkness aids considerably when your business is sneaking up on, and attaching to, other fish), sea lamprey rely heavily on their extremely keen sense of smell. Basically a swimming nose, a sea lamprey uses this prowess not only to track down prey, but also to detect pheromones that lamprey larvae emit in order to draw adults to suitable spawning tributaries (more larvae in a stream means more of *your* young are likely to survive here), and that adult females emit in order to attract mates within a tributary (more females translates into more mating opportunities for males). Smell is also used to pick up alarm cues emanating from dead or injured lampreys (warnings of impending danger). In fact, the odor of dead larvae elicits what might be termed a freakout-level flight response in migrating adult sea lamprey, which is one reason that lampricides have driven adults to spawn in untreated streams (and so further complicated the problem of control). Luhring and Wagner were investigating how pheromones and alarm cues might be used strategically, in "push-pull" tandem, to concentrate breeding lamprey in places where they would be easier to control. This could involve using positive olfactory cues (pheromones) to lure adults into traps, into unsuitable spawning habitat, or into areas that were easy and cheaper to treat with lampricides; and using negative cues (alarm odors) to repel them from areas with good spawning habitat or where control was more difficult or expensive to implement.

The idea's utility rests on whether both types of cues mediate habitat selection at discrete decision-making points along lamprey migratory routes from open water into rivers, and from there into upstream spawning reaches. Positive and negative cues might be simultaneously present in different upstream branches of a dendritic river system and only overlap downstream. This means that confluences of those upstream branches may be stark olfactory decision points between opportunity or risk that could be used to influence lamprey migration routes. The strategy was tested by manipulating the presence of risk cues for three decision points in the spawning migration: river mouth, river-to-stream junctions, and merging stream branches. Manipulations at merging stream branches showed the highest degree of avoidance to alarm-cue scent, such that you could even dictate on which side of a stream lampreys traveled; conversely, when presented with a plume of positive pheremones, lampreys happily entered the stream and accelerated.

Overall, the work demonstrated how those interested in sea-lamprey management might manipulate lamprey decision-making and so guide their movements. Cost could be an issue, and, as the inscrutable Mandrak pointed out, so could the addition of more infrastructure around streams. Mandrak had done some work on different lamprey species, including coauthoring a report on freshwater lamprey species-at-risk, and he cautioned that . . . well, that this was ecology, dammit, and ecology was never simple. A major issue, for instance, was that huge amounts of lampricide used on sea lamprey also put native freshwater lamprey at risk, and Mandrak had some beta on a fascinating eco-evolutionary phenomenon in this group that not only might be lost before it was understood, but also presented another possible outside-the-box route to get at sea lamprey.

"Both chestnut and silver lamprey are also ectoparasites," he began, "while the northern brook lamprey stops feeding after it transforms from an ammocoete, immediately breeds, then dies. However, DNA analysis seems to show that 'silver' and 'northern brook' lamprey are simply parasitic/nonparasitic ecomorphs of the *same* species. And it's now thought *all* lamprey 'species' are actually members of a parasitic/non-parasitic pair. The only one *without* a nonparasitic partner is sea lamprey, so from a control standpoint, if you could figure out what triggers the switch, you could perhaps turn sea lamprey into a nonparasitic ecomorph."

Mandrak meted out more tidbits. Another nonparasitic ecomorph, American brook lamprey, occurred in the Lake Ontario basin while its parasitic alter-ego, Allegheny lamprey, was present elsewhere in the northeast United States, suggesting that different ecomorphs of the same species could live in areas not inhabited by the other. Coincidentally, in its native range along the Atlantic seaboard, populations of sea lamprey were crashing, raising the tragic irony of a last stronghold being its invasive range in the Great Lakes. What did all this mean? "It means there's a hell of a lot more going on out there than one imagines," said Mandrak.

Given lampreys' extremely conserved morphology over a 250-million-year span, it also suggests there may have been more going on back in the Devonian than one imagines, a wholly interesting line of thought that paints lampreys in a new evolutionary light, a window into the breadth of ancient aquatic ecosystems—which, it now seems, had always been complex.

Despite concern about native lamprey in official circles, the bottom line was that a new methodology like olfactory cueing—which might increase the margin of control on a damaging invasive like sea lamprey—would be welcome.

If only it were possible to fool nuisance terrestrial invasive vertebrates as easily.

## Jug o' Rum

If you grew up in eastern North America where it's native, there's something decidedly romantic about the bullfrog. Its throaty *jug-o'-rum* was the bass track of steamy summer nights from Ontario to Florida, its size and ubiquity further enshrined in children's stories—wise, sedate, perched on a log, overseeing a swampy kingdom of dragonflies, swallows, and lily pads until alarm sent it leaping into numinous depths. Kid-lit and the mind's eye variously paint the bullfrog as a lover of water and land, a prince and pariah, a symbol of good luck and bad, a familiar but friendly enigma.

If you live in a growing list of elsewheres, however, the American bullfrog (*Lithobates catesbeianus*) is writing itself a different fable. Another of the villainous World's 100 Worst Alien Invasive Species, bullfrogs—yet again, both literally and figuratively—are a monstrous problem in places as far-flung as Belgium, China, Colombia, Cuba, France, Germany, Haiti, Italy, Jamaica, Japan, Mexico, the Netherlands, Puerto Rico, South Korea, Taiwan, Uruguay, Venezuela, and the western United States (it's in the Yellowstone River, for god sakes!). Oh, and of course, British Columbia.

From a handful loosed in various B.C. locations decades ago, populations ballooned in the wetlands of lower Vancouver Island and the highly modified agricultural floodplain of the Fraser Valley east of Vancouver. The trouble begins with the bullfrog's favored food—other frogs, particularly young ones—though, as with other generalist invaders like black spiny-tailed iguana, it will happily stuff anything that fits into its cavernous mouth. That means crayfish, salamanders, snails, snakes, turtles, small mammals, birds—even other bullfrogs. It is, of course, also prolific: a single female can deposit 20,000 eggs, a high proportion of which survive, in part because most animals find the taste of bullfrog eggs and tadpoles yucky.

After establishing in a water body, bullfrogs can eliminate native frog species through both predation and competition; by the time they gleefully begin to feed on their own kind, it's usually too late for the species they've displaced to recover, and the opportunistic and fast-dispersing bullfrog will have already spread to nearby wetlands, a hop ahead of any would-be control—like a Victoria-area business that actually guarantees bullfrog eradication. Not everyone, however, is convinced of the efficacy of such initiatives. Purnima Govindarajulu, a herpetofauna and small mammal specialist with the B.C. Ministry of the Environment, studied bullfrog control for her Ph.D. and believed the problem to be too far gone on Vancouver Island.

"In most areas there's no hope of eradication—it's either too costly or would require control forever—so it's better to put money and effort into habitat restoration for native species and hope for some balance of co-existence," said the soft-spoken biologist. "Then you stop bullfrogs from spreading further by increasing public education."

In Victoria and environs, where the amount of time, energy, and planning focused on controlling invasive bullfrogs by commercial, scientific, and community groups is flirting with the unrealistic, education and culling by both professional and citizen conservationists are indeed the measures encouraged.

## Mr. Big

As if it were dropping into a giant slot machine, a blood-red sun inserted itself into a cleft between two hills on the Saanich Peninsula just outside Victoria. At Trevlac Pond, where a clutter of cars and canoes filled a small driveway beside a woodland cottage, the mid-July gloaming buzzed with mosquitoes as equipment was sorted into the boats: one would carry Govindarajulu, along with local naturalist and invasive-species journeyman Christian Engelstoft (who'd also worked on deer in Haida Gwaii) and Nanaimo biologist Hitomi Kimura; a second would contain Neville Grigg and Pattie Whitehouse of the Highlands Stewardship Foundation; and a third would bear non-technical interlopers—a photographer and myself.

As we shuttled loads to the heavily vegetated shoreline, Grigg, the last to arrive, uncrated his gear: a headlamp like the rest of us, a white restaurant bucket like Govindarajulu's crew, and, more ominously, a copper-tube spear topped with four devilish prongs, each sporting a menacing barb.

"I got it from a supply company in the Okeefenokee Swamp," he bubbled. Upturned spear in hand, headlamps on, Grigg and Whitehouse stood side by side. In this rural setting, cabin in the background, the tableau channeled the Grant Wood painting *American Gothic.* But this was no dour couple: the pair were oddly animated, certainly more bloodthirsty. When a sonorous chorus erupted around the pond—a beaver-flooded former peat mine now but an inky blot beneath the looming forest—Whitehouse grinned. "Let's go massacre some bullfrogs!"

In the dark, the moonlit gunmetal we floated on seemed almost sinister. Dive-bombing bats, wings beating in our ears, exacerbated the mood. Ditto beavers that met our incursion with loud, tail-slapping displeasure. The convoluted shoreline variously comprised weedy shallows, downed timber, beaver lodges, and floating peat islands. And from every quarter sounded a bullfrog, the depth and resonance of its call indicative of size. The largest males—"boomers"—spaced themselves at prime calling sites around the lake's perimeter. Between them lurked smaller "satellite males" seeking to intercept females lured by boomers, and even more insidious "sneaker males," which surreptitiously released their own sperm over eggs extruded by a female *in flagrante delicto* with another male. That is to say, a pair in the throes of "amplexus," the two-armed death grip that males employ on females in order to facilitate and ensure external fertilization; once locked in amplexus, frogs are too hormone-addled to bother fighting off small-time genetic challenges.

Frozen in our headlamps were the glowing eyes of the players in this primeval drama, and soon, frogs the size of puppies were accumu-lating in an anesthetic solution sloshing in our buckets (anesthetized frogs would later be frozen and distributed to various university projects and high-school biology classes). When the two hunting canoes split up to cover more ground, we followed the Highlanders first. If a frog was spotted, Whitehouse maneuvered in slowly while Grigg held the spear out front until ready to impale the animal, then unceremoniously dump it in the bucket. When I noted he was quite good at this, Grigg shrugged.

"Sadly, it's a job that needs to be done. I'm just an engineer who runs a furniture company and got sick of bullfrog soup," he said, referencing the lake on which he lived. In fact, despite her feigned en-thusiasm, neither Whitehouse nor Grigg—nor anyone for that matter—enjoyed killing these creatures. After all, it wasn't their fault they were invasive, a destructive pest outside its natural habitat—it was ours.

We'd heard the tuba serenade of a huge boomer near a large floating island. "That's gotta be the granddaddy of all time," Grigg croaked. "He's mine!" But it wasn't to be: the Barry White of bullfrogs eluded detection, even after Grigg stepped from the boat to trod gingerly on peat that shift-ed beneath his boots like a waterbed. When our canoe peeled off to join the biologists, we discovered that Govindarajulu, too, was in hot pursuit of Mr. Big. "He's so sexy," she said, peering through a tangle of shoreline wood and vegetation. "I'd happily get all scratched up for *that* guy," she ribbed, distancing herself from the reality of extermination.

But even the practiced biologists couldn't find Mr. Big, so they bore down instead on another boomer squatting on a muddy bank. Govindarajulu eschewed a spear, catching the frogs by hand (you had to be *extremely* experienced to pull this off), which also meant the canoe had to be a lot closer. With Engelstoft paddling at the speed of Zen, Govindarajulu snaked her motionless hand through reeds and small branches and then, like lightning, made the grab. The frog, as long as her forearm, made us wonder about the size of the one we *didn't* find. Govindarajulu admired the beast for a second before passing it to Kimura, who fastidiously swabbed a few skin cells from its body into sample tubes before dropping the animal into the bucket. The swabbing procedure was related to an entirely different problem likely triggered by the unbridled global transport of nonnative amphibians.

Let's talk about that for a bit.

## Vector's Secret

The invasive bullfrog's proclivity for mass consumption of fellow amphibians was especially troublesome given the number of endangered native forms, but a related issue could prove even more deadly: American bullfrog is a natural vector of the highly contagious, freakishly pathogenic chytrid fungus (*Batrachochytrium dendrobatidis*—or *Bd*), the organism largely responsible for a worldwide decline of amphibians in recent decades. Over the span of a few years of swabbing amphibians around Vancouver Island to test for *Bd*, Govindarajulu detected it on about 50 percent of individual bullfrogs. This means that *Bd* is *always* present in ponds that contain bullfrogs; by contrast, the fungus may or may not be present in ponds without bullfrogs.

"We're still figuring out where *Bd* is and isn't present in the province," cautioned Govindarajulu about leaping to overarching conclusions about a chytrid outbreak in British Columbia, "but we're definitely telling people 'if you see lots of dead frogs somewhere let us know.' "

Although the hypothesis that an invasive species like bullfrog could move *Bd* around had yet to be studied comprehensively, what had been gleaned was highly suggestive. *Bd* in the Americas, for instance, was strongly linked by epidemiologists to the importation and escape/

invasion of the African clawed frog (*Xenopus laevis*), which had been used globally in human pregnancy tests in the 1930s and 1940s. Clawed frogs were known carriers of *Bd* resistant to its ravages; bullfrogs appeared similarly immune.

As with all of ecology, other factors were at play in the chytrid story. Numerous studies suggested that the unstoppable chytrid epidemic likely had no single cause—likewise a growing raft of infectious diseases posing significant threats: outbreaks of ranavirus in amphibians and reptiles in the Americas, Europe, and Asia featured death rates exceeding 90 percent. No wonder Govindarajulu and Kimura were so methodical.

Back on shore we checked the take—thirty-five breeding adults between us—and swapped a few stories. Grigg recalled the time he took sixty-five frogs in a four-hour haul and spoke of the need not only to control your *own* wetland, but those arrayed on the landscape around you—especially hidden, off-radar ponds like Trevlac, whose bullfrog population would continue re-seeding larger surrounding lakes after those had supposedly been cleared of frogs. He also talked up the Merville Frogfest, an annual community affair held in August that took bullfrog control to new—and tastier—levels. Frog legs and other gustatory novelties were a hallmark. I could hear George Cera and Jackson Landers cheering.

"Merville's bullfrog control program is a grassroots effort—non-biologists taking on invasive species to preserve their environment. And they've tried to make it fun to keep the enthusiasm up because they understand bullfrog control is a long-term effort," said Govindarajulu. "I attend every few years to make sure the biology and methods are fine because, well . . . things can go wrong."

There was a moment of silence and then, as if to mock her, a loud, low, thrumming erupted from the middle of the pond. Mr. Big was still out there.


## Victoria's Secret

With its massive, herbivorous tadpoles grazing ponds like cattle and highly carnivorous adults eating everything in sight including each other, the bullfrog, in the words of Govindarajulu, is "its own food web." A nifty analogy, but it was important to keep perspective: bullfrogs were just one extra-large, extra-obvious organism in a legion of Vancouver Island invaders, that, in turn, represented but a microcosm of a global problem.

Checking more ponds the next day we'd wound further inland, the roadside lousy with Scotch broom, meters tall and all of it in seed. The plant was out of control all over the island, and many communities were holding "broom bashes" each May with the aim of cutting it before it seeded. Engelstoft wanted to show me something, so we eventually pulled up to an ornate metal gate fronting a mansion with immaculate landscaping. Across the street was a small rock-cut topped by a cedar hedge. In the shimmering heat, it took a few minutes to locate what he sought, but there they were—and plenty of them. Scuttling beneath shady cedars or lodged in cool rock crevices were dozens of Italian wall lizards (*Podarcis muralis*), a startling emerald green with salt-and-pepper mottling. They were all over the neighborhood. And the next one, and the one after that, in a widening circle of dozens of square kilometers with satellite populations in downtown Victoria and on a sprawling community farm north of town where the highly active predators swarmed greenhouses for invertebrates—the farmers happy to have them. Already there were apocryphal tales of urban microhabitats with *Podarcis* populations being virtually devoid of insects, and there was a doctoral thesis in waiting on the potential competitive interactions between these newcomers and the indigenous northern alligator lizard. Introduced *Podarcis* had also established in places as geographically disparate as the United Kingdom, California, and Ohio. On various islands off their Mediterranean homelands, introduced *Podarcis* were studied for the rapid evolution they'd shown in their dietary shifts to herbivory, concomitant changes to head and gut morphology, and even mode of reproduction—although *Podarcis* are typically egg layers, some populations were drifting toward ovoviviparity in which embryos underwent most of their development in the oviducts—a strategy likely attributable to suboptimal egg-laying conditions in their adopted home. With a maximum of forty generations since introduction, the lizards were demonstrating the effects of natural selection in a fascinating way, quickly adapting to their new environments through modification of various physical and physiological characteristics. This plasticity is one reason that researchers find invasives instructive for studying broader ecological and evolutionary questions around dispersal and colonization.

But let's back up a bit: how did an Italian lizard land in British Columbia? Engelstoft related how he'd once done a TV interview where, ever-circumspect, he explained how it was *believed* that the lizards had originated from a small zoo that went broke and released all its animals. After the piece aired, a man phoned him to proudly claim this wasn't true, that in fact *he* had released two breeding pairs of lizards nearby. It was a scenario eerily reminiscent of multiple individuals or groups taking credit for a single act of terrorism. But that was the last Engelstoft heard from the eco-bomber, so the origin(s) of *Podarcis* remained as mysterious—and moot—as those of the reed canary grass, Scotch broom, Japanese knotweed, Himalayan blackberry, and California poppies where we now stood. They were, simply and irrevocably, here.

The origins of Vancouver Island's invasive bullfrogs, however, were no secret: waves of releases and escapees from deluded schemers who dreamed of a fortune in frog's legs had seeded wetlands enough for the prodigious frogs to do the rest. The same was true in the Fraser Valley— with the bonus there of regular floods to move them around.

I'd spent an evening earlier in the summer with Rylee Murray, a graduate student at Simon Fraser University doing ecological niche modeling (ENM) for nonnative invasive frogs on some seventy sites spread over a vast stretch of the Fraser Valley. His interest in using bioclimatic ENM to predict occupancy could possibly be applied to strategies for containing frogs' spread. One concern with bullfrogs in the Fraser Valley was their potential effect on the last remaining populations of highly endangered Oregon spotted frog (*Rana pretiosa*), which is on the cusp of extirpation in both British Columbia and Washington State.

After a half-hour drive from Vancouver, Murray and I exited the Trans-Canada Highway at Abbotsford, parking on a quiet street next to a hedge-lined laneway that led to a park. Walking silently in the dark, headlamps off, fifty strides landed us at the edge of a small lake. Lily pads glowed in the moonlight and geese honked obnoxiously on the far shore. We could hear the deep thrum of at least one boomer to our left, and another to our right. Bullfrogs had infiltrated here, too, deep in pseudurbia. "We don't often get big choruses out here," Rylee whispered. "There's usually only one or a few calling."

Sure enough, at our next stop the few discernable bullfrog calls were drowned out by a chugging chorus of northern green frogs (*Lithobates clamitans melanota*)—another invasive easterner. Indeed at many sites there were *only* green frogs, which seemed extraordinarily abundant, much like Anticosti's mink frogs. "Makes you wonder why we're not more concerned about them," Rylee said, echoing my thoughts.

Though the green frog was also large-*ish*, big males were still only a third the size of a boomer, and the reason for low concern was that so far, unlike bullfrogs, greens didn't seem to eat much more than insects, and they cohabitated with Oregon spotted frogs—though ecological partitioning studies to understand this had yet to be done. Murray's study, later published with Govindarajulu as a coauthor, found a higher correlation between ENM and occupancy for the climate-driven, valley-bottom-wetland-generalist green frog than for the habitat-driven, permanent-water-specialist bullfrog. It also found that four sites beyond the bullfrog invasion front were, due to habitat suitability, at risk of being occupied pending a successful dispersal. That was useful information if you had any capacity or authority to protect those ponds, especially given what was made abundantly clear on that evening's tour: eradication of *anything* was impossible in this floodplain of annually connecting and disconnecting wetlands, and in the event of an attempt, you'd have to take out green frogs—now far too numerous—in order to get to the bulls.

"Only in places where populations are so far contained do you have a good chance of eradication worth putting money into," Govindarajulu had told me during our night at Trevlac. And she'd done just that. In collaboration with the University of Waterloo, Environment Canada, multiple private landowners, nonprofit organizations like The Land Conservancy BC-South Okanagan Stewardship Program, and countless volunteers and field assistants, her ministry had squelched a small invasion near Osoyoos in British Columbia's Okanagan Valley. Even that had been costly. Master's student Natasha Lukey quantified for me the overall human effort involved in bullfrog management since the species had been first detected there in 2003: nighttime canoe searches for adults required five hours per bullfrog; active day searches for egg masses and tadpoles, eighty-five hours per bullfrog; live-trapping took eight days per tadpole, juvenile, or adult. In addition, auditory surveys for calling males required 552 man-hours per detection. "We've installed permanent auditory recording devices at high-risk locations," said Lukey by way of explaining how they'd made remote monitoring more efficient. "We have 3,140 hours of recordings."

According to Govindarajulu, five years with no frog sightings (or sounds) were required before you could declare an eradication successful. Though there was always the chance of reintroduction from another source, this one passed the milestone as a win.

## Frog-Shocker

Wherever there were significant numbers of people on Vancouver Island there were significant numbers of bullfrogs. Which is how biologist-entrepreneur Stan Orchard ran a profitable business based on bullfrog extermination. Unlike Govindarajulu, who advocated for organized control and facultative eradication, Orchard preached a gospel of permanent extermination, which he believed could be accomplished anywhere with enough effort (that is, money). Feasibility notwithstanding, it was hard to argue results: as of 2012, Orchard's BullfrogControl.com Inc. had removed 30,000 bullfrogs from Victoria-area waterways since 2007, with much of that effort funded by the Capital Region Commission responsible for the watershed. Whatever the politics, that was significant biomass—and a huge dent in the population that could have been.

The first time Orchard was called to Cordova Bay Golf Course north of Victoria, where I met him and assistant Kevin Jancowski on yet another sultry evening, it was because someone had phoned in a municipal noise violation. Owners of expensive condos overlooking one of the course's ponds had repeatedly complained of "foghorns" outside; some reputedly had put their units up for sale. "We took out fourteen *big* males on our first visit," said Orchard as he inflated a small Zodiac, loading in powerful homemade lamps that ran off motorcycle batteries packed in waterproof pelican cases.

To stun frogs, Orchard used a self-modified electrofishing rig; instead of the backpack version employed by fisheries scientists, the power unit sat upright in a plastic tub in the Zodiac. Also unlike the two-paddled fish version, both positive and negative terminals resided on a single, long-handled wand that also featured a small, boomer-sized net. Orchard held a patent on this "Frogshocker," having built and sold several to U.S. wildlife agencies that had failed in their own feeble efforts with high-powered rifles. The French had also used rifles in a *ne-inspirant-pas* campaign that landed them just 120 frogs over ten months, poor numbers compared to what Orchard removed from May through September each year—5,500 in 2011 and 3,000 in 2012. According to Orchard, the declining totals could be misleading; yes, bullfrogs had been vanquished from some areas around the city and prevented from entering others, but despite a full decade of effort, they continued to spread further afield, presenting an ever-distant frontline.

Battles had clearly been won, but without legions of Orchards and millions of available dollars, the war, as Govindarajulu held, was lost.

Returning to Trevlac next morning with the biologists, a swarm of perky gartersnakes completed their morning rounds underfoot while Govindarajulu, in chestwaders, checked minnow traps that had been left out for tadpoles. Not a great haul—only two—but both were large and would metamorphose that summer: one was a plum-sized bullfrog, the other a native red-legged frog sprouting rear limbs. Threatened red-legged frogs, yet another worry when it came to bullfrogs, bred as early as February, which should have conferred an ecological advantage over the July-breeding invaders. Bullfrog tadpoles, however, remained in the water for two years, competing with the red-legged frog during its entire aquatic stage; if the red-legged frog survived long enough to metamorphose, as this one appeared it would, upon emergence it would have to run a ravenous gauntlet of sub-adult bullfrogs patrolling the pond's edges. Furthermore, bullfrog tadpoles in Trevlac got help from—no surprise—another invader. Native dragonfly larvae that typically preyed on tadpoles were being snapped up by introduced pumpkinseed sunfish, and because centrarchids (bass and relatives) like sunfish coevolved with bullfrogs east of the hundredth meridian, they were likely wired to avoid the frog's foul-tasting spawn.

Checking for egg masses, we circled the pond in canoes, forcing the boats up every bum-wiggle channel crisscrossing the peat. Kimura and Govindarajulu tip-toed across floating islands, scouring the edges. Nothing was found; our night assault seemed to have impacted breeding, and the remaining frogs seemed especially skittish. Kimura insisted it was because bullfrogs learned fast; if you missed catching one on a first try, each subsequent attempt became harder.

Packing up, the crew demonstrated true dedication to the invasives cause, taking a half-hour to carefully spray down canoes, paddles, and waders with a weak bleach solution. Fastidious cleaning ensured that propagules like seeds, viruses, bacteria, and fungal spores weren't accidentally transported to other water bodies in which they might work. It was a lot of effort, but freed them from worry over perpetuating invasions or spreading disease. Nevertheless, the problem was inescapable. As we pulled away, the car radio droned with a news report about a lake being drained in the Vancouver suburb of Burnaby in order to find a single invasive Asian snakehead fish, followed by another item on an American tourist in the B.C. Interior found transporting highly invasive quagga mussels—a species yet to establish in British Columbia—on the hull of his boat. And then there was the stark reality of Govindarajulu's job: while we were out frogging, two eastern snapping turtles, unknown west of the Mississippi basin, had been found on Vancouver Island, and she and Engelstoft were off to investigate the nearby site where one of them was discovered . . . laying eggs in a roadside nest.

## Eradicat-er

If eradication is ugly, then control is ugliness in perpetuity.

The unpleasantness of ongoing culls is an unspoken fact (unless you're a PETA member) that reinforces, along with impact and cost, the urgent need to build up EDDR capacity. Sea lampreys may have habits we revile, but they're a masterpiece of evolutionary resilience that say much about the 3.5 billion years of life on this planet. Bug-eyed bullfrogs are equally unlikely to win a popularity contest in today's panda-dominated view of nature, but they, too, are ecological marvels. Eradicating large numbers of any damaging animal represents a cruel flipside to the same speciesism responsible for many intentional introductions. Yet people are wont to accept it so long as there is a measure of ecological restoration—seen by some as a form of environmental justice. In many ways, the question boils down not to what we should or shouldn't do about a particular invasive species, but what's possible and what isn't. And ergo, what's worthwhile to undertake and why.

"My undergraduate supervisor, Henry Regier, hammered home how important it was to differentiate between *restoration*—trying to return a system to how it was before—and *rehabilitation,* which is improving the function of a perturbed ecosystem," Mandrak mentioned during one of our talks.

Regier's foresight points to an increasingly important distinction. In either case, tolerance for terminal action plummets when the process involves warm-blooded creatures like birds and mammals, and people become noticeably queasier when the subject is a familiar furry friend— regardless of its impact. Such is the case globally with the domestic cat (*Felis catus*), a highly effective carnivorous hunter that most often depredates prey under two hundred grams but can kill animals up to two kilos. Cats have been shown to both contribute to, and directly perpetrate, wildlife extinctions on islands both pint- and continent-sized. On larger mainland areas where extinction isn't the measure, cat-caused mortalities are more difficult to quantify, with most large-scale estimates based on non-systematic analyses that tend to skimp on scientific data. In 2013, however, a group of scientists conducted a systematic review from which they were able to quantitatively estimate cat-caused mortality to small animals in the United States. The results were stunning. A conservative estimate showed that free-ranging domestic cats killed anywhere from 1.4 to 3.7 billion birds and 6.9 to 20.7 billion mammals *annually.* Ownerless (feral) cats were responsible for the majority, but pet cats allowed to roam outdoors also contributed. "Our findings suggest that free-ranging cats cause substantially greater wildlife mortality than previously thought and are likely the single greatest source of anthropogenic mortality for U.S. birds and mammals," wrote the authors, calling for scientifically sound conservation and policy intervention to reduce this impact.

If that sounds dire, Australians can empathize. Abundant populations of feral cats on the continent and its forty-plus coastal islands are linked to both the extinction of seven mainland marsupial species, as well as island and regional extinctions of native mammals and birds.

Cats have also foiled reintroduction attempts for several threatened species and were, as of 2015, thought to be adversely affecting thirty-five vulnerable and endangered bird species, thirty-six mammals, seven reptiles, and three amphibians. With other species further impacted by cat-transmitted infectious disease, it's literally impossible to calculate the true environmental and economic cost of feral cats to Australia.

Though most Australian states and territories have legislation aimed at restricting the reproductive and predatory potential of pet cats, a nationally coordinated program of feral-cat control isn't as feasible as for other invasives, and control remains targeted at protecting specific threatened species or habitats. Thus, cat eradication has been successful only on islands or within areas bounded by predator-proof fencing, though most eradications also required multiple modalities (integrated pest management). Problems associated with cat control include time, cost, and the social impacts of enforcing legislation; the status of cats as agents of pest control in some jurisdictions; variable cat densities between habitats (and lack of information on this); and relatively low bait acceptance by skittish cats. This last issue has proved particularly vexing, but may have recently been solved.

In Western Australia, conservationists hope a new bait targeting feral cats will aid recovery of fifty-three threatened species in 3.9 million hectares of national park and conservation areas. A decade spent developing baits to appeal to the notoriously fussy cat palate by the West Australian Department of Parks and Wildlife finally led to success: a sausage of minced kangaroo meat and chicken fat mixed with a cat-calling flavor enhancer. It sounds like the latest television pet-food fad until you get to the part where the sausage is laced with 1080 poison (sodium fluoroacetate). "Eradicat," which killed 70 to 80 percent of feral cats in field tests, would deploy in areas where fox baiting has been used since 1996 to save endangered species like the woylie (or brush-tailed bettong). The woylie had recovered so rapidly once its nonnative invasive fox predator was removed that it was actually delisted as an endangered animal. That lasted only as long as it took for cats to fill the predatory vacuum and reduce woylie populations to a tenth of their post-fox high point. In 2013, woylies were relisted as critically endangered. Double jeopardy: for the woylie, as well as for whomever was footing the bill for its recovery.

"Once again," said Mandrak when I shared the example, "prevention is key. After that you've essentially given up; I think it's actually possible that on detailed analysis it would prove cost-ineffective to do *anything* about an established invasive. The only reason to do something after an invasion is usually economic, but it's usually not ecologically sound. Until we get much better at carrying out true rapid responses we should just put all that money into preventing the next species coming down the line. True early detection means you're actively out there monitoring, all the time, because even when you get a hit you don't know what part of the invasion process that is—that first propagule is the hardest one to detect."

Government bureaucracy, as Mandrak frequently noted (often unintentionally), was wont to impede rapid response even where a plan was in place. "We find something, have to jump through all these hoops to initiate a response, then wonder why it failed."

The Asian Carp Program addressed this, pushing the detection point way down the Invasive Curve by utilizing more targeted sampling and efficiencies to raise the probability of success, and it had already proven itself. "What we can do with invasive species is only as good as the weakest link at the government level, whatever that might be—municipal, state, provincial, federal," said Mandrak, who remained cynical on the matter of legislative tools. "Australia and New Zealand have the best invasive species laws, but they also have enforcement. Partly because they've had big problems, but also because they're islands that can control entry points better than a country like the United States that's contiguous with two other large countries."

As always, the role of islands in the global shuffling and convergence of ecosystems demonstrated why control and eradication remained important despite the gloomy forecast.

"Homogenization might be stayed by the fact people have recognized it. But the *trigger* to actually doing something about it will likely come from recognizing its effects on the global economy . . . luckily, ecosystems can still benefit from that," Tony Ricciardi had told me, acknowledging the economic driver behind any real action. But he was okay with that, providing it also freed up money for research that might inform wiser future decisions. His was no blanket indictment against all introduced species, he was careful to note; only those that were truly problematic. "All I'm interested in is how ecosystems work and how their functions are changed by perturbations like invasive species. There are selfish reasons for looking into this—personal and societal," he'd said. "Take bees. No one is saying 'OK, let's get rid of introduced European honey bees,' because they're far too important to agriculture and as replacements for lost native pollinators, but there's no *way* the invasive *Varroa* mite would have jumped from Asian beehives to European beehives unless we [had] brought them together."

With *Varroa destructor* fingered as a contributor to the "colony collapse disorder" annually killing up to 80 percent of European honeybees in North American hives, the enormous cost of this blunder was clear incentive for study into increasing the efficacy of mite control. Mandrak might prove correct about certain eradication or control efforts being ultimately cost-ineffective, but, as he also pointed out, no one had yet done the math, and that wasn't really the point: much ecological restoration work had intrinsic value and utility in the context of a living laboratory. "Invasive species removal on islands, for instance, can reveal a lot," said Ricciardi. "When you have a native species just barely hanging on, and you remove the invading stressor and see those species suddenly become common again, it's pretty clear cut."

## Rat-Whacker

With its blue-collar feel, fishing-fleet docks, grimy DFO trucks, and a glut of funky guesthouses garlanded in maritime kitsch, Queen Charlotte City could have been just another fishing town in British Columbia if it weren't for the mop-haired kid clutching a ceremonial eagle feather in the Sea Raven restaurant. He was a reminder that the first fishermen on Haida Gwaii were, in fact, the Haida people, and that their current influence over resource, culture, and land issues was happily greater than at any time since Europeans began their systematic exploitation of these islands. But there was one thing the Haida would never gain control of, a troubling trickle turned threatening flood that raised grave uncertainties about the future. And it was more than deer.

The first hint of trouble in paradise was a billboard at the ferry terminal on Graham Island. The title, "Haida Gwaii Invaders," seemed to portend ruminations on historic raiding parties of First Nations rivals, or a litany of logging-company misdeeds. Instead, the text profiled three invasive insects and three invasive plants whose introduction the government and the Haida Nation were partnering to prevent; these organisms weren't here yet, but they'd caused extensive damage on the mainland, only a ferry ride away.

It's telling that even *potential* interlopers merited an XXL warning—though this may also have been because a more comprehensive list, including alien species already adversely affecting indigenous plants, animals, and humans on Haida Gwaii, would be so long as to be unreadable on any billboard. Were such a complete register rendered, however, with the culprits set out in descending order of most pernicious to least, and you were able to scan it with binoculars, you'd find the organism responsible for my latest visit to the archipelago somewhere near the top. An animal so commonplace it is rarely construed as exotic, which is itself a problem: the rat.

I'd bunked in at a Queen Charlotte City B&B with a sweeping view to the northern reach of Moresby Island, explaining to the proprietor that I was waiting for a plane to the Bischof Islands in Gwaii Haanas National Park Reserve and Haida Heritage Site. "Oh, you must be with the *rat people!*" she'd exclaimed a bit too joyfully, as if it were some third clan behind the traditional Haida kinship houses of Eagle and Raven.

"Not *with* so much as interested in what they are doing," I'd downplayed, hoping a dopey smile would obviate any chance of direct association.

"They" was Parks Canada, and what they were doing—killing rats—was hardly unique in the annals of human history, certainly not enough for someone to travel 700 km from Vancouver to see how the whacking was going. Unless you were interested in how such exterminations reflected a new wave of conservation—ecological restoration—that spoke to the very idea of national parks, and so held significance for not only visitors to Gwaii Haanas, but parkgoers the world over. As terrestrial ecologist Carita Bergman had explained, the federal Action on the Ground program kicked off in 2009, with millions earmarked for projects within national parks—from controlled burns to reinvigorate native grasslands, to salmon stream rehabilitation, to invasive species management. In part, it was also hoped the initiative would reconnect Canadians to their abundant diverse landscapes and help them embrace the value of ecological restoration. When that came to killing rats, it didn't take much convincing.

Rats—be they Polynesian, Norwegian, or black—were the world's most notorious hitchhikers, a ubiquitous feature anywhere human watercraft had ever nosed in. But it took a dozen millennia of express rodent delivery via seafaring for mankind to finally declare war on rats in the places it mattered most: islands. William Stolzenburg's 2011 book, *Rat Island,* tracks the global battle to develop a safe, effective way to clear small islands of these destructive pests, its title referencing an eponymous, uninhabited rock that boils up from the icy waters of the Bering Sea between Alaska and Siberia. Here, a costly experiment to exterminate a large population of shipwrecked rats that had devastated native seabird colonies was, with hiccups, ultimately successful. A growing list of such victories in the South Pacific, Caribbean, and elsewhere had encouraged Bergman to tackle similarly rat-infested islands in Gwaii Haanas.

After their eighteenth-century introduction by whalers, rats spread widely throughout the archipelago to do what rats did best—occupy and multiply in the shelter- and waste-rich netherworld created by human habitation. Where there was no human-supplied food chain or controls in the form of either cats or native predators—most of Haida Gwaii's wilderness—the ecological impact of rats was noticeably greater, in some cases overwhelming. As on Rat Island and other oceanic outposts, rats had decimated seabirds on many small islands in Gwaii Haanas. Surveys in the 1970s on tiny Arichika and the equally minuscule Bischof archipelago had turned up 1,600 breeding pairs of ancient murrelet and 700 pairs of Cassin's auklet. By the millennium, however, the Canadian Wildlife Service could find neither, with the pigeon-sized murrelet—whose Gwaii Haanas colonies represented fully 50 percent of the global population—teetering atop Canada's endangered species list. Parks Canada and the Haida, joint administrators of Gwaii Haanas, hoped that their Action on the Ground contribution—dubbed the tongue-twisting *SGin Xaana Sdiihltl'Ixa* (Night Birds Returning) for the birds' habit of holing up at night in shoreline burrows where rats easily attacked them—would turn things around.

Resourceful allies are key to any siege, and for this one Parks partnered with California-based Island Conservation, an adept sanitizer of islands with high-profile successes for invasive-mammal-removals in places like the Galapagos. A first order of business was to study the rats' habits via sensor-equipped cameras, particularly around bait stations pre-deployed to sit empty for a year so rats could get used to them. "In general, rats are wary of traps. When we put out live traps the success rate for detection is generally 0–1 percent in wild areas where we know rats to be, and only as high as 20 percent in urban areas. Cameras, however, have an 80 to 90 percent detection rate," Bergman told me.

On August 1, 2011, crews finally placed poison bait into some four hundred previously distributed stations on Arichika and the Bischofs. Hexagonal, black-plastic boxes made it almost impossible for non-target species like eagles to reach the bait; rats, however, could enter at will, ferret delicious bits laced with brodifacoum—a lethal vitamin K antagonist and anticoagulant—back to their burrows, share it with family and friends, then quietly bleed to death. The plan was to continue replacing the bait and removing rat carcasses discovered on the surface for eight weeks, after which, were no more found, success would be . . . contemplated.

My September arrival coincided with the project's denouement, where I hoped to catch the end of the work and join in the victory dance. But that meant hopping a float plane south, and with the mother of all Pacific storms looming, it didn't look promising. I'd gone to sleep amid blinking-red-light weather warnings. At breakfast, a crusty old dude who'd been on Haida Gwaii since the last Ice Age burst into the dining room brandishing a wooden barometer, his face ashen. "It's the lowest I've *ever* seen," he whispered, pointing to a needle buried well off the scale.

Outside, rain lashed, and trees—*big frickin' trees!*—bent double. We weren't going anywhere, illustrating the difficulty of even attempting such an ambitious project in a storm-wracked area. Being grounded, however, offered time to visit with Parks Canada's Haida cultural liaison, Barb Wilson. Although Haida carving enterprises—totem and mortuary poles, longhouse entrances, canoes, argillite rock—leaned heavily on animal iconography, nowhere would you find a *kuggin* (rat). Ergo, the Haida were also firm partners in Parks' campaign. "We're all interested in reestablishing an intact ecosystem," Wilson explained.

"Ancient murrelets and their eggs were food when I was a little girl, and though no one eats them now, Night Birds Returning is important for cultural reasons."

Bergman had, in fact, selected Arichika because Haida elders had identified it as historically one of their best seabird sites. The appeal of these islands to rats was clear when, during a weather window the next day, I flew to the Bischofs with Laurie Wein, project manager and de facto head of Parks' rat patrol. Low tide delivered an aerial appreciation of how the intertidal zone effectively doubled the area of some of the smaller islands. "Rats thrive here because each time the tide goes out it's like someone dumping bags of food on the beach," Wilson had told me. "A buffet restocked every twelve hours."

We'd circled the float camp—a cluster of linked concrete barges moored in a lagoon—before landing on the island's leeward side. As we taxied across the water, a small bear scuttled off a cobbled beach into the bush. Offloading supplies onto the patrol boat *Yo-Dang*, we heard how the bear had showed up a couple days prior—on an island where bears had never been recorded—to trash a few bait stations and consume their contents. Not enough to harm it, apparently, but as precaution the bear was left a tasty deer haunch containing vitamin K—the antidote to brodifacoum—which it happily scarfed. The plan now was to snare, tranquilize, and move the interloping bruin to a plush, far-off salmon stream. This, however, was a side note to a more gratifying bigger picture. Victory in the War of the Rats now seemed all but certain: bait uptake had ceased on both Arichika and the Bischofs, with only seven carcasses found—proof of the plan's efficacy in confining poisoned rodents to burrows where they couldn't be consumed by unintended targets like eagles, as had happened in Alaska.

After a quick lunch the group had split. Haida Gwaii's lone conservation officer, James Hilgemann, summoned to tackle the bear problem, remained at camp with Parks' Peter Dyment and Clint Johnson Kendrick. Meanwhile, Roger Packham, Lexi Forbes, and Ainsley Brow of B.C.-based Coastal Conservation, another project partner, joined myself and a Parks crew of Wein, Tysen Husband, and affable Debby Gardiner, who piloted the launch *Adelita* a half-hour south through swelling seas to Arichika while reciting from the wheelhouse rat-killing poetry she'd concocted.

On Arichika, bait stations had been laid out in fifty-by-fifty-meter grids. (An adult male rat's territory is about 10,000 square meters and contains several females, so there was built-in overlap—or over*kill,* as it were: "We don't establish rat densities beforehand so it's always over-engineered," Bergman told me. "Inevitably there's enough bait."). Wein, Forbes, and I headed up one gridline; Packham and Brow each took another. Hiking slick intertidal rock and scrambling up vertical embankments aided by fixed ropes, it was clear that the work's physical demands matched the project's lengthy research, design, and deployment components. And that wasn't the only labor-intensive part: I watched Forbes unlock each bait box, make notes, enter data on a portable device to be computer-uploaded later, bag brodifacoum in a small Ziploc, pin it inside the box, then shut and lock the box again. With some one hundred stations to service in dense bush and rugged terrain, Arichika suddenly didn't seem so small. Indeed by this point Packham and troops had walked some 1,500 km, taking over 822 person-hours to check baits, cameras, and carcasses.

This impressive and thorough campaign also came with some interesting lessons to further inform management efforts. On Arichika, for instance, when stations were first loaded, seventy-five bait blocks were immediately taken over two days and then … nothing, which didn't seem right. Most of the uptake occurred in a cluster on the island's northeast reach, but Bergman had live-trapped rats on the rocky southern tip—surely there were more rats than could be accounted for by the low uptake of bait blocks? Indeed when Bergman sat down to back-calculate rat energetics based on the amount of food consumed, results suggested there were only 250 rats on the island. Though hard to believe, it proved true, and the only hypothesis to account for this followed observations that rat detection on islands was always high around shorelines, tapering off as you moved inland. "It's like people living on islands—most of the food is around the edges," said Bergman, bolstering Wilson's buffet analogy.

## Bear-Trapper

A second storm broke in the night. Tied in the lagoon's protective hug, the camp moved gently with the waves, like a baby in a womb. By morning the rain was again torrential, the wind a howling fury, and breakers appeared to tower over trees along the island's forested rim. The tempest combined with a rapidly ebbing tide to spin our barge, which strained mightily against the anchor ropes, like it was resisting the flush of a giant cosmic toilet.

After breakfast, Hilgemann, Kendrick, and Dyment headed out in *Yo-Dang* to check on some bear traps. In the strange half-light entering the cramped cooking/dining/mission-control center, cook Kris Leach cleared dishes while next to her, stationed at computers, Forbes entered data and Packham reviewed photos. *Yo-Dang* returned within the hour, thwarted by weather. "Nasty out there," said Hilgemann, shaking his head. "Couldn't get anywhere near shore."

"It's blowing fifty-plus knots—spray, waterspouts, everything," added a wide-eyed and usually laconic Dyment. The largest swells that day would reach nine meters, the strongest wind gusts 170 kph. Kendrick settled into a chair and picked up a logbook. There wasn't much to do save what every other living creature out there was doing: hunker down.

Later, a satellite conference call was made to headquarters on the mainland to discuss battle strategy. Over speakers, the conversation, like the entire operation, seemed so facilitative and professional that it was like some weird hologram projecting an optimal model of people working together, so contra everyday life—let alone government—as to be surreal. Perhaps more surreal was that the reason we were bobbing in the angry North Pacific—surrounded by strategic maps, wall charts, radio and battery chargers, computers displaying images from dozens of remote cameras, bottomless coffee, and a plate of enormous cinnamon buns—was because we were effectively at war. With this much effort and this many resources marshaled against them in what should be a deci-sive battle, one imagined the rats having little chance of prevailing. Yet all it would take would be one pregnant female to escape the dragnet.

When the weather cleared the next day, I accompanied Packham on a mission to restore bait stations trashed by the bear. It was tough going, the tangle and humidity equivalent to any jungle. It was also treacherous. As on Arichika, we ascended steep, mossy embankments, clambering over and under massive logs and other obstacles. Packham re-anchored upturned boxes with rebar and zip-ties to the piercing soundtrack of gulls wheeling in pointillist clouds over feeding sea lions and whales—the great natural cycle that night birds once belonged to.

Behind camp, Packham and I traipsed across a doubtless rat-friendly cabin site since reclaimed by forest. It had belonged to a draft-dodging hippie named Butterfly who'd died in a woodstove fire in the 1970s, his charred body reputedly found on the beach clutching a guitar as if he'd been struggling toward the water. There was a story about how Gardiner and another female staffer had been spooked while surveying here a couple years back. Retreating to Parks' permanent encampment on nearby Huxley Island, the other woman had gone directly to her room and shut the door. Gardiner, strangely quiet, sat down beside Kendrick.

"What's wrong?" Kendrick asked.

"Well . . ." Gardiner hesitated, "we heard someone crying for help."

"Then what are you doing *here?*" cried Kendrick, leaping up in alarm. "Let's go!"

"Um . . . You don't understand. There was nobody there."

The idyll of these far-flung islets was often usurped by the unexplained. On their first outing of the day, Hilgemann, Kendricks, and Dyment discovered that the bear had inexplicably managed to clean out the pot snares they'd hoped to corral it with, so they'd dismantled them and returned in defeat. But surrender didn't sit easily. The animal might move off the island on its own when the berries were gone, but Wein and Packham were nervous about gambling on that; this final mop-up phase of the rat project was its most important, and the risk of a bear messing it up wasn't worth it. Though the bear hadn't touched any bait stations in days, removing it safely would bring peace of mind.

Kendrick loaded a photo card on the computer and the group gathered to watch the bear's actions over the past few days. Photos revealed that it always returned to places it had been rewarded—like the pot snares, or the deer haunch. With this information, Hilgemann decided on one last attempt before our outgoing plane arrived at 5 p.m. Kendrick, Dyment, and I helped lug the pot snares to an area above the same beach where Wein and I had seen the bear when we first arrived.

To encourage it to reach up and into a pot—where its paw would be held in a soft snare until it was tranquilized—Hilgemann attached them to trees at thigh height. Then in went the high-graded bait: chicken, crabmeat, gravy from last night's prime rib, cinnamon buns, frozen blueberries, sausages fried lovingly for the occasion, and dollops of molasses. Hell, if it weren't for the can of cat food Hilgemann added, *I'd* have eaten it.

An hour before our float plane extraction, the problem took care of itself in a way as inexplicable as the bear's origin. Overexcited, the bear had managed to squeeze its head through a snare meant only for a paw and, panicking when the mechanism tripped, hung itself over a massive, two-meter log. Hilgemann had never seen anything like it.

Animal deaths of any kind weren't fun, rats included, but it was beyond sad to lose even a single member of an abundant species to random accident when superhuman provisions were being made to protect them. Before the rat-baiting had begun, fifty invasive deer carcasses were laid out on points and islands surrounding the Bischofs as mitigation for resident ravens and eagles; each morning the islands' birds happily flew off to feed on free deer meat while the crew fanned out to look for anything that might harm these or other indigenous species. The bear's appearance was unaccountable, as was its death, but the unfortunate mishap took a distant backseat to what by any measure had been a superhuman act of hands-on conservation: by the time Hilgemann, Wein, and I departed, rats, it appeared, had been eliminated from the Bischofs and Arichika.

As relayed by Wein via email, checks that November and in the spring of 2012 also revealed no sign of rats. Good news. If Parks could, as intended, eliminate rats from the larger nearby islands of Faraday and Murchison, it would create a significant buffer for Hotspring, House, and Ramsay Islands, on which rats had never been documented. This meant that a good chunk of this damaged archipelago could take a step closer to ecological recovery, a small but important victory along a much larger battlefront. If night birds indeed returned to nest on the Bischofs and Arichika, not only would it be worth the time and effort; it would also prove an important, perhaps motivating, point: not all our environmental mistakes are irreversible.

## Rat-Whacker Redux

Returning to Haida Gwaii in the summer of 2014, I looked forward to being debriefed on the Bischofs/Arichika effort, especially given that the standard timeframe for officially declaring anything rat-free was two years. But I was also interested to hear about that summer's big initiative on Murchison and Farady Islands. Wein was happy to fill me in.

Arichika had passed the test and they were confident it was rat-free. The Bischofs, however, presented a more complex case. With no rats detected since November of 2011, out of the blue came a hit on May 20, 2013. An emergency crew deployed to set and rebait traps logged three camera detections in different areas and eventually live-trapped an adult female Norway rat. This wasn't good given the animal's breeding potential, but it didn't disparage the original work.

"If the 2011 eradication had failed, we would almost certainly have found rats sooner given the number of traps and cameras, so this was a reinvasion scenario," said Wein about a somewhat inevitable development. "We always knew the distance from Lyle [the area's largest island] was short—five hundred meters—and there have been many, many storm events that blew kelp mats back and forth through that channel. When we get the DNA info back on the rat we trapped we'll know whether it's a match with the Lyle population or maybe came off a boat—another possibility. But ultimately, if we can't keep rats off the Bischofs, the question becomes do we try and figure out a control regimen—say every two years—or do we just let those islands go?"

A precedent for the Bischofs scenario is New Zealand's Ulva Island, a birder's paradise and sanctuary for numerous rare species where successful rat eradication was thwarted by the short distance to a source population on a nearby island. There's little information about the long-distance swimming ability of rats, but what does exist also comes from New Zealand: a biologist radio-tracking rats lost one only to have it turn up on an island five hundred meters away. Boom.

Given the years of labor that had been invested in making Ulva rat-free, and that afterward several rare and endangered birds prone to rat predation had been reintroduced to the island, the Kiwis viewed the re-ratting as a genuine disaster. To salvage what they could, the New Zealand Department of Conservation opted for aerial spread of poison bait by helicopter, carried out in 2011. It would require years of monitoring before Ulva could again be declared rat-free (rats subsequently picked up in traps so far appear to have swum to the island), but in the meantime, the aerial drop also affected some bird populations due to direct and secondary poisoning, requiring even *more* assisted colonization—a slippery slope as Simberloff and Ricciardi have pointed out.

Obviating secondary effects to other wildlife was the biggest challenge to ridding larger islands of rats, an endeavor that now appeared to be morphing into an international game of ecological one-upmanship: first New Zealand's 113-square-kilometer sub-Antarctic Campbell Island was acknowledged as the world's largest landmass ridded of the creatures; not to be outdone, Australia leapt ahead with rat eradication on its own sub-Antarctic holding, the 128-square-kilometer Macquarie Island (the Australians also took out rabbits and mice, passing the two-year mark on all three). Small potatoes, said the surprisingly well-funded and ambitious South Georgia Heritage Trust, clearing South Georgia Island, which spans 3,900 square kilometers and claiming the trophy. It was hard to imagine anyone beating that milestone.

In any event, avoiding secondary effects was a top priority for Parks in its next step, clearing Farady and Murchison Islands, which required a switch to aerial tactics. "Eradication couldn't be done on either island with ground crews for a myriad of reasons," said Wein, itemizing an impressive list of prohibitions. "They're too big, and would require too much diligence—people get tired after ten weeks in the field. We costed it out as a theoretical exercise and needed something like sixty personnel in five different camps, so you can see why people use aerial bait broadcast—the chance of success is better. Also, these aren't Norway rats, they're black rats with smaller home ranges, so you might miss some with a ground array."

Black rats were also climbers, occupying a more three-dimensional habitat with all its attendant problems. Wein felt they had the technology to get around those problems, but it needed testing. In the fall of 2012 they conducted trials on three small islands off Farady and Murchison, plus an inland section.

"To mimic a helicopter, we hand-broadcast a placebo bait which had no toxicant and included a biomarker so rats could be scanned under UV light to see if they'd taken it. It was a pretty massive exercise. We had to ring the islands with traps, scan any rats we got, then euthanize them for DNA samples. From that we figured out which application rate gave the highest efficacy. We also learned that you needed to have the bait on the ground, with full access to the pellets, for three nights," said Wein, sounding exhausted just talking about it. "That's when we discovered that deer would be a problem—they were attracted to the bait as well."

Aerial rat eradication had been accomplished where there was a de facto bait competitor in their niche like squirrels, but never deer, whose potential interference here was compounded by hyperabundance. Though no one worried about losing invasive deer to brodifacoum, the cervid swarm would likely remove the bait completely, compromising rat eradication and posing a much greater risk for nontarget species that might scavenge deer carcasses, which, unlike dead rats, would be in the open. The answer was to mount a pre-rat-kill deer cull. "We took about two hundred off the island over nine months. It wasn't easy; you had to pick them off one by one; if there were three to four in a group you couldn't shoot just one because you'd increase wariness in the population. We had to adopt all sorts of techniques but it yielded good information on how to cull deer. There were only a handful left."

As always, the project was overengineered to ensure success, upping the amount of bait broadcast to account for potential deer uptake. Wein ran helicopter bait-drop trials to see how much made it to the ground versus getting hung up on trees, again adjusting the application rate upward. After the initial broadcast in the fall of 2013, Wein dispatched a crew to collect any dead nontargets to reduce the risk to scavengers. This task alone comprised five hundred hours of searching over three weeks, yielding only three deer carcasses. Knowing that 85 percent of baited rats should die in their burrows, they radio-collared forty-five and tracked them until they expired. A high number perished in the right time frame, confirming what was known from other eradications. This fastidious follow-up was all to one end: the Parks staff wanted to demonstrate the efficacy, low-risk, and short time periods involved; and they wanted to show that a rigorous environmental assessment had been done and that although there may be minimal short-term impacts to other wildlife, there would be no long-term negative effects, and affected populations would have the added benefit of subsequent rat-free breeding to rebound.

"We're still in the two-year waiting game but so far it looks good," Wein reported. "In the monitoring phase we're also looking for ecological response to rat removal—monitoring seabirds, as in are they starting to prosper, can we get an index of abundance, that sort of thing. And how are shorebird species responding to rat absence? We chose black oystercatcher because it's a good indicator of coastal ecosystem health, a keystone species, and a proxy for many others. So we're looking at its abundance and breeding success. We're also measuring responses in small mammals—a native shrew and an archipelagic mouse; the mouse could barely be found before, and shrews were in extremely low densities, but now, on rat-free islands like Arichika and the Bischofs it's like night and day, with shrews rebounding in a big way. We're also looking at intertidal community composition; where there's no rats, there's food, and shorebirds are feeding in the intertidal zone. Tangentially related is bald eagle reproduction. On islands with no rats, eagles have a higher nesting and fledging success. Not because rats were consuming eggs or hatchlings, but because of food availability—rat-occupied islands lack seabirds for eagles to predate. As for night birds, we figure ten to twenty years to see measurable nesting and numbers increases. To speed it up there's been some thought to recruiting birds via acoustic playback or visual cues with egg shells."

All this success and Wein still found it hard to get people to care about the impact of invasives. And what she found craziest was the difficulty of getting them to actually give a rat's ass about little-known species like ancient murrelets teetering on the brink of extinction. "We got more public attention through the media talking about rats than through talking about seabirds—it's the fear and fascination appeal. If people are grossed out, well, they're interested. In essence, playing on old human fears works: using the rat angle galvanized more attention to the birds."

Restoration ecology, Wein firmly believed, should play an increasingly large role in public agencies like Parks Canada. And allocating adequate monies to restoration was a key imperative for government. "The time for monitoring is years past," she'd summed. "From a conservation point of view, we need to think more about restoring ecosystem health and less about saving individual species. Conversation needs to shift to managing invasives on the land base in support of ecosystem resilience."

This was averring of a greater truth, one that especially mattered with invasive plants.

## Tied Up in Knots

On a late-July morning in 2014, high on a hillside above Britannia Beach, the view across Howe Sound was, once again, nothing short of spectacular. Suspended in a cloudless sky, the sun infused an ocean turquoised by snowmelt and glacial flour brought down in the latest heat wave. The day looked to be another smoker, and deep within a jungle-like stand of Bohemian knotweed, Rob Hughes was already sweating.

Wielding machetes against the woody, leaf-heavy, three-meter stems, Hughes and Breanne Johnson, one of the SSISC's two field crews, were busy knocking knotweed over to make it easier to spray with herbicide. I'd learned enough about this plant in previous outings to understand why it was such a focus of the group's control work and a problem here: knotweed had already claimed considerable ground on this dead-end street, pushing up through gardens, lawns, gravel, driveways, and anywhere construction had taken place. It had arrived along a right-of-way for the power company BC Hydro, following steadily behind a row of houses to make incursions into every property. At the bottom of a steep backyard, Hughes and Johnson earlier had crossed a fence line, disappearing into a bind of knotweed that strained in every direction. Only now, an hour later, were their torsos rising above the foliar fray.

A Lower Mainlander with a degree in ecology and environmental biology from the University of British Columbia, this was Johnson's first year on the crew. Prior to heading into the tangle, she'd spent time carefully mixing chemicals and readying sprayers and injection guns. The guns delivered a preparation of 48 percent glyphosate (N-phosphonomethyl-glycine—the controversial metabolic enzyme-inhibiting herbicide of numerous commercial formulations best known as Monsanto's infamous Roundup) directly into knotweed stems, while the leaf spray comprised the same active ingredient diluted to 5 percent and mixed with blue dye so you could see what you'd sprayed. This was science, and, as in the laboratory, careful preparation was critical not only for protecting the crew from the substances they were handling, but also for getting things right: the goal is to kill as many plants as possible by the most efficacious means and, with the cost of chemicals, not to spray money indiscriminately. Actually, it's best not to spray at all if you can help it, which is why guns were used to inject the chemical directly into larger stems. Sometimes, as now, a patch was so large and dense that the process had to be done in stages, starting with knock-and-spray followed by stem injection in late summer when the plant was winding down metabolically and sending most of its energy—along with the glyphosate—to its considerable root system; the plant would then die over the fall and winter. The job on this particular street was more complicated: the crew was here to treat knotweed in a few private yards where only injection could be used because of pets and children. But the ultimate source of that knotweed was over the fence in the right-of-way, hence the jungle warfare overture.

Every time Hughes emerged he looked as if he'd just gone for a dunk. This was hot work. Tough work. And potentially depressing when you lavished so much effort on a patch of knotweed surrounded, as here, by other patches that were part of the same individual super-organism. Looking either direction down the right-of-way, knotweed stretched as far as the eye could see; even Sisyphus would have quit by now.

"I try to keep focused on the long term," said Hughes, a veteran crew supervisor with a forestry background. "This guy is paying to have his yard cleared. He's making an effort. So maybe the people who don't want it done will see the results and buy in at some point. Sure it seems like an uphill battle—we're working on the outside, the perimeters of the problem. But more people are getting on board. Last year the District of Squamish was behind us with a little funding, but after we did a small project for them they really came on board. So, little by little."

Earlier in the book I detailed how knotweed, an aggressive, semi-woody, perennial ornamental, was introduced in the 1800s from eastern Asia to Europe, and thence carried overseas. No one knew the precise forensic trail of knotweed in the Pacific Northwest, but there was plenty of it, with amounts expanding exponentially every year. It was hard to imagine where the saturation point lay, or if there even was one; knotweed's ramified spread and lack of any significant limiting factor to growth seemed less to follow the contours of the Invasion Curve than some kind of asymptote approaching infinity. Often mistaken for bamboo by gardeners and plant-lovers who were (usually briefly) happy to have it, knotweed was easily distinguished by its broader leaves and more vigorous root system. If you were looking for it, that is, which most weren't. Its current critical mass had been fueled by a perfect ecological storm: spreading quickly by both vegetative (rhizome, plant fragments) and sexual (seed) means, dense thickets had degraded wildlife habitat and reduced plant biodiversity everywhere by physically and chemically impeding native vegetation. As in Great Britain, knotweed in the Pacific Northwest readily moved itself down hills and along rivers by destabilizing slope and bank alike with probing roots that broke off to slide or float away and establish new populations. And both Squamish and Britannia were lousy with it.

I'd left Hughes and Johnson and headed south along the highway where another SSISC veteran, Sharon Watson, who has a degree in environmental science, was working with rookie Sam Cousins, a GIS geographer by training. They'd parked by the highway in front of a large aggregate pit to check on what was left of knotweed they'd sprayed down the previous year. Not much: as we walked, dry knotweed husks crunched underfoot like old cornstalks. Here and there new eruptions flared, snuffed out by Cousins and Watson with hand-held spray bottles. Ding-dong the knotweed's dead: now what? Well, if you couldn't rehabilitate a site after eradication—and almost invariably, regardless of how much they'd contributed to the cost of spraying, no one wanted to pay the freight on this—the answer was usually more invasives. In fact, one graying pile of knotweed debris was now crowned in an invasive salad that included clumps of thistle, burdock, comfrey, and foxglove. Surrounding it were tall, dense swaths of Scotch broom, which almost matched knotweed for persistence and difficulty of control. If it weren't for a few scrappy native alders and conifers it would have been a totally alien landscape.

"Now here's a classic jurisdictional problem: the Ministry of Transportation is an SSISC partner and this is its land we're working on. But over there," said Watson, gesturing into a line of trees separating us from the pit, "it's private. And knotweed doesn't know the difference."

Of course. Fortunately the landowner—a wealthy developer—allows SSISC access to the pit to control invasives; not so fortunately, he's chosen not to fund the work, nor any follow-up. Hence the constant eruption of knotweed along the property's frontage, providing source populations for further spread. To extend a previously trotted-out knotweed analogy, this was like putting out the main body of a wildfire but leaving the edges to smolder, eventually to re-erupt into a full-blown conflagration. It placed SSISC and other nonprofits doing similar work between a rock and a hard place: though they wouldn't be paid for their time, they had to utilize the opportunity of control in the pit as insurance against the considerable labor capital invested along the highway. Even then, in the end it might all be for naught. A month from now, a Ministry of Transportation crew would mow the roadside for parking-lot sightlines prior to a large music festival, sending knotweed fragments flying in a hundred directions.

Such scenarios have led to knotweed leap-frogging up and down the highway like a silent green monster described locally as an evil intruder, a burglar of garden space, an underground iceberg, or, as I saw it, a plant that invited analogies the way conservatives invite condemnation—through their very existence.

Still, if knotweed deserved some of the grudging admiration that Tony Ricciardi reserved for champion invaders, it could also be lauded for sheer vegetative exuberance. Where most plants aspired to an adaptively derived shape—the rounded crown of hardwoods, the spear-point of conifers, the fan of ferns—knotweed stretched in every direction simultaneously, occupying maximum airspace leaf by offset leaf, blocking out other plants (even ever-tenacious Himalayan blackberry) and reflecting that vigor even more perniciously underground, such that scions of the alien mothership—cleverly disguised as seemingly unrelated, rubicund sprites—popped up like remote mushroom rings meters away. In only a few years a single knotweed plant could obtain titanic proportions, smothering the competition, treating infrastructure in its path as mere annoyance, and appearing to transcend plants' typically sessile nature to the point of veritable animation. Let's step back to imagine the fragment from which this mass first erupted, perhaps as small as a fingernail, expanding in time-lapse. Thus visualized, a knotweed infestation resembles a slow-motion explosion, a chlorophyll Big Bang whose interminable billowing begs to be, if not outright stopped, arrested wherever and whenever possible.

Though it didn't appear in his book, knotweed presented a very reason for the militaristic lexicon erected by Elton to describe alien invasive species: it *made* you want to fight back.

## The Green Admiral

Richard Beard is a wiry, snow-haired, sixty-something Brit who emigrated to Canada in 1974. Drawn, like many Europeans, by vast spaces and all the animals, fishes, and verdure that occupied it, he settled in West Vancouver with his wife, where he put in a thirty-eight-year shift as a successful dentist. A few years back he'd returned to school, relinquishing his dental practice to attend the University of Victoria's diploma program in restoration of natural ecosystems. "I'd been involved for a long time in conservation on an uninformed level," he said—modestly by my mind, for someone who had more scientific information on the tip of his tongue than most. "Mostly a director of organizations, working on committees or fundraising. But I was really looking for something practical to do as a career change."

It didn't get more practical than this. Dressed in a gray, one-piece boilersuit, with work-worn boots and blue surgical gloves (did it seem odd to wear these for other than poking around people's mouths?), Beard fronted the bed of his white pickup, its lowered tailgate revealing a familiar bricolage of water jugs, spray backpacks, injection guns, and carboys of glyphosate. His company, Green Admiral Nature Restoration, serviced about 140 homeowner clients and a raft of jurisdictional contracts around the Lower Mainland; most involved invasive plant removal, jobs where he'd roll with one to three employees depending on what was required. "September is our busiest time," he explained, "because that's when knotweed flowers and seeds."

Of course—the demon knotweed. Because Beard had long subscribed to ecological restoration as a keen lepidopterist and occasional birder, he'd been preoccupied with the idea from the get-go with Green Admiral. "I could see where the need was. I'd seen it coming for a while and that people would eventually need to address this enormous problem. At the same time, I recognized that invasive species presented us with an opportunity as long as we embraced restoration—the presence of knotweed, hogweed, and blackberry are literally *telling us* to reclaim a place in order to reestablish native forest."

The troika he referenced was North Vancouver's de facto Axis of Invasive Evil, and his comment put me in mind of a spate of recent literature arguing the same. It also raised an interesting question. "How long would it take for native forest to reestablish if we just left that stuff alone?" I asked.

"Probably over a century," he answered. "We need to acknowledge the concept of arrested succession when it comes to invasive monocultures. A huge patch of blackberry stops succession in its tracks, so it'll be a long time before an area it grows in will be conducive to sun-loving colonizers that create the soil conditions necessary for forest to reestablish. With that comes fear that these areas provide loci for the spread of invasives into native forest elsewhere. With the same basic biogeoclimatic zone from Seattle to Alaska, why wouldn't shade-tolerant invasives spread into all these areas with the same disastrous consequences—especially in riparian areas?"

They would. With serious blackberry and knotweed infestations already plaguing north coastal British Columbia and Southeast Alaska, increasingly common early and persistent heat waves in these areas were creating opportunity aplenty for a plethora of invasives. As Bergman's plant find on the seagull nest in Haida Gwaii suggested, seeds were already taxiing up and down the coast with birds; all that colonizing invasives required was to have their propagules deposited on some seriously disturbed habitat, which was fortuitously abundant in any human-occupied outpost along the tortuous 2,000 km coastline of the world's largest temperate rainforest. Beard now recited "The Group of Six" most problematic shade-tolerant invasives: holly, yellow lamium, English ivy, periwinkle, cherry laurel, and daphne laurel. Holly, thick in the Lower Mainland and on the islands of Howe Sound, was, by an order of magnitude, the biggest garden-grown invader of neighboring intact forest because it was so easily transported by American robins that ate its berries. According to Beard and SSISC's Clare Greenberg, there was a host of similar emergent threats that no one was heeding: small-flowered touch-me-not was the fastest spreading invasive in West Vancouver; also gaining ground was sweet woodruff, which packs the same characteristics as knotweed—underground rhizomes, rapid spread, dense monocultures.

"They're *all* sneaking up on us," stated Beard. "Largely because we're too busy catching up with established problems."

Getting out in front of plant invasions—predicting problems where none yet existed—was the next horizon in prevention, much like Cudmore and Mandrak's work with Asian carps. Modeling invasions in ways that could inform prevention decisions, however, was harder for plants than for fish. Long-established plants often took decades to become invasive; rhododendron, for instance, a U.K. denizen of over a century, became invasive only in the past three decades. Likewise the lag in knotweed detonation. "Do some invasive plants need to reach a critical population mass before they can take off?" Beard wondered aloud of these examples.

"Or maybe have just one more small, adaptive mutation fall into place, like a genetic slot machine?" I added, thinking of papers I'd recently read suggesting just that. We look. We shrug. This is the science before us.

Today's mission had started with the Tsleil-Waututh First Nation, a North Vancouver band on whose unceded territory Beard and crew had mapped thirty-five knotweed patches. The map looked impressive on the iPad that Beard's assistant Wade McLeod showed me, but a mere scatter of spots at even a smallish scale belied reality: Beard estimated 1.5 hectares of the stuff comprising upward of 53,000 stems. Oy.

Our first stop was the one that precipitated the Tsleil-Waututh's interest in dealing with invasives on their land, an engagement that began, as commonly happened, with a wake-up call. Kids riding a bike trail in the nearby forest returned home with horrific burns on their hands. Investigation revealed a major infestation of giant hogweed along the trail; coincidentally, the hogweed grew amid an even larger patch of knotweed. Green Admiral had injected a thousand hogweed stems here—some as big around as a forearm—and some 1,400 of knotweed. Now, six weeks later, the hogweed had lain down, deliquescing into the forest floor. Hogweed seeds might remain viable for eight years, however, requiring a long monitoring process. Shockingly, a few small plants were already popping up. "We call this spot 'ugly patch,'" said Beard, showing me a photo on his phone of what it originally looked like.

The team—which also included Ecological Restoration grad Julia Alards-Tomalin—would take care of further patches near the trail because these were the ones most likely to spread. But they'd have to pass, for now, on a massive infestation near an outdoor hockey rink: 2,000 stems of 400 square meters meant about thirty hours of work, in part because blackberry removal was required to reach it.

"The key to keeping this stuff from growing into forests is to have forests dominated by natural conifer canopy as opposed to a purely deciduous second-growth forest whose canopy is seasonal," explained Beard. "Knotweed starts growing early in the year when the canopy is still open— same with blackberry. But as soon as you put in a few cedars or Douglas-fir, blackberry fades out. It's an emotional issue around here, though, because birds eat blackberries, and people like birds. So when you mention controlling blackberry, the first thing they think about is losing birds, without realizing how blackberry frustrates re-establishment of a natural forest that has much higher wildlife values."

This last point, which numerous individuals I'd spoken with circled back to with frustration, was nevertheless an important one: removing invasives that people had become used to (shifting baselines again) could push emotional triggers; the concepts of rehabilitation and restoration, not so much. Reduced bird traffic in the backyards of housebound seniors, for instance, might bring them to tears, while the thought of a towering native forest reappearing half a century from now would be too abstract to elicit an equally lachrymose response. This was why rehabilitating such sites immediately after invasive removal was key.

"The focus on removal is actually clouding the debate on eco-restoration," said Beard, a truth I had yet to hear articulated quite so baldly.

It was an interesting point that also smelled of money and jurisdictional politics, both of which also seemed worth exploring. And I had a friend who knew a guy who knew a girl who worked in exactly the kind of place where I could do just that.

## Nature

Fiona Steele is a professional biologist at Diamond Head Consulting, a specialized, multifaceted firm working to integrate environmental values into a range of projects, hoping, as its website touts, to "create great places that people will value for years to come." It may sound lofty and poetic, but it's a sincere ethos that leans heavily on collaboration, working with nature, and being both strategic and creative. With fourteen years of forestry, ecology, and GIS under her belt, Steele's passion for native plants led her into ecosystem restoration and urban invasive plant management. She spent years working with municipalities to create policy and strategic plans for managing invasive plants that include inventories, habitat restoration, and coordinating large-scale control and removal programs. A past board member of the Invasive Species Council of Metro Vancouver, she currently sat on its Regional Invasive Species Strategy committee.

I'd call that a career. Which I mentioned shortly after Steele and I sat down for coffee in a hip Vancouver industrial area. I also shared that I thought Diamond Head's touchy-feely brand statement was a first for managing invasives—it almost sounded sexy.

"Most of my projects are still pretty mundane," she laughed. "We have a variety of clients but the big guys are Lower Mainland municipalities we do policy-development work on invasives with—everything from how crews will operate to how to educate the public on what they're doing. These are higher-level operations, but we're also on the ground with invasive audits, removal, and restoration planting."

In particular, that meant municipalities like Surrey, Burnaby, and North Vancouver, jurisdictions comprising enormous peri-urban and urban areas, with equally enormous invasive plant problems. How did that work?

"The District of North Vancouver commissioned us to do a full inventory of invasives within their boundaries, then used it to develop an in-house strategy that they went full-steam ahead on," Steele related. "First thing they did was tackle *all* knotweed and hogweed on public lands—a brilliant move that gave them leverage to come down hard on private landowners."

While DNV was currently in the throes of concocting a comprehensive invasives policy, the City of North Vancouver, DNV's urban core where Richard Beard's Green Admiral often worked, was even further ahead. Unfortunately, knotweed traveled downstream through watersheds, and much of DNV sat above the city. Similar issues were problematic across the region; Surrey, for instance, was doing much on the invasives front while Metro Vancouver, which accommodated tens of thousands of Surrey commuters daily, was, officially at least, doing doodly-squat. "Vancouver is lucky their largest parks have active volunteer groups making measurable impact," said Steele. "Stanley Park's 'ivy busters' removed English ivy from every single affected tree in the park, and while the plant still grows there, the trees are all protected for now."

A few things Steele told me were disheartening, others uplifting, some just plain funny. "We inventoried knotweed along the City of Richmond's dike system because it was concerned that if it treated well-established knotweed, dead rhizomes might create air-pockets that could compromise the dike. So they contacted the government of Holland to see what it was doing about knotweed on its dikes, and the Dutch were like 'Oh . . . interesting—what are *you* doing about yours?'"

Richmond's dikes and agricultural canals were also plagued by aquatic invasives. Two commonly invasive aquarium plants—parrotfeather and Brazilian *Elodea*—had recently become established and were spreading. Regularly scheduled dredging of the canals would now likely serve to further spread the invasives; to counter, Diamond Head was conducting a trial with barrier membranes on the canal's surface to block sunlight and keep plants from pushing through. The municipality that Diamond Head was most invested with, however, was Burnaby.

"We have a long history with Burnaby. They got on the invasive bandwagon early. In 2009 we conducted a city-wide inventory in natural areas for their parks department. It was a *massive* job, especially because we didn't have iPads or anything and did it all on paper. But it's the way for any jurisdiction to start because then it knows what it's facing— maybe an audit doesn't catch *everything,* but there's a sense of what the priorities will be. The original intention was to take the inventory infor- mation and create an invasive plant strategy, but Burnaby didn't do that—probably because full integration between departments on the idea hasn't happened yet, which is problematic. Their approach has been a bit haphazard, but Parks was still able to follow up on some of our recommendations."

Within that follow-up, however, lay another hurdle: Burnaby's by- law against cosmetic herbicides stood in the way of eradicating invasives. To nudge higher-ups to allow managers to use them, Parks adopted a straw-man strategy, contracting Diamond Head to attempt the physical removal of knotweed for three years just to prove it didn't work. "It was silly not just because it wasn't working, but because it also cost a ton of money," said Steele.

Nevertheless, one could sympathize with Burnaby's Parks. Work- ing within governments on invasives always required prioritizing. What was an imminent problem? What did you have the capacity to work on? Who was going to help? Were there things you could do *without* the slow-motion parliament required for a plan like Johanna Niemivuo- Lahti's Finland strategy?

Melinda Yong, environmental coordinator for Parks, Recreation, and Cultural Services for the City of Burnaby, knew all too well the lim- itations alluded to by Steele, filling me in when I dropped by her office one September day.

"Right now, Parks is the *only* department actively engaged in inva- sive control, yet the edges of parks actually comprise engineering right-of- ways—sidewalks, gutters, etc. The Engineering Department is interested but they don't have a mandate. So now we're treating invasive plants in parks whose perimeters are being mowed without concern by Engineer- ing," said Yong, noting that things may be improving. "In general, staff in other departments don't know what knotweed is. But over time, because of sites where things have gone into areas outside the aegis of

Parks, they are slowly coming on board. I expect this to accelerate over the next couple years."

Having sat with Steele on the Vancouver Invasive Species Council board, Yong was also familiar with the challenge of inter-jurisdictional responsibilities and getting everyone both to the table and onto the same page. "Even *within* Parks it has been slow to get different groups asking the right questions when they're designing sites, to get contractors to report things they see, and to get everyone thinking together about issues like soil screening."

When Yong began her job, with invasive plants part of the portfolio, her first move was to commission the baseline audit by Diamond Head, which listed a top thirteen species threats. She'd been able to contract the company to initiate control measures that targeted recommended species. Internally, however, the City of Burnaby had its own priorities: enlisting volunteers to help clean out key parks with large green space and natural areas. "We're lucky to have both community and corporate groups coming forward to help with removals," said Yong. "Volunteers are usually gung-ho for the first few hours and then lose interest, *but* . . . if I have sixty people working two hours each that's 120 man hours. Plus they've made a personal connection."

As with any environmental issue, it was important to educate individuals about the short-term ecological and aesthetic costs of bad gardening choices, for example, or the monetary burden of infrastructure damage. Sometimes, however, galvanizing interest and concern was as easy as a sensationalistic news story—like the infamous "snakehead incident."

Hanging out by a lagoon in Burnaby Central Park on Mother's Day, 2012, a man had spotted a strange fish in the shallows. He filmed it, posting the footage to YouTube where it went viral—but not before several viewers inveighed that it was an Asian northern snakehead, a vicious predator that would eat everything in the pond. Officials were alerted and Burnaby's snakehead was suddenly headline news; Yong found herself both in a media spotlight, and tasked with the heroics of saving the pond's other innocents from a snakehead's rapacious jaws.

"I wish I'd saved those news articles," she recalled. "We even had stuff in the U.K. and Australia, but dealing with the media wasn't so bad because it was multijurisdictional—our communications department worked with the provincial Ministry of the Environment. In the end the exposure was good for *all* our invasive species initiatives, highlighting the importance of citizen science because it was reported by a community member."

After fruitless netting attempts, they'd ultimately drained the pond to get at what proved to be a lone fish; later analysis showed it to be a subtropical snakehead species unlikely to have survived the winter. Not only that, but the effort came with great irony: virtually everything else in the pond—the animals they were trying to *save* from the snakehead— were invasives.

A freakish exotic fish and a media circus at least offered risible counterpoint to the daily tedium of planning, managing, and planning- for-managing invasive plants. Even as only part of your portfolio, this last was a daily process. Yong walked me through a typical scenario.

"Let's say we have a park in development. We'll first see what kind of invasives are, or could be, onsite and create a plan, then, after initial devel- opment, the area goes into monitoring. That's one approach, the other is to go by species, and for that we'd jump back to the audit. Some species are at a low enough occurrence for EDRR removal—pickerel weed and butterfly bush, for instance. More widespread and problematic species are addressed site by site; for example we'd ask whether, if we removed black- berry and replaced it with native grass, someone was willing to maintain that extra greenspace. If they said *yes,* great; otherwise we'd cut it back for sightline issues. With knotweed, the challenge was our herbicide ban— we'd barely used chemicals on *anything* for twenty years."

Yong recounted Steele's fruitless knotweed work. "As a first-year trial, we started with three knotweed sites that spanned different types of habitat from riparian to dry, and attempted physical removal while going forward with requests to our environment committee for herbi- cide treatment. For instance, in the second year we got approval to treat sites where physical removal failed, in the third year we got approval to also treat sites where the plant could be spread by maintenance prac- tices. And that's about where we are now. Baby steps. There's a lot to do but it's looking better . . . No, *really,*" she emphasized, suspicious of my sudden silence. "Am I the *only* hopeful person you've spoken to?"

I didn't have the heart to say. I also didn't have the heart to tell her what I'd seen outside.

Parking on the street across from City Hall, I was agog at the lot I'd pulled in beside, a seeming laboratory-crafted demonstration of the pattern, process, and powers of knotweed. In a far corner the mother plant, densely foliate, stood four meters tall. Fanning out ninety degrees from it were satellite patches a couple of meters high and just as wide. Around these circled still smaller scions, and out past those, some five, ten, fifteen meters distant, were lone, shin-high sprouts nudging the grass aside. Along the lot's periphery, knotweed climbed houses, ringed trees, leaned on sheds, and crawled over and under trestles of blackberry, reaching across the scant boulevard lawn for ... well, anything. Which it apparently found in front of my car, where knotweed had entered the base of a metal-sheathed conduit for an electric cable that ran up a utility pole. About two meters aloft, where the conduit was clearly being pried from the pole, a spray of knotweed exited in a surreal, conquering bouquet.

Time to call Engineering. And the power company. And maybe the phone company, too.

## Nature*ish*

Another November day of heavy rain, another day spent driving south. There was flash-flooding in North Vancouver—perfect conditions for spreading invasive plants around—although that wasn't what I was thinking about. Instead, I'd noticed how, despite the downpour, traffic spray, and mist-shrouded highway margins blotting out detail, a familiar sight was obvious to me by shape alone. A sculpted green molding cascading from the forest edge at a uniform height of two meters for as far as the eye could see, it resembled a long trough of dark syrup being tipped, the viscous liquid first pushing up over the trough's edge in an upward surge before slumping toward the ground. Running for tens of kilometers along woods and across any openings, this dense wainscoting of Himalayan blackberry looked entirely manufactured, which was probably why ecologists referred to a patch of the stuff as a "berm."

Fittingly, I was heading toward the U.S. border and the seaside community of White Rock to look in on one of Diamond Head's blackberry eco-restoration projects. Easing off the highway, I made a quick right onto Crescent Road, which wound its way to an eponymous beach through large houses on big lots buried in towering forest. But it wasn't long before a leitmotif emerged—English-ivy-coated gate posts, blackberry berms, sprays of yellowing knotweed.

At a gas station rendezvous I met up with Jeff Hunter, a part-time Diamond Head foreman. Tall and easygoing, with a black baseball cap covering short-cropped salt-and-pepper hair, he sported a set of black earbuds hanging from the collar of his gray t-shirt like a necktie. Back in our vehicles, I followed Hunter seaward into Blackie Spit Park, a favorite property of the City of Surrey, located on the Pacific Flyway route for numerous migratory birds. Being a bird sanctuary, abundant avian-vectored seeds from near and far were deposited in the sandy, fertile, highly disturbed soil, leading to invasives spreading like wildfire. Prominent Surrey folk (and you know what that means) wanted something done, so its parks department called Diamond Head.

A gravel road led into the park, sandwiched between a dike and a rail line. Hunter pulled in behind a cargo trailer. In front of it a Diamond Head sandwich board read: "CAUTION Ecosystem Restoration Crews Working Ahead." Nice touch.

A train whistled, grinding past about four meters above us on the right. It was from California and had just crossed the border carrying a load of who-knew-what—plus invasive propagule stowaways of who-knew-what varieties. A marina on the other side of the tracks serviced itinerant boats, and a community garden hemorrhaging nonnatives like comfrey stood kitty-corner to the site where Hunter's crew was working. Along with the bird traffic, it was like an arrivals lounge for invasives. Even the sign for Blackie Spit Park bore a warning for four species of salt-tolerant *Spartina* grasses invading British Columbia's coastal marshes.

A year before, Diamond Head had cleaned out part of a gargantuan blackberry berm here and planted a few widely spaced conifers; in adjacent areas they'd put down a native field-grass mix to discourage the blackberry seed bank from rejuvenating. They were here now to clear out a new berm already threatening those young conifers. They'd been at it for six days of a budgeted seven, physically removing blackberry because spraying and injecting were ineffective, and it was clear they wouldn't finish in time. The patch, about 560 square meters, now looked to require, just for blackberry removal, ten days and three hundred man-hours. When Hunter's group finished, they'd be followed first by a replanting crew, then a contractor spreading weed-suppressing mulch.

Blackberry's thick growth results from the plants creating their own elevated scaffolding for each subsequent season's effort to build on, increasing the plant's surface area—and, hence, its energy-gathering capacity. The plants also spread by "layering," sending root systems down where branches touch the ground, and so further anchoring the berm. These features made removal hellish; large, woody root balls required levering out with shovels, many of which broke from the effort. And in any event a significant seed bank remained, constantly refreshed by squadrons of berry-eating birds.

When I'd asked how blackberry removal shaped up to other jobs, tall, laconic crewmember Naji Mohammed Ali said it was probably the most labor-intensive of all invasive restoration work. No doubt, I thought, watching him and a young woman wrestle out root balls the size of small dogs. It was also a job best tackled in stages. First, Ali took a gas-powered brush saw to the leading edge of the berm, topping and cutting back into it. But you could only take it down to a certain height before the cut material had to be raked out with a pitchfork, which was also used to lift the accumulated debris into the trailer. According to Ali, you could also beat back blackberry by cutting and removing, cutting and removing, so that eventually the roots were depleted of stored energy and unable to throw up anything capable of photosynthesis. But that took years. Wholesale removal seemed the better bet. Still, said Ali, "Look at all this work—and you can come back in two or three weeks and blackberry will be growing here again."

Spent poplar leaves rustled in the strong breeze of another approaching Pacific storm, twisting off their stems, blowing upward and spinning down around us like yellowed ticker tape. What about ivy removal, I asked Ali, which was also known to be tough. "It can be if there's deadfall involved or it's on a vertical rock face, but it can also be easy," he answered. "Sometimes we can just roll up a mat of it like a carpet."

With the trailer filled, Hunter was off to a special invasives dump. He could take one to three loads a day depending how far it was from a site and how much they were pulling out. The trailer was lined with a tarp so that when they arrived, they could fork out the uppermost layer then just slide the bottom pile out all at once. "It's like herding a very thorny elephant," said Hunter, whose top-line, calfskin gloves could only take a few days of this before needing replacement. Even with new ones he suffered frequent punctures: thorns would poke through and break off under the skin to immediately abscess. "It's a constant problem," he noted, shaking his head. Still, one thorny elephant was no doubt easier to deal with than the thousand barbed snakes comprising it.

A walk up the road and onto the dike offered a *tour des invasives*. Tansy intertwined with blackberry everywhere along a brackish canal, and a second variety of invasive blackberry with more palmate leaves was happy to use the trestles laid down by its Himalayan cousin. A planted field mix included aliens like reed canary grass and I thought I even saw *Phragmites,* but of course, everything looked like an invasive to me now. Waterfowl abounded on the seaward and canal sides, a bald eagle chasing flocks back and forth across the mudflats and marsh, seemingly for fun. The canal led into an upscale neighborhood where the streets were, quite literally, crazy with invasives: English ivy and some honeysuckle-smelling vine were everywhere; someone had even let ivy turn an ugly chain link fence into a hedge; at a bus stop they'd chased ivy out with a chainsaw to keep the waiting pad plant-free. A tall, ornamental grass suggestive of phrag-meets-bamboo marched out of people's yards to cross guardrails and ditches. Fittingly, a sign on the corner said Agar Street—agar being the nutrient medium for Petri dishes on which the spores of bacteria and fungus are grown in microbiological laboratories. This neighborhood was a de facto Petri dish for invasives that made Hunter's task seem not only Herculean but, once again, Sisyphean.

"You just can't think about it," he said in a familiar refrain.

Driving to a mall for lunch we clock a checkerboard of rust-painted squares on the hillsides, the fall-reddened leaves of blueberry farms. The lower Fraser Valley was berry country, with 500 or so farms and over 11,300 hectares of fertile farmland annually turning out up-ward of 54 million kilos of blueberries alone, with similar volumes in raspberries, strawberries, and cranberries. Yet the amount of land devoted to these crops paled in comparison to the acreage of invasive blackberry occupying every nook, cranny, fence row, dike, levee, right-of-way, field edge, and vacant lot. A cover story in that morning's *Vancouver Province* newspaper screamed about the dangers of vacant lots as havens for drug dealers and homeless encampments, social phenomena that, ironically, posed more tractable problems than the catalogue of invasives harbored by the typical vacant lot.

After lunch I joined another crew under head foreman Keith MacKenzie, who'd logged extensive invasive-related work. We embarked with his two-truck, three-man crew on a long, looping tour to the City of Surrey works yard, where MacKenzie was picking up plants for restoration work in a place called Hi-Knoll Park. The prescription called for alder, Douglas-fir, western red cedar, and broadleaf maple—the kind of proper native forest that Richard Beard had lamented. As we loaded trees, a Pacific chorus frog (*Pseudacris regilla*) called from an adjacent row of potted shrubs, demonstrating how easy it was to move small animals around; while not an invasive concern per se as a native species, its obvious transplanted presence here highlighted similar issues of genetic contamination and disease transport that could also have serious impact.

It was another long way under a still-brooding sky to Hi-Knoll. MacKenzie backed the truck down an access point and the crew schlepped potted trees across a tiny, muddy creek past erupting holly and clumps of canary grass. Three years ago this patch of land was cleared of blackberry, then restored in stages. The first plantings looked great: the alder was already tall and doing its job of colonizing, fixing nitrogen, and shading out blackberry; the conifers were healthy, and one small fir had even produced cones; a few broadleaf maple looked to get in on the game; and second-year plants were doing fine as well. Seeing *anything* cleared of blackberry and returned to bare soil was astounding enough, but seeing the tremblings of an actual native forest where the scourge had once resided was particularly gratifying, especially in light of Beard's estimate of how long the process would take without intervention.

"We have almost 100 percent survivorship," noted MacKenzie, surreptitiously hauling on a cigarette and kicking at the rich, sandy soil.

"Look at this stuff—it wasn't too hard to get the root balls out, and it has been really good for the trees."

In eight years of restoration work, MacKenzie had seen awareness around invasive species grow in the community mind—and his own. Now when he drove around, his eyes tended to register mostly the restored sites, undistracted by the surrounding glut of aliens.

"This place has grown rapidly since the 1980s. Early on there were yard-waste disposal problems; people didn't know they were creating issues we'd all be dealing with now," he said. "Still, among municipalities, Surrey is ahead of the game. It put a fair amount of money into removal and restoration . . . of course, it also *has* a fair amount of money."

But it also had a fair amount of invasives, and it was almost impossible to imagine the investment required to tackle a problem of this size that was riding the middle, rather than the beginning, of the Invasion Curve.

## A Bug's Life

One of the biggest reasons we should get on with eco-restoration is actually very small—insects. More precisely, the lack of native insect species in invasive monocultures, and the diversion of pollinator insects away from native plant species toward nonnative invasives.

While hundreds of insect species utilize knotweed in its native range in Asia, these notoriously specialized arthropods have been understandably slow to take to it in exotic locales like North America. Richard Beard had discovered that the generalist larvae of the California tiger moth—the black-orange-black woolly-bear caterpillars familiar to most on this continent—were more than happy to eat knotweed, finding them on about thirty different patches. Whether this was their chosen forage or merely adventitious, it still represented pretty slim pickings for birds and other native insect feeders expecting to cruise local vegetation for a meal. Knotweed's abundant, nectarous, early fall florescence was also attractive to late-season bees, which swarmed it to the exclusion of late-pollinating native plants. That problem was exacerbated when plants like knotweed out-muscled locally adapted colonizers like alder. One measure of alder's huge importance to B.C. ecosystems is that some one hundred species of moth have been documented to feed on it, which in turn feed (or once did) many neotropical birds migrating up the Pacific Flyway. Disturbed soils in the Pacific Northwest were failing to see a typical alder colonization because any number of invasives were getting there first. And that was bad for birds. As University of Toronto forestry professor Sandy Smith told me, migrating birds preferentially sought out native trees because, with far more insects on them, foraging opportunities were greater.

In Beard's estimation, the Vancouver North Shore currently had some thirty hectares of knotweed coverage, and probably double that for blackberry. This made it both a nexus for problems and a ground zero for further spread—a dubious showcase. And the place he and I next visited, in lower North Vancouver, was the failing heart of the North Shore's infested body. We'd pulled up in front of a construction site occupying a several-hectare block where, some years back, Beard had mapped ten patches encircling the site. After two years of treatment, ground was broken on a multi-use retail center. The construction company was forward thinking in addressing invasives: it had removed the upper half-meter of topsoil from the entire site as prescribed by U.K. standards, sequestering it in a fenced zone where it now languished as a chest-high pile regularly visited and treated for knotweed sprouts by the intrepid Beard.

We walked the perimeter, checking on various patches to see what, if anything, had popped up. Nothing, it seemed, according to Beard's increased pace. Halfway round, at a corner, we diverted, crossing the street to an unrelated area that Beard's crew had treated earlier in the year. A large stand of ghostly gray stems beside a phone company relay box elicited a tale from Beard of a nearby hydro substation where knotweed had grown up through the concrete base and was discovered only as it was about to short out the entire neighborhood. Beard pointed to a patch of knotweed running the length of the block beneath towering cottonwoods; these thousands of stems had required removal of a huge amount of blackberry to reach, and Green Admiral's efficacy here was about 99 percent. Impressive work with equally impressive results.

Inside the fenced yard of a landscaping company bloomed a different story: a bounty of knotweed sprouting in a dozen places around an outdoor asphalt storage area. Beard had left it to grow until it was big enough to be worth spraying and today, with sun drying dew from the leaves, it was ready to go. Beard sprayed and prowled, prowled and sprayed, scanning the ground like he was chasing cockroaches. Knotweed flared at the end of almost every row of pallets on which garden material was stored, including a large corner patch growing around small potted trees and plants. In at least one case, knotweed had entered a planter from beneath, sprouting up beside the small tree it contained. Even if someone cut this off before shipping it out, that rhizome would remain in the pot. This was a whole different take on an invasion front: a propagule clearing house.

We made our way back to the worksite's periphery, methodically squaring around it and down a long, meridian-split boulevard. Beard had earlier treated the boulevard for the city and he now mopped up the few surviving stragglers with the spray gun. A city landscape crew alerted us to a larger plant on the meridian, which he also took time to meticulously douse. On the final corner, where the biggest infestation once stood, construction had proceeded in close proximity, and now, as a voluntary precaution—at Beard's behest and again based on what U.K. law prescribed—the developers were protecting a fifty-meter stretch of foundation with thick, black geotextile (a plant-barrier fabric) that went three meters deep and extended some seven meters to either side of the original knotweed border. It was a huge and costly proposition, and as we walked, Beard related the difficulty of convincing the powers that be that it wasn't so much precaution as necessity. Here, at the very end of our circuit, a small, lone knotweed struggled toward the sidewalk from beyond the fence; as we passed, Beard misted it with glyphosate without even breaking stride. "A desk is a very dangerous place from which to view the world," he said.

For our lunch break we met up with Julia and Wade at the Wild Bird Trust of British Columbia Conservation Area in Maplewood Flats. We washed our hands in a small office and plugged in the tea kettle. As usual Wade and Julia had done far more than could be expected.

"I had my knotweed eyes on this morning," claimed Julia.

"We only did thirty knotweed and about twenty hogweed," Wade said dejectedly.

"But that's amazing!" answered Beard.

As we ate, Wade regaled us with tales of being a part-time arborist who occasionally entered professional tree-climbing competitions.

The area we'd lunched in was a covered work area next to a wooden poster board pinned with invasive plant literature, information on bird sightings, and an anise swallowtail butterfly project that had an interesting invasive-related component. As part of an ecological restoration, the conservation area had planted cow parsnip (*Heracleum maximum*), the main food of the butterfly's spectacular caterpillar. But cow parsnip looks an awful lot like a more problematic plant to the uninitiated, so an additional info sheet with a large photo had been tacked to the wall. It read: *No Worries! This plant is NOT Giant Hogweed!* And that reminded me of the day in Britannia Beach with the SSISC crew, where an after-lunch hogweed hunt had proved to be an entirely different animal than the morning's knotweed struggle.

## Promised Land

Copper was extracted in Britannia, a former mining community, until 1974. Despite the riches spirited away, decades of contamination had fouled the land, creeks, and surrounding ocean bottom, creating one of the most contaminated places in all of British Columbia. When the mine shuttered, the community fell into disrepair; despite artisan studios and gift shops, it remained a thinly veiled blight on the drive between Vancouver and Squamish. Much of the area had been reclaimed in recent years, with the old mine transformed into a museum to divert tourist dollars flowing up the highway during the 2010 Winter Olympic Games. While tailings areas adjacent to the mine remained devoid of vegetation due to heavy-metal contaminated surface soil, in other places greenery had crept back in. Few of these colonizers, early adapters, and pollution-tolerant plants, however, were native. Compounding the problem, rehabilitation had included trucking in fill from the lower mainland, none of which had been screened and, as was now apparent, included a litany of invasive hitchhikers.

On a hillside high above where we'd knocked down knotweed all morning, and past where monster homes were being constructed after a rock-bottom land grab by developers, we parked in front of low bungalows and triple-wide mobile homes, leftovers from the original mining community. Behind them, a disseminate hogweed stand had taken up residence at the junction where five backyards converged. Some plants had been cut, doused in gasoline, and placed under tarps (not advisable); others had been cut and left on the ground (ditto—I'd suffer a burn from one of these whose blister lasted weeks); still other smaller hogweeds were growing through lawns, gardens, and compost. Rob Hughes set to digging up and/or spraying what was left. The plants were small enough that it didn't require donning full-body hazmat suits that would otherwise be the norm. It didn't take long to finish the job.

While ferreting out and removing a danger like giant hogweed ranked higher on both satisfaction and good-deed scales than endless, mindless knotweed control, the crew all expressed an equal affinity for the work (Rob: "we've had some great successes"; Sharon: "I like having an impact on the environment"; Bre: "it's a trend in ecology that's easy for people to understand"; Sam: "it's a great outside job with a small team") and concerns with it (Rob: "CN Rail didn't pay us for last year and isn't returning our calls"; Sharon: "not psyched about the possible long-term effects of pesticide exposure"; Bre: "it can be super frustrating"; Sam: "if everyone in the corridor was on the same page how much more could we get done?"). So went the often yin-yang world of invasion biology.

For myself, the day had been fascinating and horrifying in equal measure; when you invested any thinking in the invasive potential of plants like knotweed it was an exercise in infinitude akin to contemplating deep time or staring into the firmament, enumerating stars and trying to divine the depth and vastness of the universe—there was literally no place for the mind to come to rest. It also galvanized a changing view of where I lived that had begun forming years before on that spring drive north from Vancouver—what I once saw only as greenery, I knew now to be endless broom stands and the beast of knotweed. It was also obvious that the lines they traced on the landscape—along road, rail, and power company right-of-ways, ditches, parks, and fields—were lines that we ourselves had drawn. It recalled something Hughes said that morning, as he'd emerged dripping from the first knotweed stand he'd treated. "Once knotweed has a route of distribution it doesn't take much for it to take off . . . but disturbance is always the key to getting it started."

I looked left across the aquamarine waters of Howe Sound, a marine environment only now recovering from the excesses of the mining and pulp-and-paper industries that did it in. Whales and dolphins were back; people were harvesting shellfish. But on the far shore, a mere wind gust away from the folly/fiasco of Britannia and Squamish, I saw new land clearances for liquefied-natural-gas terminals, clear-cut logging marching up mountainsides, and brutal slashes at all altitudes to accommodate transmission lines from myriad independent power projects. In essence, I saw a diorama of disturbance—the promised land for a new wave of aliens.

## A Lot of Clams

Every year, hundreds of invasive alien plants and animals cost the economies of North America billions in lost revenue and in direct control and eradication costs. Dutch elm disease, Japanese knotweed, Scotch broom, European milfoil, common carp, zebra mussel, European starling, gypsy moth, and emerald ash borer are but a few of those etched into public consciousness in Canada, yet invasive aliens in this sparsely populated northern landmass now included a minimum 27 percent of all vascular plants, 181 insects, 55 freshwater fish, 26 mammals, 24 birds, 2 reptiles, 4 amphibians, and dozens of mollusks, crustaceans, and fungi. A government study's estimated annual price tag for a representative cross-section of just 16 of these was $13 to $35 billion CAD; Canada's agricultural and forestry sectors alone lost $7.5 billion annually to invasive species. We Canadians felt the impact firsthand when introductions altered entire landscapes (Scotch broom), posed a danger (giant hogweed), or brought disease (West Nile virus). Yet these are the children born of globalization and consumerism, their numbers increasing in tandem with trade liberalizations that create shorter transit times, more numerous countries of origin, and a greater diversity and higher volume of propagule-bearing products. Along with climate change, habitat destruction, pollution, and ocean acidification, invasive species are now a marquee phenomenon of the Anthropocene, a driver of the Sixth Great Extinction in which we are all participants and for which we will all pay the piper.

As we've seen, the lag time between detection and decisive action dictates whether you can achieve eradication or be forced into a regimen of control. The window for success is small; from the get-go, for instance, Skip Snow suspected that the Everglades' python problem was already beyond eradication and had to be about control. After all, research and control of the destructive brown treesnake in Guam required $3 to $4 million annually in addition to related health costs and lost revenue from electrical outages, reduced tourism, and declining agriculture. By comparison, until 2015, research on Florida's python problem had cost federal, state, and local governments $20 million USD, with no real control in place other than ad hoc amateur snake hunters. Bounty hunter George Cera's contract to draw down invasive black spiny-tailed iguana populations propelled him into the ranks of those advocating that we eat our way out of invasive species problems; regardless of how many copies his black-humor cookbook sold, it remains a zeitgeist statement. The United States and Canada combine to spend $30 million USD annually on sea lamprey control, and together lavish tens of billions more on other Great Lakes invasive issues. In fact, direct and associated costs to the U.S. economy from invasive species like zebra mussel are on the order of $145 billion USD annually—that's a lot of clams.

Around the globe, the world's foremost invasive species—rats—are being eradicated island by island, to the tune of around $10 million USD each, buoyed by a growing conservation trend of ecosystem restoration/rehabilitation. Meanwhile, archipelagos like Haida Gwaii have been plagued with surreal numbers of invasive deer, to the extent that on Anticosti Island special vehicle modifications are needed and the economy is now based on sporthunting, the deer's sole and inadequate means of control. On neighboring Newfoundland, collisions with invasive moose have constituted a multimillion-dollar auto insurance, health cost, and liability hit to the province.

To wrap this screed, here's an example of the entire ball-of-wax rolled into one: the tiny Puerto Rican coquí treefrog (*Eleutherodactylus coqui*). Introduced to Hawaii as putative insect control in the state's burgeoning flower industry, the coquí would eventually reach absurd densities of up to 90,000 per hectare. It certainly consumed insects—up to

690,000 invertebrates/hectare/night—but some of those weren't pests, and instead were important to local ecosystems. In addition, the frog's piercing 85 dB call was dangerous to the human ear at close range and impeded the intra-species communication of key pollinators like birds and bees. With its negative impacts on the environment, the economy, and human health, this small invader became one of the most costly pests the state had ever seen—which was saying something considering Hawaii is *the* most invaded archipelago on the planet. Years of research uncovered few options among means for managing frog populations: mechanical (traps, barriers, stream treatments, vegetation management, hand capture), biological (disease, parasites, predators), and chemical (caffeine, hydrated lime, citric acid) methods were deemed unlikely to do the job. Moreover, once coquí was established on Hawaii and Maui, little effort went into controlling its spread, eradicating populations, or studying the problem despite warnings from scientists about potential impacts. When public calls for coquí control finally brought funding, the point of no return had long been passed. Eradication on Hawaii failed (at up to $10,000 USD/acre extrapolated to all of the big island, that would have been $96 million). Further spread was slowed, however, and small populations that had made it to other islands were successfully eradicated. Tens of millions have nevertheless been spent and millions more will be required annually to suppress populations. Indian mongoose eat coquí, though it will never be a principal food source for the invasive carnivore. And that's a good exit segue.

Neither the globally introduced mongoose nor cane toad had much impact on their target pest species, but they did effectively devastate populations of small native vertebrates in each of their new homes. Current work on the cane toad in Australia, in fact, shows how it fundamentally changed the food web by altering energy flow in the continent's apex predators, a reverse echo of the yet-to-be-quantified effects of Burmese pythons' near-removal of small mammals from the Everglades.

When I'd first started looking into invasive species, I'd come across a Daniel Simberloff paper whose title—"How Much Information on Population Biology Is Needed to Manage Introduced Species?"—made the point that research-in-progress should never be casually invoked as an excuse for inaction on an invasive species. I also encountered a crude, tongue-in-cheek answer to that question that precisely captures the reactionary gestalt regarding invasives. Though I can no longer find the source, it went something like this: "Stop studying them and start fucking killing them!"

# PART VI

# Sunrise on Homogena
## *A New Planetary Paradigm*

It is interesting to contemplate an entangled bank, clothed with many plants of many kinds, with birds singing on the bushes, with various insects flitting about, and with worms crawling through the damp earth, and to reflect that these elaborately constructed forms, so different from each other, and dependent upon each other in so complex a manner, have all been produced by laws acting around us . . . Thus, from the war of nature, from famine and death, the most exalted object which we are capable of conceiving, namely, the production of the higher animals, directly follows. There is grandeur in this view of life, with its several powers, having been originally breathed into a few forms or into one; and that, whilst this planet has gone circling on according to the fixed law of gravity, from so simple a beginning endless forms most beautiful and most wonderful have been, and are being evolved.

*Charles Darwin,* On the Origin of Species

# Disaster

Life on our planet is changing, of that there can be no doubt. That alien invasive species are a measurable component of this is also clear. The questions raised, then, are simple, and essentially those we began with: do we care about this? And if so, what are our roles and responsibilities?

To care at all about the impacts of invasive species you have to care about biodiversity. And to care about biodiversity you have to believe in the interconnectedness of the entire biosphere, and, in turn, its interdependencies with the inanimate atmosphere, lithosphere, and hydrosphere. Acknowledging the intimacy of these relationships going both forward and back in time is not only a shout out to the Earth-as-organism Gaia theory promulgated by planetary systematists James Lovelock and Lynn Margulis, but also an idea implicit in Darwin's famous "tangled bank" passage in the epigraph, perhaps his most subtle but important gift to our big-picture thinking about life.

Earth's rich biodiversity, the result of a 3.5-billion-year game of tectonic Rubik's cube, has, in only a few human generations, attained a rate of unraveling matched only by previous catastrophic extinction events in our planet's history. As with much else occurring today, ecologist Charles Elton saw this profound alteration to the planet's biological landscape coming. "When contemplating the invasion of continents and islands and seas by plants and animals and their microscopic parasites, one's impression is of dislocation, unexpected consequences, and increase in the complexity of ecosystems already difficult enough to understand let alone control, and the piling up of new human difficulties," he wrote of how our influence was even then reshaping animal distributions.

This acknowledgment is key. Lest anyone adopt the mindset that a host of invasive aliens are remodeling Earth into a planet one might whimsically re-label Homogena, Elton reminds us that it is, in fact, mankind who is perpetrating the transition. In one form or another this thought echoes endlessly through literature devoted to invasive species, from Quammen's "Planet of Weeds" to a growing list of academic texts and popular books.

"This place isn't truly natural . . . most of the plants around here are Central American," a scientist who worked on the infamous brown

treesnake problem in Guam told author Alan Burdick in *Out of Eden.*
"Introduced species do very well here. It's a McDonald's ecosystem, and
these are real fast-food animals. If people don't watch out we're going to
have the same metropolitan animals around the globe. I see the same
birds in Copenhagen as I see in Toledo and Sydney. In the St. Lawrence
Seaway, forty to sixty percent of the animals are introduced . . . What's
wrong with cosmopolitan fauna? There may be people who would like
to just get it over with because it would make it simpler to get on with
business."

The ranks of such people—a small but vocal minority—will surely grow in the face of an issue that increasingly seems both too overwhelming and too intractable to deal with, but for now they remain
thankfully scattered about the margins. At the other end of the spectrum, devoted to upholding biodiversity and ecosystem integrity, tens of
thousands of dedicated individuals in myriad organizations and disciplines are engaged in the study, analysis, understanding, management,
and, most importantly, prevention of invasive species, a shift in global
mindset that parallels the changes to our planet's surface.

The Guam scientist's allusion to McDonald's is apt in another
way—these same folks are working to prevent a bio-homogenization
that would be in lock-step with current trends to cultural uniformity
wrought by globalization, in which the same food, automotive, electronic, and warehouse-shopping franchises will bookend every town
no matter the continent. We've met a number of these people over
the course of this book, and while all were happy to discuss their work
in the moment, there was always a subtext, another simple question:
what next?

It would be easier to answer if our species was a future-focused lot.
We are, however, famously not, as Kathryn Schulz pointed out in "The
Really Big One," a *New Yorker* exposé on the inevitability of a devastating Pacific Northwest earthquake. "On the face of it, earthquakes seem
to present us with problems of space: the way we live along fault lines,
in brick buildings, in homes made valuable by their proximity to the
sea. But, covertly, they also present us with problems of time. The earth
is 4.5 billion years old, but we are a young species, relatively speaking,
with an average individual allotment of three score years and ten. The

brevity of our lives breeds a kind of temporal parochialism—an ignorance of or an indifference to those planetary gears which turn more slowly than our own."

In noting that the problem is bidirectional, Schulz opens up the thinking on earthquakes to other natural-disaster analogies. "Invasive species," in fact, might replace "earthquake" in many of her sentences without any loss of relevance. Extending the analogy via paraphrase, invasive species have often remained hidden—not from sight, but from *notice*—simply because we can't see far enough into the time and space components of an ecosystem; they pose a danger now because "we refuse to think deeply enough about the future." Similarly borrowing from Schulz, this is no longer a problem of information; we now well understand the potential impact of many invasive species. Nor is it a problem of imagination. There is no shortage of movie treatments of the world threatening to succumb to a host of plagues and apocalyptic scenarios that Schulz fingers as "a form of escapism, not a moral summons, and still less a plan of action. Where we stumble is in conjuring up grim futures in a way that helps to avert them. That problem is not specific to earthquakes, of course. The Cascadia situation, a calamity in its own right, is also a parable for this age of ecological reckoning, and the questions it raises are ones that we all now face. How should a society respond to a looming crisis of uncertain timing but of catastrophic proportions? How can it begin to right itself when its entire infrastructure and culture developed in a way that leaves it profoundly vulnerable to natural disaster?"

## Apoplectica

"It's probably fair to say, in fact, that without alien species there wouldn't be any human civilization at all."

Working at home one Saturday morning, radio gurgling in the background, I'd long since tuned out the discussion on a weekly science newsjournal when this statement shocked my subconscious. The host was interviewing Ken Thompson, the British botanist whose book, *Where Do Camels Belong? Why Invasive Species Aren't All Bad,* I've previously mentioned.

As the title suggests, Thompson's core thesis turns on the "natural" intercontinental movement of various biological taxa—in this case camels, once native to North America but in absentia for some 8,000 years. It's a premise that immediately seems as sophist as it is facile.

There's no argument that the idea of belonging lies at the heart of the invasive species issue, whether from a human or ecological perspective. And although Thompson admits that our own species is moving organisms around faster than they could have achieved on their own, he wonders when this became "unnatural" (his word), making much of the fact that we've been a vector of dispersal forever (on which all invasion scientists agree)—while simultaneously overlooking that such agency is true of many organisms. On the show, he entrained a number of similarly perplexing statements that, while garnering attention on the book-flogging circuit, were at odds with both the received scientific view and what I was seeing in the field.

"Quite honestly, I have to say if zebra mussels were native we'd probably think they were wonderful things," he quipped, lauding the filter-feeder's turbo-cleanup of dirty ol' Lake Erie and the food it now provided native birds and fish. Here he was completely ignoring the select nature of benefits and the obvious cost of invasional meltdown scenarios (disease outbreaks, food web shifts, extirpation of native mollusks). Not to mention the multi-billion-dollar infrastructure problems collectively inflicted across North America by dreissenid mussels. The host, a genial, usually more responsible type, egged him on, leaving even the most preposterous claims unchallenged. "And it's really quite difficult to figure out whether an alien species is really a problem and how big a problem it is," Thompson continued in a droll, acerbic tone.

Nonsense. Mandrak had told me the book was considered "a joke" in his circles. For a more thorough thumping, he'd suggested, I should bring it up when I spoke with Ricciardi.

"I find it hard to even know where to start," said Ricciardi, clearly incensed. "I've *never* made a more reluctant purchase in my life. I read most of it in the bookstore, but I knew I had to write all over the margins and I couldn't do it there. So I go to buy it, and the checkout person says 'Oh, great choice!'—you know how they try to encourage your purchases—and I said 'No, actually it's a piece of garbage.' And they were like, 'Really—why?' So I went through a whole long list with them. I'm not sure what they thought."

Unless they had any previous knowledge of invasive species, what could they think? And herein lay the problem according to Ricciardi: if Thompson's book was the only thing people read on the subject, they'd be delivered little more than a crackpot view of a biological tempest in a teapot.

Various invasion scientists pointed to other efforts they felt to be on similarly soggy ground, or advertised with "equally misleading odes to legitimacy." These included Tao Orion's *Beyond the War on Invasive Species: A Permaculture Approach to Ecosystem Restoration*, a justified screed against the overuse of pesticides, but erroneously touted by self-proclaimed reviewers as part of a "rapidly growing literature about the futile and destructive attempts to eradicate nonnative species" (the pesticide thread is plucked more succinctly by Andrew Cockburn in his 2015 Harper's essay "Weed Whackers: Monsanto, Glyphosate, and the War on Invasive Species"). There was also *Invasive Plant Medicine: The Ecological Benefits and Healing Abilities of Invasives* by Timothy Lee Scott; surely, expressed the scientists, the moot point of this title is no excuse for planting invasives? And finally *The New Wild: Why Invasive Species Will Be Nature's Salvation*, in which environmental journalist Fred Pearce erected the rickety scaffolding that conservationists had become enemies of nature. Though said to challenge "the long-held belief that keeping out nonnative species and returning ecosystems to a pre-human state are the only ways to save nature as we know it," no one I'd spoken with advocated returning any ecosystem to its pre-human condition—only to functionality.

"The whole thing is an example of shallow thinking," said Ricciardi. "Like the link some critics make to 'xenophobia.' I suppose [invasion biologists'] use of the word 'aliens' stimulates the connection, but the value judgments are completely imagined. As for conservation biologists, there's no prejudice—they're as concerned about local species being transported elsewhere as they are about aliens moving in where they live and work."

In Ricciardi's assessment, Thompson's position was that it had become orthodoxy to see the transport of alien species by humans as unnatural, and that all of invasion science was thus burdened. But rarely was the word "unnatural" used by serious invasion scientists; "unique" and "unprecedented" were commonly employed in referencing the ac-celerated speed of introductions due to globalization, yet these were clearly observational and process-oriented, never value-laden. When I suggested that this made Thompson's example of slow-motion, tectoni-cally driven continental faunal exchange a spurious comparison, Ric-ciardi practically exploded in agreement. "What exactly are these people apologists *for*—unrestricted human-mediated dispersal?"

It was an entirely reasonable question that strongly suggested something fundamental is missing from these arguments. Critics ap-peared to accuse invasion scientists of removing themselves from nature to conduct their work, but my experience suggested the opposite—particularly on the philosophical front. Assuming that the lamentably rapid movement of alien species by human vectors *is* natural, would not our definitions and categorizations—for example, those recognizing certain alien species as potentially invasive and some of those invasions as problematic—be equally natural? Likewise natural, then, would be that as the only species aware of, and concerned with, such processes, we could construct an argument that it was our natural responsibility to meliorate their impacts. And that line of reasoning, it seemed, actually fit Thompson's notion of *Don't blame the invasives, blame ourselves for the invasive-friendly modifications we have made to the environment*—without presupposing that it was okay to leave unattended the mess we'd "naturally" created. Obviously we don't blame garbage or toxic waste for the conditions we've nurtured for its ill-advised disposal, and we know damn well that there is a moral imperative not only to clean it up, but also to actively curb its creation. Ricciardi agreed.

"I'm one of those people who thinks we should clean up our mess," he said, "but I don't like to get bogged down by philosophical discus-sions when I'm wearing my researcher's hat. I use the word 'natural' as a comparative tool to say 'here's what things were like without our influ-ence' and 'here's what they're like *with* our influence' ... Thompson likes to point to how few introductions succeed in establishing and the vanishingly small number of those that go on to be problems. That's both overstated and not really the point. Again, most storms are rela-tively benign, but given the costs of clean-up, few see it as a waste of time and money to be prepared for the occasional large destructive one."

Most of the arguments advanced to buttress the notion that the threat posed by invasive species was overblown were equally specious, but a particularly hollow example was the idea that native species could sometimes also become "invasive" (that is, irruptive, as with whitetail deer). This was deftly swept aside by Daniel Simberloff and colleagues in a 2012 paper, "The Natives Are Restless, But Not Often and Mostly When Disturbed." The authors began by noting that scientific observations of pestilential irruptions of North American native plant and animal species, due specifically to large-scale ecological disturbances, could be traced back to 1874, and that no less a natural philosopher than Aldo Leopold held forth on these in 1924. Examination of the plant-invasion literature for the United States showed that nonnative species had been termed invasive sixfold more often than native species, and that a naturalized nonnative was forty times more likely than a native to be perceived as invasive. In an overwhelming majority of cases where a native plant became invasive, it was associated with an anthropogenic disturbance such as altered fire or hydrological regimens, livestock overgrazing, and changes wrought by introduced nonnative species. The implication: native plants were significantly less likely than nonnatives to be problematic for local ecosystems. Furthermore, "the question of whether nonnative species are more likely than natives to become invasive seems not to have been asked. Some data suggest this might indeed be the case. For instance, approximately half of the freshwater fish, mammal, and bird species introduced from Europe to North America or vice versa have established populations, and more than half of these became invasive . . . We know of no research on what fractions of native species of any of these taxa have become invasive, but surely it would be less than this."

It is—as dozens of published studies have now shown. Simberloff's 2014 book *Invasive Species: What Everyone Needs to Know* does a yeoman's job of parsing the foolishness addressed by Richardson and Ricciardi in their 2013 paper "Misleading Criticisms of Invasion Science: A Field Guide." Yet I remained intrigued by a subhead in Richardson and Ricciardi's paper: "A Cottage Industry of Criticisms," which accurately summed up the amateurish assailing of invasion science. As an erstwhile researcher in a decidedly esoteric area of biology (multispecies complexes of polyploid unisexual salamanders), I had

found that being forced to clash with a highly critical, seemingly illogi-
cal, often territorial alternative school of thought was always the case,
and operated as an economy of scale. So I'd put it to Ricciardi: given the
large and expanding critical mass of invasion science, wouldn't he consider this cottage industry normal? Objective criticism was, of course,
welcome, he answered, but most of what was circulating seemed unfounded, indefensible, and wrapped around a vaguely articulated death
wish for invasion science. Ricciardi insisted that what many detractors
wrote about was *not* invasion science as understood and practiced by
almost all biogeographers, conservation biologists, and ecologists, but a
self-constructed caricature or parody of the discipline.

"I'm not sure what's motivating them. For instance, ignoring time
lags and all the evidence of nonnative species as a major cause of global
animal extinctions is a big problem. So what else are they missing? You
rarely see the word 'evolution' in their rants and yet it's at the heart of
it—the native versus nonnative dichotomy is a continuum, in some cases the status of a species is bloody obvious, and in other cases fuzzy—
like the species concept itself. The best arguments I can put forth about
why they're misguided are in my most recent papers."

To that end, Richardson and Ricciardi had summarized the most-
often parroted criticisms under six main groupings: (1) modern invasions were nothing new (per Thompson), (2) impacts of nonnative
species on biodiversity and ecosystems were exaggerated, (3) increased
species introductions actually increase biodiversity, (4) positive (desirable) impacts of nonnative species were understated and at least as important as their negative (undesirable) impacts, (5) invasion science was
biased and xenophobic, and (6) biogeographic origin of an invasive species had no bearing on its impact.

How any of these persisted was a mystery given that none stood up
to even the mildest scientific scrutiny. That raised a troubling question:
were arguments by this faction less about science and more about ideol-
ogy, human nature, even professional bitterness? Certainly Thompson's
take on how invasion science was funded—not to mention parentheti-
cal references to "the tireless D. Simberloff"—seemed positively churl-
ish. "There's a whiff of climate-change-denial to this in that you see the
same arguments made after they've already been disproven, like the au-

thor hasn't noticed or is purposely ignoring the challenges," stormed Ricciardi. "And yes, there are all these weird, snide little insinuations . . . about conspirators and funding and editorial control. It's ridiculous. If so, who's in on the conspiracy—the *best* and *brightest* scientists, the most celebrated journals in the world, the science councils that fund them? I mean, c'mon."

Of the many analogies in both modern and historical science where a discipline—born either of a new understanding or an emerging phenomenon—had drawn sharp-fanged detractors, climate change was the most recent and obvious to come to mind. Despite its well-funded deniers, the field was growing not only in size and importance across a range of other disciplines, but in heuristic value as well. Ditto invasion science.

"If invasion science *was* a waste of time or headed in the wrong direction, you'd expect to see the number of papers published on the subject go down," noted Ricciardi. "Instead, it's going up—and steeply, meaning that they're recognized *by peer review* as being of increasing importance. In other words, we're clearly learning from them."

Richardson and Ricciardi's conclusion, in fact, sounded a stark urgency: "It is well accepted that pragmatic approaches to dealing with nonnative species are needed to ensure that limited resources are applied to the most important problems. Indeed, one of the principal goals of the field—to predict which introduced species will become disruptive—is of increasing societal importance, given the enormous rates of invasions driven by globalization, the synergistic interactions of nonnative species with one another and with multiple stressors including climate change and the potential flood of future novel organisms (e.g., GMOs, synthetic cells, products of nanotechnology) into the natural environment."

In the end, however, unfounded criticisms posed more of a problem than the need to simply shoot down unsubstantiated arguments, point to the comical re-murdering of straw men, or, as with creationism, simply ignore the nonexistent "debate" that naysayers believed they were engaged in. Like the false pulpits occupied by climate-change deniers, there was clear and present danger in giving equal voice to what, with no rigorous scientific review, amounted to mere opinion, as in the unwarranted airtime afforded Thompson.

"You have to wonder how a book like that gets the same kind of treatment as real science," lamented a clearly frustrated Ricciardi. "When media attention is given to people like Thompson, it hinders the galvanizing of political will required to deal with serious problems that require legislation—like vectors and pathways."

## Carp, Repeating

In August 2015 I was back at CCIW, visiting Becky Cudmore's new office, a move merited by her transfer to another section of DFO that, as usual, came with more responsibility and little else. On a wall I noticed a Japanese tapestry of stylized carp that had once adorned Mandrak's office.

"He didn't have room for it at U of T so he gave it to me when he left," she said, before we got down to catching up on the Asian Carp Program's work since the previous November.

It was, according to Cudmore, "chugging along." During the summer of 2015 another field crew was added at the Burlington base. "Having three crews in the southern lakes means we're able to both visit higher-risk sites more often, and be very thorough at any given site," said Cudmore, noting that should something be found, crews had also undergone more training in emergency response. "We're in good shape and I think the only change for next summer will be to prepare for *larger* efforts. Until now, we've only had to mount small responses, but if we capture several fish in an area or discover a breeding population, the response will involve more manpower and operational tools, a larger geographic area, and more agencies."

With no such talk the last time we'd spoken, it sounded ominous— what had they been finding out there?

Twice that summer, crews visited the Grand River site on Lake Erie where triploid grass carp had been found but turned up nothing; there likewise had been no further news out of Sandusky, on the Ohio shore. But there *was* a reason that Cudmore wanted an overdrive gear on her response engine: in late July, a pair of meter-long grass carp had been found in Lake Ontario at Toronto's Tommy Thompson Park, a landfill spit on the city's eastern waterfront. Partway along the spit an embayment was being converted to a wetland, and after it was blocked off from the lake a salvage was initiated to remove fish that could be affected by construction. The biologist on site became immediately suspicious of the two unfamiliar behemoths found during this process. The Asian carp unit was alerted.

"When the news came in, I called my biologist, caught him on the highway heading west and had him turn around," recalled Cudmore. "The whole response went smoothly. Even though we were working with the Toronto and Region Conservation Authority, an agency we'd never partnered with, we were able to bring everything we'd learned to bear on the situation."

Blood samples rushed to the CCIW lab showed the two male carp to both be diploid. "Male" was a relief; "diploid" the opposite. Though diploid grass carp were part of both legal and illegal releases in the United States, as well as the live food trade, grass carp found in Lake Erie to date had all been triploid—incapable of reproducing and thus, establishing a population. Mandrak happened to be cleaning out his office at DFO when the fish arrived, and was asked if he'd dissect them to assess reproductive status.

"Each had one really well-developed testis in breeding condition and my guess is they were looking for a place to spawn," Mandrak later told me. "The big question is why they were *together*. One reason might be they found each other while searching the shoreline for the same thing—though where they were discovered wasn't good breeding habitat."

The Asian carp unit nevertheless learned plenty from these fish and was about to learn more: as Cudmore and I spoke, stable isotope analysis of the carps' otoliths was under way that would determine whether the animals had been in the area a long time, or were captive-bred elsewhere—perhaps as aquarium or food-trade fish—and more recently loosed in the lake. (On November 9, 2015, Cudmore would tell Toronto's CTV news that various chemical signatures in the otoliths showed the fish to be eleven to fifteen years old and likely hatched at U.S. fish farms.)

While not as worrisome as a fecund diploid female, the two male carp nevertheless merited the largest response to date: seven boats, twenty people, and 550 man-hours. Media response was proportional—insane, according to Cudmore—with national print, Internet, radio, and

TV coverage. At one point, news helicopters hovered over the spit where biologists were manning nets and cruising the area in electrofishing boats. Despite the circus, and like Melinda Yong's snakehead incident, Cudmore deemed the incident positive in terms of public awareness. Nevertheless, she had to deal with a lot, including dozens of interviews to field—and some to fend off. Furthermore, increased attention on the Asian carp threat had at least one equivocal result: a tenfold spike in reports of Asian carp.

Intrigued by this psycho-social phenomenon, Cudmore fired off an email to the regional Invading Species Hotline to see if she could use data collected by their system over the past few years to quantify the uptick. "Prior to this we were getting two to three reports a week from people saying 'I think I caught an Asian Carp.' The week of the Toronto find we got twenty to thirty reports. None panned out, but it shows the public is more aware, engaged and ready to pitch in. It's a lot to manage, but we'd rather they report erroneous finds than not. It only takes two seconds to look at a photo and say, 'No, that's actually a common carp . . . or a fallfish.'"

While no one would mourn a common carp killed as a suspected Asian carp, native fallfish—a cyprinid that might be confused with a juvenile Asian carp by a non-expert—exemplifies the hazards of misidentification by a lay person, who might not only, in a knee-jerk reaction, kill a fish or three, but also impulsively kill everything in a stretch or body of water. Technology, however, was helping obviate such mistakes.

"Biologists tend to fail at outreach because what appeals to them are biological issues. So we've been working with non-government groups to help target messages to the public. Now people can call in sightings, report on a website, or use a phone app where you submit a photo whose location is logged into a database," said Cudmore.

The digital infrastructure to support this involved collaboration by federal and provincial governments, which together funded an initiative by the indefatigable OFHA to log sightings, receive and compile reports, and circulate this information among a prescribed group of federal and provincial biologists. "Whoever opens the email first gets to reply," said Cudmore. "Even with a crappy photo we can usually get to 'It's not an Asian carp' pretty quickly."

The latest finds were an excellent demonstration of prophylactic effectiveness on the part of both informed workers in the field and Cudmore's team. If an Asian carp species *was* to become established in one of the Great Lakes, however, what would be the prognosis given the situation in the Mississippi basin?

"The main population front of bighead and silver carp in the U.S. hasn't moved since 2007, so frontline work there is buying time," Cudmore responded. "That involves frequent removal of metric tons of fish and research into integrated pest management techniques. The USGS is doing a lot of really neat research into control—like poison pills that react to gut flora which would be unique to bighead or silver carp. Still, although it takes lots of manpower, our best tool is monitoring. In the past we pretty much haven't noticed invasives until they're established, so any amount of proactivity is better than doing nothing—and far cheaper."

Cudmore, who knew better than most the cost-benefit picture, was repeating the invasion biologist's mantra: *An ounce of invasive species prevention is worth a pound of perpetual management.* To that end, there were more positives on the prevention front in the beleaguered Great Lakes: DFO scientists charged with ship ballast issues were finally addressing the potential movement of aquatic invasive species between basins, research that included different on-ship technologies—such as chemicals or UV light—to mitigate their transport via ballast.

Beyond the concerns on her own plate, I'd wanted Cudmore's take on where she thought the global reshuffling of ecosystems was headed, and why the kind of work she did was important despite an increasingly gloomy bigger picture. "I've spent a lot of time working on the live-organism trade and when you see what we're purposely moving around the world it's pretty overwhelming," she said of the strategic risk assessments she'd undertaken with Mandrak. "None of this stuff is meant to be released into an exotic environment, but it's not going to stop. You can never stem the tide of introductions 100 percent, so the best way to maintain the diversity of native ecosystems is through education, awareness, and policy. Unless it's something so dangerous it requires an outright ban—like a northern snakehead."

Noting that the Burnaby snakehead proved to be a tropical species that would have died over winter, Cudmore wondered whether DFO would have bothered to undertake the expense of draining a lake to look for a northern snakehead if an importation ban had existed on the species. Importation bans would auger well for the Great Lakes, whose story over the past century was one of relentless invasion, a narrative into which the basin's other issues—water levels, industrial base, shipping, agricultural pollution, and degradation of habitat and environment—readily fit. As an undergraduate, Cudmore had started down what now seemed a path with no end; how did she see everything that had happened or was continuing to happen fitting together? "Humans have had a lot of influence on how the lakes have come to be; they would have been much different otherwise. Going forward it would be best to try and minimize our impact and influence to support ecosystem restoration. But as for active intervention, I'm still not sure we fully understand nature enough. Tinkering is Russian roulette and maybe we haven't shot ourselves in the head yet, but we will. So I don't like that game. I've seen too many surprises."

Later that fall five more grass carp would be found in scattered locations around Lake Ontario, from St. Catharines at the mouth of the Welland Canal in the lake's western basin, to the Toronto Islands, to the Bay of Quinte near the eastern egress of the St. Lawrence River. At least one was a diploid female capable of reproduction—a first real warning that the waters of the Great Lakes were poised to go the same low-diversity, monoculture ways of the land base surrounding them.

## One-Dimensional

Sandy Smith has been talking about invasives for twenty years, or, as the University of Toronto forestry professor puts it, "since we called them exotics." Although she can't help herself, Smith doesn't enjoy peering into Toronto's ravines these days. "Everyone thinks they look great but they're full of Norway maple—both overstory *and* understory because the young trees have evolved to grow under the shade of a dense Norway maple canopy."

Norway maple was yet another invasive hiding in plain sight. In an all-too-familiar story, the tree was imported in the eighteenth century as a garden adornment by a Philadelphia merchant. With few native insect or fungal challenges it spread furiously, crowding out less-competitive indigenous species. Fast growth and a prodigious number of seeds—borne in those familiar, twirling keys that lodge in car windshields each autumn to be transported all over a city—allowed it to take off across the continent's northeast. In Canada, its spread was greatly aided in the nineteenth century when the maple leaf became a national symbol (the country's first anthem, "The Maple Leaf Forever," was penned in 1867, the year the British North America Act created the Canadian Confederation)—to most, the tree was just another stately maple to plant with abandon. Even today, although 98 percent of Canadians can identify the tree as a maple, the same number likely assume it to be a native silver, sugar, black, or red maple.

Norway maple's rapid invasion helped it to leap from fringe to familiar, eventually occupying cultural pied-à-terre, including the official logos of Wilfrid Laurier University, the FIFA under-twenty World Cup of Soccer, the Canadian Television Fund, and even the Alliance of Natural History Museums of Canada. Though the museums' logo was particularly galling, those in the know merely "tsk-tsked" in displeasure, crying full foul only when its leaf turned up on a twenty-dollar bill issued in 2013 as the first in a series of new polymer banknotes. At that point, botanists publicly and loudly decried having an alien invasive species gracing Canada's national currency. The Bank of Canada pleaded botanical license (or ignorance), claiming it was a stylized amalgam of *all* Canadian maples . . . that just happened to look exactly like the stout palmate leaf of a Norway maple. By 2015, almost two billion new $5, $10, $20, $50, and $100 bills were in circulation, each paying inadvertent homage to the Norway maple's ability to hog space not only in Toronto's ravines, but also in the national consciousness.

It would be funny if it weren't so sad, said Smith, who saw it as part of a broader environmental apathy. "If you look where humans have been in the last 20,000 years it's not a pretty sight. Lebanon was once full of native trees. Ditto Easter Island and the south of Spain," she said of a few notoriously bare landscapes. "So where's the tipping point that moves us into the permanent dysfunction of only a few widespread plants and the limited microbial and invertebrate worlds that go with them?"

Human landscapes, of course, include cities, whose one-dimensional insect faunas reflect an environment of invasive plants and the monocultures common to landscaping. "In urban areas, the diversity of birds and insects in native trees is higher than in nonnatives. So if most of humanity is going to live in cities, as seems to be the case, then we need to make cities better," she said. "Currently we're reducing things to something that either looks good or is easy to manage. It's this kind of homogenization that worries me most because we're creating a vulnerable world."

No one knows where that tipping point might be, and we seem not only content to find out the hard way, but then, once the penny drops on the latest anthropogenic folly, to expend uncountable amounts in a bid to remedy situations that are now only within the grasp of suppression. As controllers of nature, we have also become controllers of the unwanted outcomes of our various interferences. Smith knew this all too well. Her area of expertise, biocontrol—something Charles Elton expended abundant ink on—has long been the poster child for creation of unforeseen invasive problems. Having recently completed a book chapter summarizing 125 years of biocontrol in Canada, however, Smith understood both its lessons and its perils, and was bullish on its potential.

Most problems, she noted, stemmed from "classical biocontrol"— the phrase that describes going back to the country of an invader's origin for a control organism specific to it. Early attempts at classical biocontrol often loosed new problems on the land because they were fraught with political and economic urgency, and as a result were poorly implemented. As Smith explained it, well-thought-out, fastidiously tested, successful biocontrol "is silent."

(A much-cited example of negative, cascading effects in classical biocontrol involves European spotted knapweed, now a scourge across North America. To control it, a European fly was released that lays eggs in knapweed flowers where hatching larvae then consumed seeds. These same larvae, however, proved irresistibly tasty to mice, who ate the flower heads with abandon, spreading seeds around in their poop; predators that ate the mice dispersed seeds even further. Not only that, but the burgeoning knapweed food source made mice more abundant and widespread—along with the deadly-to-humans hantavirus for which they act as reservoir.)

Smith's interests lay in the emerging area of conservation biocontrol and augmentation, an ecosystem-level approach where you strive to shore up exotic and native enemies of an invader with indigenous forms that lend resilience to landscapes. "There's a risk to everything you do, especially in pest management. The do-nothing option is, of course, also a management choice, which has risks and consequences as well. But in terms of biocontrol, let's see what nature has provided us first before we go tampering again."

Waiting for our meeting in the hall outside Smith's office, I'd sat on a wooden bench crafted from beetle-addled wood, fascinated that such beautiful cellulose tracings could actually destroy a tree. Once inside, my eyes made their own tracings across an eclectic mix of maps, public outreach posters, and fridge magnets warning about various insects and plants.

"West of me is buckthorn, east of me dog-strangling vine, and everywhere *Phragmites*," said Smith, laying out the waypoints of what would be a wide-ranging, non-linear discussion, something I'd come to expect with invasion scientists.

"Parasitoid wasps have already been brought into the U.S. and Canada to control emerald ash borer. The loss of millions of trees is a huge, expensive problem and people are screaming for answers. Governments are under great pressure to do something, but they have few tools to use and each comes with some level of risk—like someone insisting they come up with a cure for cancer. The province screams at the feds saying 'You let it in!' The feds defend themselves by saying they don't have the resources to control everything. It's states and provinces that deal with the consequences in loss of resources, but it's ultimately municipalities that suffer with immediate impacts. If invasions are a disease, then munis are the patients waiting for the cure."

By that measure, Ontario and New England were waiting rooms brimming with "sick" municipalities suffering the ravages of various afflictions—emerald ash borer, hemlock woolly adelgid, dog-strangling vine, and, increasingly, *Phragmites*. Smith had spent two years in the south of France at the USDA biocontrol lab surrounded by native *P. australis* that posed no problem there. With everyone in North America thinking phrag was a battle we were losing, however, the public policy question, according to Smith, was clear: "Do we *want* that to happen, or are we willing to risk bringing in a moth that has a phrag-loving caterpillar? So far we haven't found agents that *won't* attack native phrag, so we're gambling with the remaining 3 percent of phrag that *is* native. You could look around and say, 'Well it's almost gone anyway . . .' but that's not the right approach. If you want to know where we're at in that thinking, you'll need to talk to Robert Bourchier—he's Mr. Weed Biocontrol."

## Million-Dollar Question

I would eventually speak with Robert Bourchier, a research scientist in biological control and insect ecology for Agriculture Canada based in Lethbridge, Alberta. After so much time spent immersed with knotweed, my top interest was discussing imminent biocontrol trials involving the Japanese knotweed psyllid, *Aphalara itadori*. Bourchier was happy to fill me in, but we'd first veered into the *Phragmites* thicket, whose spread far outpaced that of knotweed. Fortuitously, he'd picked up right where Smith left off.

"The invasive European genotype is a huge problem in North America," leveled Bourchier. "A few isolated patches of native North American *Phragmites* still exist on islands and elsewhere, but it's being outcompeted and there's a real possibility it will be extirpated."

Or absorbed. Per Bob Brett's voiced concerns, it had now been conclusively demonstrated that the invasive and native genotypes of *Phragmites* were actively hybridizing, though the extent and ecological outcomes weren't yet clear. This development only added to the current feeling of defeat being experienced by many in eastern North America.

"I was at a workshop in Ontario where they were reviewing *Phragmites* problems and I was disheartened to see the extent and nature of the infestations," said Bourchier. "Right now they're trying to get permission to spray it with herbicide right down to the water. That's a serious and dramatic decision to make, but when you see the size of rootbeds and the biomass represented—I mean, it's literally like an iceberg—there aren't many other options."

I had to agree. A couple months earlier, kayaking the lower reaches of the historic Rideau Canal where it enters Lake Ontario near Kingston, one-time capital of a nascent Canada, the view from water level was almost indescribable as I paddled past kilometer after kilometer of dense, four-meter-high *Phragmites.* Unlike the few patches of native cattail it towered over, inevitably surrounded, and doomed, the phrag didn't simply occupy shoreline, but pushed further into the current, creating rafts from which it could spread further. Per Mandrak's shifting-baseline argument, this was indeed the new normal for people who grew up here, and I was old enough to remember when Southern Ontario's marshes, lakeshores, and ditches were phrag-free. What I'd observed in periodic trips east over the past decade was a time-lapse of complete takeover, an ever-moving invasion front in New York State, Quebec, and Ontario, and a march along any number of highways as far west as Nebraska and Saskatchewan. The only limit to the spread of *Phragmites* would ultimately be climate—which was warming and so moving that goalpost ever farther.

Bourchier was working on a moth whose larvae would feed in the *Phragmites* stem and then travel down to the root. The insect was considered a pest in reed beds in Europe so it had the potential to cut down the plant's biomass and fecundity on this side of the Atlantic. The agent had so far been test-screened against other North American plant species to assess their vulnerability; bowing to political pressure didn't happen anymore, with long-term testing required to ensure nothing was released that would harm nontarget plants. The major concern, as Smith hinted, would be if the moth was also able to develop on native phrag, accelerating its decline and eventual loss. According to Bourchier, at least two natural limitations might obviate that. First, the moth oviposits in leaf sheaves where the eggs then overwinter before hatching; but unlike its European counterpart, North American phrag drops those sheaves for winter, with anything on them unlikely to survive. Second, in oviposition choice tests, the moth preferentially chose introduced phrag. Promising results, but they wouldn't be put into action anytime soon—given the due diligence required, the regulatory road for approval to deploy any biocontrol was long and arduous. And Bourchier had been through it many times.

"We're petitioning for the agent's use in 2016. The regulatory bodies might want more testing or justification, and that will slow things down. Here it's the Canadian Food Inspection Agency and in the states the USDA's Animal and Plant Health Inspection Service. How long will it take? Hard to say. The U.S. regulatory process has a couple more steps, but even in Canada we're conservatively looking at three years to approval. And then once the agent is out there, it'll take a few more years to figure out how best to establish and monitor it. That lag doesn't necessarily mean we're falling further behind—remember that most of these weeds had a hundred-year head start on us and already have a huge seed bank out there."

Bourchier further made the point that biocontrol, at least for weeds, had a good record in Canada. "About seventy-five different biocontrol agents have been released targeting twenty-one different weeds. There have been some huge successes, while others have been ineffective, but in no case has the insect either gotten out of control or done something unpredicted. *Phragmites* is a bit different in having a native genotype that the insect could theoretically wreck, but with the ecological filters in egg deposition we think the risk can, in this case, be justified."

Some researchers, however, are leery of "preference for the invader" being an acceptable criterion for biocontrol, in part because a shift to the most available food source is an easy evolutionary switch, and once an invading plant has been sufficiently reduced, a native might represent that most available source. Furthermore, even given preference by an agent for the invader, sufficiently large biocontrol populations supported by similarly large invasive populations could still exert competitive harm on less-preferred natives.

Petitions for approval include information on the impacts of the plant in question and the justification for releasing a new insect. It's always a question of risk versus risk management: the more societal, ecological, and economic costs of the invader, the more justified the risk. And, as Bourchier noted, biocontrol was a last resort when nothing else was working. "When you *do* find a successful agent, though, you're looking at a long-term solution for exerting reduction in invasive populations that you acknowledge you'll never be entirely rid of."

There were, of course, always unanticipated challenges even after approval, including context-dependent effects unanticipated in a lab setting.

"The insect released on purple loosestrife, for instance, worked well in some areas but not in others, and we haven't figured that out yet. It's really a research question: how do we best do this?" said Bourchier. "There's still a lot of work to be done once you take an agent into the field."

And that work was onerous. As Bourchier outlined, each biocontrol project was different and largely dependent on an insect's life cycle. The insect chosen was usually the one they had the best chance of rearing, either in laboratory quarantine in North America, or where the plant originated. Figuring out how to grow the insect starts with the challenge of growing host plants; every plant has different vulnerable stages from seed to flowering, with insects chosen to target a particular stage.

"The ultimate filter is feeding specificity," noted Bourchier. "That starts with no-choice testing—you present a plant, and the insect either feeds on it or dies. We learn a lot about rearing the insect before we petition for a permit. Once you have permission, you might do open releases of adults because that's what you can produce in the lab—with the bonus that they'll actively seek out the target plant. Or you might place larvae directly onto plants in the field if defoliating is what you want—though manually transferring individual larvae onto shoots is very labor intensive. With phrag the insect is a stem feeder, and it's hard to captive-rear anything in stems—you need the right diameter stem and have to keep the plant alive while the insect feeds, so with this one we'll probably release adults and hope they do our work for us . . . maybe we'll put them into a field cage so they don't disperse and then you know where to go back and look."

Bourchier took the opportunity of discussing release tactics as a segue to talk about *Aphalara itadori,* the agent on which was now pinned the salvation of many a knotweed-infested nation. The 2010 release of the psyllid for non-target field testing in England was auspicious not only for the hope it represented, but for the anxiety created—it was, amazingly, the first time anyone had tried biocontrol on a weed in Europe.

"We have a permit to do cage releases of *A. itadori.* In December 2014 we released diapausing adults while the plants were senescing, hoping the insects would go to sleep and then come up synched with the plants in spring. Then this summer, we released non-diapausing adults to try and build up a population that we would see next spring. If either of those works to establish the insect, we still have the problem of growing its population. Which is what happened in the U.K.—they're just now moving to releases in their more problematic watershed areas."

Field cages retain humidity, which *A. itadori* likes, so cages are useful both in this respect and in their ease of monitoring. By using them, researchers have learned that the psyllid does best on Japanese and Bohemian genotypes, observations they've had a while to make.

"We've had the insect here eight years. It's a bit of a double-edged sword when you raise something in captivity that long; you learn about it, but without challenging it in a natural environment you may accidentally select for a wimp. That was also thought to have been a problem in the U.K., so they went back to Japan in summer 2014 to get new insects from the same place and mix them back into their populations. At any rate, we're in a position to rear a lot quickly if we get approval for more widespread release."

For over a decade the Centre for Agricultural Bioscience International (CABI), based at Wallingford in the United Kingdom, sought a suitable biological knotweed killer. Although CABI was doing the work, the project was backed by a consortium that included Britain's Environment Agency and stakeholders/blameholders like Network Rail and British Waterways (which risked infrastructure damage and incurring control costs, but also had helped spread the plant). This in contrast to Canada, where outfits like the power company BC Hydro and CN Rail either refused to address problems and pay their fair share, or threw up their hands with lip-service coins to collaborative efforts.

"*Phragmites* is a political issue in eastern North America because people are aware of it and there's pressure everywhere to do something," Bourchier told me, echoing Smith's assessment. "People there aren't as aware of knotweed yet, but it's creeping in all over and will be a huge political issue as well. In my first talks in Ontario back in 2010 and 2011, I always mentioned knotweed, but no one really paid attention. They were all talking about garlic mustard and dog-strangling vine. Even phrag wasn't mentioned back then. It'll be better to get in with biocontrol at the stage knotweed is at in the East now—sporadic infestations—than the stage phrag is at, which is ubiquitous."

Despite the overwhelming *Phragmites* threat, garlic mustard and dog-strangling vine were still concerns. Dog-strangling vine (*Cynanchum rossicum*), which forms dense, impenetrable thickets, was particularly problematic in peri-urban and rural areas around Toronto. Here was another biocontrol project fifteen years in the making. "We have an insect approved and did open releases of larvae, and cage releases for adults that seem to have overwintered. The strategy is to get them established then open the cage and let 'em go," said Bourchier, adding a poignant afterthought that further addressed the urgency of control. "Monarch butterflies lay up to a third of their eggs on dog-strangling vine now—and those eggs all die. It's a dead end. That's one of its impacts."

And extra-troubling for the monarch, whose plunging populations across North America were already under siege from a host of problems that included a precipitous decline in its primary breeding plant, milkweed, due to a massive increase in glyphosate use in agricultural areas where milkweed traditionally grew. And now dog-strangling vine was providing a deadly diversion to this dwindling icon. The situation reminded me of something Sandy Smith had said.

"How healthy is any particular ravine if this vine stays? It will degrade the soil, there will be spring and fall erosion because it has shallow roots. So . . . what's the forest of the future here going to be?"

That was the million-dollar question.

## Lot 69

In early September 2015, SSISC crew boss Rob Hughes invited me to join him and Sharon Watson for a day at the "Knotweed Farm." Though that sounds like a fun and innocent outing to a managed plantation of some type, I knew it to be pretty much the opposite—a mission to destroy something intentionally grown for just that purpose.

As Clare Greenberg had explained during a sit-down a few weeks prior, the Knotweed Farm had been inaugurated by a District of Squamish decision the previous winter to prohibit invasive plant-contaminated soils at its municipal landfill. Developers already removing contaminated soils from construction sites were caught out, with no place to dump it. With an alternative dumping place needed stat, DOS and an environmental consultant settled on an old log-sort plot leased from the Crown known as Lot 69. Here developers could either dispose of contaminated soils, or conduct off-site soil rehabilitation, letting plants sprout then spraying them down—a lengthy process that was bogging down projects like the one I'd seen in North Vancouver with Richard Beard.

At Lot 69, heavy machinery cleared and leveled large areas, centimeter-thick geotextile was laid down, and the fill was spread on top. This had occurred in winter and early spring, so the luxuriant growth now on display was only six months old. An impressive biomass completely covered the "Wilson patch," so named for the street it was removed from. Most was knotweed, albeit small, easily treated plants. On the "Bailey patch," a slightly larger, more recent area of fill, growth was both more sporadic and less vigorous. In a few areas of bare soil, heavily fissured by summer drought, the *only* thing growing was knot-weed, emerging from the shadowed depths of cracks where buried frag-ments had been exposed to air. Still, the plants were relatively small, and Watson, used to dealing with towering overhead thickets, welcomed the knee-high growth. "These are ideal conditions for treatment," she said, "because you're not battling through anything else."

The idea was to "farm" the knotweed on these soil pans, hit them hard with foliar spray, and treat any regrowth regularly until the seeds and regenerative capabilities of any plant bits were spent by attrition. Hughes estimated that it would take three to five years to decontaminate the soil, acknowledging that even then you couldn't be sure. "It'll be interesting because all this came from infestations of two-meter-tall plants," he said, sweeping his arm some 270 degrees, "dug up and churned around and put into trucks so there aren't likely any massive rhizomes, only broken fragments. Without a well-established rhizome system, and by treating it in the first year of growth, it might be easier to kill but who's to say? It's *so* different between sites; we get lucky with some that we treat once and never see another plant, but have others that go knotweed-free for five years and then *boom*, it's back. The plant is well-adapted to weathering troubled times."

Touring the lot to see what else was going on, I noted one corner of the Bailey patch densely covered in small shoots. They could all have been from different fragments, or a single large rhizome; certainly there was one very large stem nearby, about three centimeters wide and head high—was this a mother plant that had arrived quasi-intact? As I rounded that same corner, beyond the edge of the geotextile mat, I also spotted a *ménage à trois* of invasive monocultures where patches of knotweed, blackberry, and morning glory met. Only a couple meters away stood a swath of enormous Scotch broom, with burdock and spotted knapweed constellating the remaining areas. There seemed nothing good about any of this and I couldn't help but think again of those who would promote such congregations as vistas of salvation.

While I walked, Watson staked warning signs around the perimeter, took weather readings and photos, and logged data onto an iPad. Then she and Hughes readied backpack sprayers while hoping for a breeze to stir the humid air. The previous night's rain had been slower to clear out than forecasts suggested; patches of blue were now spreading in from the Pacific, but the last foggy tendrils of cloud clung stubbornly to the mountainsides. A surfactant in the glyphosate mix would penetrate dust or debris to help it stick to leaves, but these needed first to be dry.

The two struggled into one-piece white hazmat suits, Hughes leaving his partially unzipped to offset the sudden heat. "There's always a bit of leakage near the pump valve, so this at least keeps the stuff off your back," he said of his Elvis-meets-Weed-Man look. He'd shoulder the four-liter rig and tackle the 1,600-square-meter Wilson patch, while Watson wielded the smaller unit on the larger but less-vegetated Bailey patch. Glyphosate hitting the air and wafting downwind was my cue to leave.

It seemed like lonely work as I watched them disperse blue showers from a distance. Hughes worked his way along an outside embankment, raising his wand from the bottom upward in what looked like a bye-bye wave to each plant; Watson waded more methodically, arms moving side-to-side, offering what looked like a benediction. White suits against a dark, verdurous forest reminded me of something . . . but what? Oh, yes—the 1995 Hollywood movie *Outbreak*.

# A Brief History of Future Disease

Toronto General Hospital, the city's largest medical center, sprawls several blocks along University Avenue at the heart of a cluster of patient care and research facilities associated with the University of Toronto's medical faculty. When I pulled up curbside on a steamy morning in August 2015, at 9:27 a.m., people with appointments to see a wide range of specialists were already lined up at pay machines for street parking that began at 9:30. I joined them because I was also there to see a specialist: Isaac Bogoch, the epidemiologist who'd schooled me on chikungunya.

Tall, lean, and moving with purpose, Bogoch met me at a busy Starbucks in the lobby of one of the two-hundred-year-old institution's newer edifices. I grabbed a coffee, he a green tea, and we rode an elevator to the "penthouse"—as he jokingly called his office—on the fourteenth floor. Neither spartan nor cluttered, the office had one wall featuring customary diplomas while elsewhere were distributed artifacts from his multitudinous travels. A desk and enclosed shelves, both of modern, faux-wood Danish design, were noticeably devoid of surface clutter; books and papers had been filed fastidiously away—the complete opposite of the info-hemorrhaging environs of most science professors. When Bogoch apologized for the sparse decor, however, pointing to boxes he'd yet to empty despite having been there a year, it was absolutely in line with these other professionals, all of whom instinctively offer various "office apologia" as you would for a home. This was a universal behavior (personally recalled from when I'd welcomed students into my own sanctum) because you likewise viewed an office as an extension and reflection of yourself, while simultaneously imagining that its topography did *not* reflect your mental landscape.

For a guy who claimed natural hyperactivity, Bogoch seemed remarkably relaxed. Or at least more so than during our phone chats, shoehorned as they'd been between a litany of clinical responsibilities, which had elicited an eagerness on his part to immediately dig into the disease du jour. This time, however, he leaned back in his chair, an ankle resting atop the opposite knee, unbuttoned labcoat splayed in a "Where do you want to start?" pose. The question soon answered itself: my eyes had been drawn to arrows and scribbles on a whiteboard beside the door—a dry-marker sketch of the HIV infection cycle depicting how different drugs, for example, interfere with viral entry to a cell, or block the retroviral RNA genome from being reverse-transcribed into DNA for gene-expression. Tracking my gaze, Bogoch jumped in.

"We're in the golden age of HIV management right now," he began. "Think about it. When the virus was first identified, you were gonna die—that was it. Drug therapy became available, and you took a fistful of pills every day. Then it was fewer drugs that still made you feel shitty. But now you just take a drug combination in a single daily pill with minimal side effects that keeps the virus at incredibly low levels in your system."

A benefit of this advent, explained Bogoch, was that an immune system exposed to low levels of HIV over longer periods of time was better able to mount its own response, with a twofold epidemiological result: people now living with HIV into old age meant a higher prevalence of the virus in the population, but lower levels of transmission and infection. As a self-contained ecosystem, the human body under a maintenance drug regimen presents a model for broader invasive scenarios: one can similarly imagine the effect of long-term control of a destructive invasive as allowing time for a native ecosystem to evolve a response that makes further spread of an organism less likely and/or more difficult; cumulatively, occurrence of the invasive may cover a wider area, but patches would be smaller, less destructive, and spreading at nowhere near their previous rate. Also similar to invasive scenarios, prevention remained *the* most important strategy in HIV epidemiology—though even the riskiest encounters could now be mitigated to an extent by a type of EDRR. "People who think they may have been exposed to HIV can be given a twenty-eight-day drug regimen that will prevent transmission 99 percent of the time," said Bogoch. "And soon we'll have a prophylactic drug to take *against* becoming infected."

My eyes widened at the thought of how this last might undercut the safe-sex ethos relentlessly inculcated by health professionals for the past four decades. "And yes," said Bogoch, reading my expression, "it will have huge social implications."

Emergent diseases, of course, are de facto invasive anytime the pathogens that cause them are moved outside an endemic area of prevalence to take hold in a naïve population (think Dutch elm disease and chestnut blight landing in North America, and the ongoing extent of destruction a century later). But HIV was also *zoonotic*, meaning that it jumped to humans from another species—likely a chimpanzee somewhere in central Africa, as fastidiously documented in *Spillover*, David Quammen's riveting 2012 book that also eerily anticipates the 2014–2015 West Africa Ebola outbreak. Most of us are familiar with the story from there: transmitted in the human population by the exchange of bodily fluids through sexual contact, shared needle use, and blood products and transfusions, HIV circled the globe, aided by politics (war, famine, colonialism), social and cultural forces (recreational drug use, sexual liberation, human predilections for spontaneous reckless behavior), and burgeoning global travel and commerce (recall my mention in Part 2 of the increase in HIV infection associated with the invasive Nile perch fishery on Lake Victoria). The advances in antiretroviral therapy now being outlined by Bogoch, however, had also created a kind of public amnesia concerning continued impact of the AIDS pandemic. Lest we forget, March 2015 statistics still painted a stark picture:

~ 80 million people had contracted HIV and some 35 million had died of AIDS-related causes since the first cases were recognized in 1981

~ 2.1 million people were newly infected with HIV in 2015, 150,000 under age fifteen

~ 5,600 people contracted HIV each day in 2014—more than 230 per hour

~ 1.1 million people died from AIDS in 2015

~ 36.7 million people were living with HIV in 2015, 1.8 million under age fifteen

~ 25.6 million (70 percent) of those known to be positive for HIV live in sub-Saharan Africa; prevalence in some villages is 60 percent

~ 17 million living with HIV (46 percent) have access to antiretroviral therapy

Wrapping your head around HIV *distribution* was chore enough, but when you considered how many of the hardest-hit countries also suffered from additional infectious disease, food insecurity, and a range of other issues, the real magnitude of its effects sank in. Could we have done a better job given the tools at our disposal at the time? Perhaps made an effort to get in front of HIV spread with modeling once the endemic reservoirs and transmission routes had been identified? Sure. But history will show that politics and prejudice interfered, and by the time these now-routine epidemiological measures were enacted it was too late. The bigger question was this: had we learned *anything?*

"Let's talk Ebola," I said, recalling Bogoch's modeling work during the recent epidemic. In a January 2015 paper in *The Lancet,* he and seventeen coauthors had analyzed both International Air Transport Association data for worldwide flight schedules over the last four months of 2014, and historic traveler flight itineraries of the previous year, to describe expected movements, via air, out of the three countries most affected by Ebola—Guinea, Liberia, and Sierra Leone. Pairing this with virus surveillance data, they modeled the expected number of exported Ebola infections, the potential effect of air travel restrictions, and the efficiency of airport-based screening at international ports of entry and exit. Based on epidemic conditions and international flight restrictions as of September 1, 2014 (which resulted in 51 percent fewer passenger seats for Liberia; 66 percent fewer for Guinea; and 85 percent fewer for Sierra Leone), their model projected an average of 2.8 Ebola-infected travelers collectively departing this equatorial troika each month, 64 percent of whom had destinations in low-income and lower-middle-income countries. They concluded that exit screening in affected countries was the most efficient way to assess the health status of travelers who may have been at risk of Ebola exposure, but that such an onerous task might require international support for effective implementation. The takeaway for decision-makers was to more carefully balance the harm of imposing economically devastating travel restrictions on countries with Ebola activity against the benefit of reducing the risk of infecting populations of other countries.

"The outbreak had potential to be much worse, and was thankfully contained in those three countries," said Bogoch. "But [the world] could have done a better job up front to prevent the extent of its spread and resulting death. That happened because of bad policy."

Policy that saw the international community react with a collective shrug to news of another Ebola outbreak and the World Health Organization eventually apologize for its slow response—a delay that took on the familiar profile of the bottom of the Invasion Curve (Figure 6), and showed clearly that the WHO had bypassed a chance for rapid response.

Maybe we hadn't learned anything from the HIV pandemic after all.

Policy aside, containment was further frustrated because the outbreak didn't follow Ebola's usual scenario of restriction to isolated forest villages; instead, it erupted with little warning into urban settings with many more moving parts. "The scariest aspect was when it landed in Nigeria," said Bogoch of the shudders that went through the global health community when Ebola appeared in Lagos, capital of Africa's most populous country, a teeming center of continental commerce and home to 21 million. "But Nigeria's public health administration did an amazing job of shutting it down. Still, looking at the overall outbreak, 25,000 died. What about long-term economic fallout from orphaned kids in these already poor countries? On top of financial issues related to reduced travel and business, that's really frickin' sad for something that didn't have to happen."

As we talked, an alert popped up on Bogoch's computer, and he briefly turned to peruse a notice from Nepal of a cholera outbreak in a Kathmandu jail. "*That's* gonna be a problem," he said, swiveling back. "Fecal-oral transmission. It'll rip right through a prison."

He would know, having just returned from Nepal himself. His group had secured funding to set up outbreak teams after the April 25, 2015, earthquake that had killed 9,000 and injured 23,000. Maps are primary tools for medical geography modelers like Bogoch, and he quickly pulled one up of Nepal depicting outbreaks of long-simmering endemic diseases like typhoid and cholera, colored blots arranged along roads like the spokes of a wheel whose hub was Kathmandu. "Earthquakes disrupt both clean water and sanitation infrastructure, and in a place where such things are already limited you almost have to expect problems," he said.

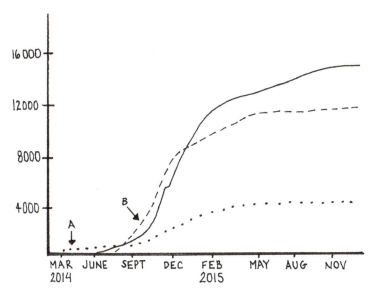

Figure 6. Assessment of the international community's failure in the 2014–2015 West Africa Ebola outbreak. The figure depicts total number of cases over time: the dotted line represents Guinea, the dashed line Liberia, and the solid line Sierra Leone. A is when the international community could and should have intervened; B marks when it finally did. (Based on case count data from U.S. Centers for Disease Control.)

Major earthquakes prepare the ground not only for reemergence of latent indigenous pathogens, but for the sowing of invasive infectious diseases from elsewhere—as in Haiti after a powerful January 2010 earthquake devastated the country's south, including the capital, Port-au-Prince. In this telling example, a massive, ongoing cholera outbreak that hit ten months after the quake had, as of August 2015, killed 9,000 and infected some 800,000 more. Disease had inevitably spread to Haiti's neighbor, the Dominican Republic, and leapt to Cuba and Venezuela.

During the nineteenth and twentieth centuries, six global cholera pandemics occurred, each originating on the Indian subcontinent, and we were a half-century through a seventh at the time of the Haiti outbreak. Pandemic number seven owes its drawn-out nature to a less-virulent strain of the *Vibrio cholerae* bacterium known as "El Tor," which, by dint of its sub-lethality, had proven an intrepid world traveler. El Tor first surfaced in the Middle East in 1905, but wasn't behind a major outbreak until 1961 in Indonesia, whence it spread rapidly through Asia, India, Bangladesh, Iran, Iraq, and the USSR. By 1973 El Tor was in southern Europe and had jumped to various points in the Americas—including an infamous 1991 outbreak that sparked in Peru and flamed across South America, sickening a million and killing 10,000. Traced to a Peruvian harbor, the outbreak was a cautionary tale of disease translocation: subsequent testing showed that the oceanic freighters that had picked up ballast water in the suspect harbor had routinely discharged *V. cholerae* into other South American ports—and even the Great Lakes.

Haiti, however—somewhat miraculously—hadn't experienced a cholera outbreak in recorded history. Thus, not only was its populace immunologically naïve, but as a country already ravaged by HIV, the prospect of yet another epidemic triggered panic, confusion, and rioting that complicated relief efforts. Politics had also interfered.

As it turned out, routine tests and sophisticated DNA fingerprinting by the U.S. Centers for Disease Control (CDC) found that the El Tor sub-strain responsible for Haiti's outbreak (serogroup O1, serotype Ogawa) had not, as imagined, merely been vacationing in the Caribbean for a few decades waiting for a chance to show its stuff, but had recently arrived from Asia. Within the country, the outbreak was traced to the Artibonite River, 100 km north of the capital and a major source of drinking water. Local suspicions led epidemiologists to the likely source of contamination—a UN military base on a tributary of the Artibonite near the town of Mirebalais that was home to a group of Nepali peacekeepers. In October 2010 a new group of soldiers had rotated in, and five days later, someone downstream died of cholera; within a month the disease had reached every corner of the country. Though news reporters dug up photographic, video, and verbal proof that mismanaged waste from the base had found its way into the river, Haitian, Nepalese, and UN officials all denied links that were obvious even to schoolchildren. In October 2013, a multi-billion-dollar lawsuit was filed in a Manhattan courtroom, accusing the United Nations of covering up its role in what many call the worst cholera outbreak in modern history. "Shameful," was how Bogoch, who had no truck for politics, summarized the situation.

Eventually we circled back to the topic I'd first contacted him about: chikungunya. According to the CDC "Chikungunya Tracker," between December 2013, when Asian CHKV first appeared on the Caribbean island of Saint Martin, and July 2015, over 1.5 million cases had been reported, and it continued to spread, particularly into the austral reaches of South America. Autochthonous cases had also been reported in (of course) Florida, but the Caribbean countries of Dominican Republic and Jamaica were hardest hit; both declared public health emergencies.

"This has really been the perfect storm," said Bogoch of what would likely become the quintessential textbook example of an invasive vector of transmission being present when a novel invasive disease arrives. "The virus landed where *Aedes* mosquitoes had already been introduced, and there was a non-immune human population. The climate profile for temperature and precipitation was appropriate for these species, plus they're urban mosquitoes, and there were lots of peri-domicile areas where they could breed in a tire at the side of the road, for instance, while malarial mosquitoes would need clean, fresh water. The disease will grumble along at a lower level for years to come. And it'll spread farther because health authorities can't tell people to stay home as you would with the flu, because when you're symptomatic—the two-to-five days-*ish* that you're actually viremic—you might feel crummy, but not *too* crummy to go out and either get bitten and transmit, or get on a plane to take it somewhere else. So within a year or two most people will have had it, gained immunity, and won't get it a second time."

"Like West Nile?" I wonder.

"Exactly. You and I and many other people have probably had or been exposed to West Nile virus and not known it. No one really cares about it now, but I get a call a week from people going on vacation asking 'What are my chances of getting chikungunya?' And I have to say, 'Well, they're not zero . . .'"

## Ticker Tape of the Damned

There's a virtual rabbit hole where the interests of disease specialists and invasion scientists converge—the website ProMED-mail, which Bogoch and I spent several minutes perusing before I forced myself to disconnect. For science geeks and fans of macabre biology, the one-to-two-line disease reports from around the globe are like a bottomless bag of Doritos to a pot smoker—you can't stop consuming them.

To highlight just a fraction of the pathogenic exotica, over the past three days, in addition to Nepal, Asia was awash in cholera outbreaks, along with dysentery and various hemorrhagic fevers. Likewise the Americas were lousy with chikungunya, hantavirus, bubonic plague, anthrax, encephalitis, West Nile, and numerous tick-borne illnesses. There was an outbreak of anthrax in Canadian bison. Invasive tiger mosquitoes were now being tracked in Australia, Belgium, and France, with France also reporting autochthonous dengue in the south. Fatalities from toxic algae had occurred in the United States and Britain, and there were updates galore from Africa's waning Ebola frontier. But the report that really caught my eye was from Ireland: the deaths of over six hundred white-clawed crayfish (*Austropotamobius pallipes*)—the country's only freshwater species—from crayfish fungal plague in a County Cavan river. The occurrence was significant from several standpoints. First and foremost, should the highly virulent fungal plague pathogen (*Aphanomyces astaci*) become established in Ireland, its impact would be severe and irreversible, with a high probability of white-clawed crayfish being extirpated from the island. Second, though it *could* have been accidentally introduced from contaminated equipment used previously in affected waters in another country, the disease was most often seeded through a single pathway: direct introduction of nonnative crayfish—typically the much-larger American red swamp crayfish (*Procambarus clarkii*), a natural carrier of the fungus that is immune to its ravages. Introduction of larger, nonnative crayfish had already eliminated native crayfish from much of their European ranges, both through spread of disease and direct competition (recall the story of American signal crayfish replacing noble crayfish in Finland). Until this moment, Ireland had been free of crayfish plague and was the *only* European country without an established invasive crayfish species. And though direct evidence of such an invasion had yet to be found, the plague deaths were highly suggestive that Ireland's prodigious effort to prevent introduction of nonnative crayfish had, as of August 17, 2015, finally been overridden by carelessness or purposeful malevolence.

Reports like this made scrolling ProMED-mail like watching a live broadcast of dozens of different natural disasters unfolding, with a similarly potent "train wreck" draw. Or perhaps it was more like a ticker tape of the damned. And yet such facile comparisons belie its utility.

As an exemplary use of modern technology and communications, ProMED (Program for Monitoring Emerging Diseases) is an Internet-based reporting system dedicated to rapid global dissemination of information on infectious disease outbreaks and acute toxin exposures—including animals and plants grown for food. That means 24/7, up-to-the-minute reliable news about health threats to humans, animals, and relevant plants around the world. "By providing early warning of outbreaks of emerging and re-emerging diseases, public health precautions at all levels can be taken in a timely manner to prevent epidemic transmission and to save lives," advertises the site, which is free of political constraints and open to all avenues of reporting. Because it is just as important not to raise false flags in a panic-prone world as it is to reveal suppressed information, a global team of human-, plant-, and animal-disease experts work behind the scenes to screen, review, and investigate reports pouring in from myriad sources. Legitimate information is immediately posted, as well as distributed by email to over 70,000 direct subscribers in some 185 countries. Monitoring the ProMED-mail feed can help researchers like Bogoch get ahead of problems—as with another little bug currently coming out of Africa. When you ask disease-outbreak specialists what they're excited about, you almost don't want to know the answer. But I'd asked anyway.

"There's a flavivirus called Zika or ZIKV. It's similar to dengue, and has the same *Aedes* mosquito vectors. It was previously thought to only be in Africa. Then there were a few little outbreaks in Pakistan and Southeast Asia in the 70s; next it went to the Pacific Islands and lots of people got sick—fever, rash, joint pains, red eye, but, like CHKV, ultimately mild and self-limited. No one really cared about it, but I wanted to model its potential spread anyway. I suspected it would go to South America because they had the right mosquito vectors and climate conditions and, like CHKV, the only thing required for it to take off and propagate through a non-immune population was one viremic person to land in the right spot."

Discovered during routine screening of monkeys for yellow fever in Uganda's Zika forest in 1948, the virus was not reported in humans until 1952. Its slow movement out of Africa nevertheless closely followed the spread of competent invasive mosquito vectors. Indeed because the *Aedes* mosquitoes that spread ZIKV are now found throughout the world, it was, as Bogoch surmised, very likely that outbreaks would reach new countries. Already a handful of cases were being reported in the United States among returning travelers. As with chikungunya, there was no vaccine to prevent Zika, no medicine to treat it.

"So we're looking at this, thinking about modeling it, and then *boom*—in May 2015 Zika is reported in several northeast Brazilian states and we thought 'Oh boy, here comes the storm—this is going to be bad.' If you have the infrastructure to make a single diagnosis, there's likely many more cases. Now maybe forty have been reported but we suspect hundreds if not thousands more. It's the same setup as chikungunya and so it could spread just as crazily, which makes it a big deal that public health officials need to know about. This is going to happen *for sure*," he finished, pulling up a color-coded map of South America.

"Red is where you could have year-round transmission of ZIKV; yellow shows where part of the year works, and blue is where it won't happen. These maps would look different twenty years ago because current risk is based on climate-change trajectories. We'll have to redo maps periodically as that accelerates. Anyway, it's important to use the best models available to figure out where it is and where it's going so you can inform Public Health Officials, clinicians, and the general public: PHO teams can do screening and monitoring in local areas and if it looks like Zika is going to be a problem they enact mosquito control; clinicians *have to* suspect this virus, because if they don't consider it they won't make the diagnosis—even if you live in northern Finland where there's no chance of transmission you'll still need to treat returning travelers; and, lastly, the public needs to make informed decisions about vacation plans to protect themselves."

This was heady, cutting-edge stuff. Knowing how disease has changed the course of human history, if you needed an example of why modeling Invasive Species + Nonendemic Pathogen "systems" might be important to public health, you need only look at how the Zika story would play out over the next year: in December 2015, spikes in the incidence of birth defects like microcephaly and the immuno-neurological disorder Guillain-Barré syndrome in Brazil would be linked alternately to the virus and to the pesticides being used covertly to dampen its mosquito vector; the virus would prove to be the culprit. By January 2016 ZIKV would be confirmed in the American territory of Puerto Rico; by spring there was autochthonous transmission in Florida, and sexual transmission of the virus in Texas and Saskatchewan. The World Health Organization called Zika's spread "explosive" and by February had declared it an international emergency. The Brazilian epidemic spiraled into summer and fear of the contagion—particularly by those who were, or planning to become, pregnant—had an enormous economic impact on tourism and the 2016 Summer Olympic Games in Rio de Janeiro. Zika raged in fifty countries in the Americas, infecting millions, including tens of thousands of pregnant women. With some 4,000 cases in the continental United States, and a burgeoning outbreak in Miami Beach in late summer, South Carolina officials panicked and began an aerial spraying program for *Aedes* mosquitoes that killed billions of other insects, including honeybees.

The virus's true latent extent and subsequent spread perfectly mirrored the predictions of the model that Bogoch had showed me a year earlier. Yet this was only the tip of an epidemiological iceberg that he and others were trying to wrap their minds and models around. As ProMED-mail shows, moreover, the many threats aren't confined to humans. Fellow mammals as well as birds, reptiles, amphibians, and fishes are all under siege by new diseases circling the globe at rates that make the 1911 influenza pandemic seem like a norovirus outbreak on a cruise ship. And animal communities have no public health authorities looking out for them.

## Gone Baby, Gone

You know things have reached a tipping point when a headline like "North American Bats in Precipitous Decline" is best categorized as understatement.

In 2013, a study in the *Proceedings of the National Academy of Science* confirmed that the epidemic of white-nose syndrome (WNS) that had killed almost six million bats in Eastern North America was caused by the fungus *Geomyces destructans*, and likely introduced from Europe by humans. The study, led by Lisa Warnecke and Craig Willis at the University of Winnipeg, also involved researchers from Saskatchewan, the United States, and Germany. Since its publication and this writing, the death toll has almost tripled.

White-nose syndrome was first documented in New York State in 2006, and subsequently spread to thirty states and six Canadian provinces. Because bats are essential to most ecosystems—some consuming a third of their body weight in insects each summer night—fear of far-reaching ecological ripples were not unfounded. WNS is transmitted mostly from bat to bat, but the far-ranging animals were spreading the staggeringly lethal disease hundreds of kilometers a year. Its toll was truly alarming: in one cave with an estimated population of 132,000 bats, only 2,000 remained a year after exposure, and subsequent census showed that number to have plummeted to almost zero. Coupled with the many fatalities inflicted by wind turbine farms (it's now known that bats, not birds, are the main victims of this rapidly growing energy sector), the cumulative effect on ecosystems in some areas may well be exponential, and wildlife biologists and governments were scrambling to deal with the threat.

While this highest-level biological emergency made headlines in eastern areas of Canada and the United States, the arrival of WNS in the west of the continent was being treated as "likely to inevitable," according to ever-vigilant Purnima Govindarajulu of British Columbia's Ministry of the Environment, who reckoned the disease would make its way to the Pacific coast in five to ten years.

Taking on a different kind of fight from that waged against invasive American bullfrogs, her agency in 2013 set in motion five priority actions to manage WNS in Canada: national coordination, population monitoring, surveillance, mitigation where present, and promotion of research. While it was assumed that WNS would eventually arrive on the Pacific coast, the timing couldn't be predicted with the linear distance models used in the east due to a lack of data on the extent of bat-to-bat contact across mountain ranges, the potential for human-vectored transport, the large knowledge gaps in bat winter range and behavior, and the role of climate change in the establishment and spread of WNS.

Some help came in December 2014, when a research expedition netted seven individuals of the hitherto elusive spotted bat near Lillooet, about 100 km north of Whistler in the arid Fraser River Canyon. The bats were fitted with small radio transmitters to track their movements and yield data on roosting and foraging habits, information that could help researchers better prepare for WNS.

"We're feeling this urgency to learn more about our bats before this devastation hits," Govindarajulu told Whistler's *Pique Newsmagazine*. "We know [the disease] kills in the winter, but if we don't have background baseline knowledge, we can't formulate management strategy."

With British Columbia home to at least sixteen bat species—the most of any Canadian province—losing an entire tier from local ecosystems would be devastating. "The ecological and economic value of bats is enormous. A small brown bat weighing less than ten grams can eat a thousand mosquito-sized insects per hour," explained Govindarajulu. "That's a huge amount of pest control we get for free from a healthy ecosystem. So if bats disappear, it's going to have an impact through insect destruction of plants in forestry and agriculture—everyone's going to feel it."

By 2016, in fact, it was calculated that a loss of bats was causing U.S. agriculture almost $3 billion a year. But while the ravages of WNS were indeed shocking, you could hear similar doomsday scenarios at virtually any scientific meeting you cared to attend. For my own part, being of the herpetological persuasion, I'd already spent a decade sitting through such sessions, only to be continually shocked—but never surprised—by the new data, interpretations, and implications found at every turn.

## New Normal

First there was the doom: a raft of presentations documenting the impacts of invasive species, clear-cutting, toxic chemicals, parasites, disease, and the time-bomb of climate change. Then the gloom: a collective realization among 1,700 scientists from forty-eight countries that, acting in concert, these forces had produced an apocalyptic symphony, and the decades-long mission to document, understand, and preserve biodiversity among the world's 17,000 or so species of amphibian and reptile was now an act of global triage.

Such was the mood blackening the hallways at Vancouver's University of British Columbia in August 2012 during the Seventh World Congress of Herpetology. Having attended the very first World Congress held in 1988 at Canterbury, England, as a wide-eyed doctoral student, and similar gatherings over subsequent decades as researcher and journalist, this week-long affair was remarkable for its melancholy—contra the typical *joie de vivre* of an oddball fraternity that had happily served as a model for the nerdy kids and pith-helmeted scientists famously sent up in Gary Larson's *Far Side*.

Missing was the puffing, posturing camaraderie that stems from studying and sometimes surviving animals that are the stuff of other people's nightmares, a badge that herpetologists wear as proudly as they do their custodianship of two separate animal classes (a unique conflation that goes back to Aristotle's low opinion of these creatures). But all of this masks more legitimate honors: as a driving force behind the molecular methodologies of modern taxonomy and population analyses, herpetologists have long claimed the forefront of conservation genetics and ecology. It was an unfortunate irony given the theme dominating the 2012 World Congress: collectively, their study organisms were now the most endangered vertebrates on earth.

The lugubrious tone was set by Tyrone Hayes of the University of California at Berkeley, in an opening plenary address on the effects and impact of the herbicide atrazine—80 million tons of which is dumped on U.S. corn, sugar cane, and golf courses each year. Dogged research on the endocrine-disrupting chemical that was now widely present in both rain and tapwater had rendered Hayes both an industry enemy and a genuine celebrity, with a day named in his honor in Minneapolis and a film collaboration with Erin Brockovich. Studies show that atrazine depresses reproductive function—sterilizing, demasculinizing, and even feminizing males across all vertebrate classes, including humans. Like many toxins, atrazine exposure also depresses immune function, increasing susceptibility to disease. Hayes's central thesis was that "a" cause of amphibian declines did not exist; rather, there is a collective culprit, with chemical contaminants like atrazine interacting with a host of other factors.

"It's not like we've done just one thing here, but far too many—and now interactions are the problem," Hayes offered. "Pathogens, parasites, habitat modification, invasive species, and climate change all serve to increase the effects of pesticide exposure."

These findings emerged just as ecotoxicology—a difficult field requiring foundations in ecology, physiology, and chemistry—was being hampered by a lack of funding. "There are serious declines in support and interest," noted Donald Sparling of Southern Illinois University in his own talk on the global impacts of contaminants, "and less money means less research." Translation: when it comes to attracting money, feminized frogs aren't as sexy as dead ones.

Funding cuts and an anti-science political climate again loomed large, but there was more. The congress's numerous symposia, comprising over a thousand studies, were dominated not by the usual ecological and evolutionary topics, but by reports of battles with invasive species, conservation skirmishes over captive propagation, translocations and reintroductions, and all-out wars on disease. This litany added up to the phrase "New Normal," which was bandied at every coffee break: in other words, these were no longer ephemeral perturbations, but ongoing problems accelerating rapidly to an unknowable end game. Herpetologists had ceased believing that they could safeguard biodiversity and were, instead, merely coping with the disasters at hand, hoping, like Govindarajulu, to learn enough to deal with an imminent ecological implosion.

Various invasive species fires that had long smoldered—for example, worldwide incursions of cane toad and bullfrog that had decimated native amphibians and altered local food chains—had fanned into full-blown conflagrations by the time of the World Congress. Ditto the swarm of alien Burmese pythons in South Florida; more novelty than threat a decade ago, they were now demonstrably responsible for the virtual disappearance of small mammals from the Everglades and a proven predator of several critically endangered taxa. Other talks focused on how invasive fire ants, crayfish, fishes, rats, treefrogs, and lizards were spreading disease and radically changing ecologies everywhere.

On the disease ledger, chytrid still led the way. The microscopic fungus plowing through the world's frogs—primarily and perplexingly in remote and pristine areas—had been a topic at such meetings since its identification in 1998. Although this plague was the primary actor behind the decline and disappearance of around 40 percent of the planet's 7,500ish amphibian species, the New Normal turned the received view on its head. As Sparling demonstrated with a meticulous meta-analysis, pesticides applied in lowlands like California's Central Valley—which grows 50 percent of America's food—were likely being volatilized and carried aloft to be deposited into otherwise pristine mountain ponds. High-altitude areas in the Sierras had indeed experienced massive frog die-offs, and this "montane paradigm" also pertained to Central America, the Andes, and eastern Australia, areas of the worst chytrid disasters. "Most amphibians currently in trouble live in an area that fits a model of oceanic effects from lowland agriculture meeting a mountainous region," Sparling concluded.

Furthermore, forensic DNA studies had revealed that the *Bd* pathogen has long co-existed with many amphibians, but new strains had emerged and/or been moved around by invasive species to increase susceptibility to it by already immune-compromised and chemically stressed frogs. As Hayes suggested, the seemingly unstoppable epidemic likely didn't have a single cause—nor so other emerging infectious diseases posing significant new threats: ranavirus and chelonian herpesvirus outbreaks in amphibians, reptiles, and fishes in the Americas, Europe, and Asia featured mortality rates that could exceed 90 percent.

The theme was picked up again at a September 2015 meeting of the Canadian Herpetological Society in Saint John, New Brunswick. Maria Forzán, a wildlife pathologist with the Canadian Wildlife Health Cooperative, began her overview with a look at the effects that ranavirus (a DNA virus of the family Iridoviridae) was having on North American frogs, which featured various gruesome pathologies in adults and tadpole mortalities of 98 to 100 percent that could lead quickly to potential extirpation/extinction. Vectored by various carriers, some of them invasive, ranavirus was cropping up everywhere, transmitted by everything from cannibalism, to infected water and substrates, to skin injuries. When Forzán's talk turned to chytrid, she hadn't minced words: its impact has been the most spectacular loss of vertebrate biodiversity in recorded history. Unlike ranavirus, which couldn't survive long on its own, *Bd* happily lived both in the skin of amphibians *and* in the environment; it liked wet organic matter, and that was a huge problem.

A poster displayed outside the hall documented work by a German graduate student on both ranavirus and chytrid at the northern limits of frog ranges in the sub-Arctic environs of Canada's Northwest Territory: all die-offs occurred in the warmest water, suggesting possible environmental mediation. It meant that climatic warming models—like those that Bogoch crafts for humans—could play a crucial role in tracking and preventing the future spread of these diseases. The student wasn't sure if the outbreaks he'd monitored were irruptions from indigenous pathogens or introduced forms, though he suspected the introduced forms, given the "*huge* propagule pressure from bird transport." The area was located in the main North American Flyway for waterbirds, evoking a sort of malevolent air-mail special delivery.

Forzán finished by discussing a new threat: *Batrachochytrium salamandrivorans* (*Bsal*), an equally lethal relative of *Bd* uncovered by Dutch scientists investigating a sudden, widespread mortality event in the already endangered European fire salamander (*Salamandra salamandra*). It was a familiar refrain: stable and endemic in Southeast Asia, the fungus had traveled and been released into an exotic environment through the pet trade. With all European salamanders threatened, measures were being taken to stop the spread of *Bsal,* including bans on the importation of Asian salamanders and other potential amphibian carriers. When the fungus was tested on North American members of the family Salamandridae (genera *Notophthalmus* and *Taricha*), they proved extremely susceptible. Furthermore, North America contained the bulk of the world's salamander diversity in five other families representing some 190 species—almost 40 percent of all known salamanders.

Given the threat level posed by *Bsal,* a group of scientists who'd been immersed in the *Bd* fight and witnessed many a frog extinction went to the extraordinary length of penning an open letter to Congress in the *New York Times* calling for an immediate ban on salamander imports from both Europe and Asia. Nothing happened. They wrote another letter. And though Canada's national pet-trade organization rallied members to endorse a temporary self-ban on the importation of Asian salamanders and newts, nothing seemed to be moving in the United

States, where economic interests had a tendency to outweigh common sense (from 2004 to 2014, an estimated 2.5 million live salamanders representing fifty-nine species were imported into the United States). Then welcome news: on January 12, 2016, the U.S. Fish and Wildlife Service announced an impending official ban—under the still-powerful Lacey Act of 1900—on the importation or movement across state lines of 201 salamander species that could potentially serve as *Bsal* carriers. It was a bold move, lauded by those who'd been calling for a ban. "This is pretty impressive," the University of Maryland's Karen Lips—front and center in the *Bd* fight and leading the *Bsal* charge—told the *New York Times*. "We're looking at this as a bit of breathing room."

As Forzán had pointed out in Saint John, *Bsal* seemed happily absent from North America. Researchers had even gone back to check swabs done on salamanders for *Bd* years before to analyze them for *Bsal*, and all came back negative. Still, without an official ban in place, optimism was in short supply. "Because *Bsal* likes muck, it will be just like *Bd* was for frogs in susceptible places; if it gets loose in an area you won't be able to stop it—it will stay there forever," she said.

Things couldn't have been gloomier when Matt Allender of the University of Illinois took the podium, but they quickly got worse. "My life these days," he began, "is ruled by snake fungal disease."

*Ophidiomyces ophiodiicola*, an emergent, debilitating, and often deadly disease, had now been seen in some fifteen species of snake from various habitat types all over the eastern United States, with the first clinical cases in Ontario and Europe recently reported. It seemed present in all months but particularly in the spring and fall. After museum specimens going back to the 1800s were looked at, it seemed that snake fungal disease had emerged in the year 2000. The transmission route was unknown but thought to be through direct environmental contamination of soil. It was possibly contagious. There were many questions. Too many.

While the audience reeled, Allender moved onto the more troubling ravages of ranavirus in wild and captive chelonians (turtles, tortoises, and terrapins), already one of the most threatened animal groups on earth. Ranavirus, an emerging worldwide disease going back ten to fifteen years, was expanding its range, and human-induced spread had been documented. It was transmitted through ingestion of infected material and by blood-feeding parasites like mosquitoes. Mortality was greater than 80 percent in a transmission study in red-eared sliders, and a particularly sensitive species, the eastern box turtle, was being virtually wiped out by it; specimens could present with sudden onset of severe illness, followed by death within hours or days. Allender cited recent work that showed how one strain—RV3—easily jumped classes, from fish to turtles to frogs and back to fish. Terrifying.

Allender kept his presentation entirely descriptive, but I knew biologists and, faced with such a crisis, imagined that other questions were doubtless being asked; that is, the clinical information he had on offer was unlikely to be the only data available. During the ensuing Q&A, I asked if, given the invasive nature of both chytrid and ranavirus, he was willing to speculate on the origin of snake fungal disease.

"Well, I believe the *organism* is indigenous," he began. "It has probably always been here."

"Can you go back and look at soil samples from before 2000 to see if there was a non-pathogenic strain?" I softballed.

"When we get the soil eDNA working," he batted back.

"And what do you think you'll find?" I'd pressed, making a harder pitch to telegraph my thoughts: nothing became that virulent without a serious trigger. He caught my drift.

"Well ... if I answer that it means introducing something else entirely."

"That's why I asked."

"*Ophidiomyces ophiodiicola* seems to be resistant to the two most common agricultural fungicides in use in the U.S. ... and it shouldn't be."

The statement cast a pall of silence because the implication was clear: in the same way that overuse of antibiotics leads to antibiotic-resistant superstrains of bacteria they target, antifungals might have here turned an otherwise harmless soil fungus into a deadly pathogen—by repeatedly challenging the organism to produce resistant mutants. This, as fungal invasions were increasingly recognized as a significant component of global change that threatened species survival, ecosystem health, and food production.

Fungal invasions have long been overlooked both because they tend to be inconspicuous, and because inappropriate methods are often employed to identify them. But they'd now reared up on numerous fronts, leading to dire prognostications for the future.

## Metastasis

Invasion. Spread. Uncontrollable growth. Self-limitation. When you look at it that way, it seems inevitable that someone would eventually investigate parallels between invasive species and systems that follow similar first-order growth functions—virus proliferation in a cell, disease through a population, cultural memes through social media. It's less intuitive that anyone would zero in on long-distance dispersal and colonization paradigms to compare the spread of invasive species with metastatic tumor growth, but indeed someone has and the information is compelling. In the first study to demonstrate that metastasis is an ecological process, researchers used computer modeling, photomicrography, and aerial photography to compare the spread of an invasive tree with that of glioma, an aggressive brain tumor.

The premise was simple: ecology meets oncology in the seed-in-the-wind dispersal of cancerous cells over long distances within a body. In computer models and growth simulations, malignant tumors and invasive species both spread according to scale-invariant power laws in which objects or populations of vastly different size may be directly compared (and here, you might consider a parallel utility of the good ol' Invasion Curve), with their similar processes occurring at different spatial scales (for example, kilometers for trees versus microns for cancer cells).

Evidence from crop pathogen studies showing that such power laws govern the spread of invasive species led the authors to suspect these same laws might govern tumor growth. Certainly tumor-modelers had, in the past, used mathematics to demonstrate how fractal geometry was both characteristic and predictive of metastatic tumors. In this study, the same fractal signature was observed both for uncontrolled cell growth in digital photomicrographs of stained tumor sections and for the gradual spread of English elm (*Ulmus minor*)—an ornamental introduced to Ar-

gentina in the mid-twentieth century—as visualized through aerial pho-
tographs covering a forty-year span (in the period 1987–1997, for instance,
74 elm patches blossomed into 189 patches scattered across a much larger area). The shared spatiotemporal signature includes a unique patch-
work growth pattern, a distinct geometry along growth boundaries, and
a comparable distribution of propagules (seeds or cells), indicating that
the early spread of each is similar; calculations yielded an almost perfect
model fit between the dispersal dynamics of both.

Beyond simply representing analogies between different systems,
mathematicians who research cancer have found several ecological
models useful for generating testable hypotheses and intervention strategies. The invasive species comparison likewise generates a different
ecological model for metastatic cancers, eliciting hope that imaging
technologies might one day be able to use the underlying rules that drive
the geometry in both to enable the direct observation of early spreading
tumor cells. Reversing the idea for the utility of invasion science, those
same geometric signatures should be easy to divine on a landscape. Especially with the photographic capabilities of a new generation of small,
affordable, hand-controlled drones, a tool increasingly employed in a
range of studies—everything from mapping the spread of invasive
earthworms via the proxy of tree-crown physiology, to examining the
distribution and microhabitat preferences of invasive American bullfrogs in ponds shared with the highly endangered Oregon spotted frog.

Finally, as both invasive species managers and oncologists can at-
test, once either system reaches the exponential growth phase (middle
of the Invasion Curve), it's far too late to address the problem simply by
taking out the source—whether that means a primary tumor or parent
tree(s). But making a threatened habitat more *resistant* has, in some
cases, worked with invasive plants and animals, and may do so for cancer. The authors suggest that making metastasis-targeted tissues more
invasion-resistant, or less likely to foster drug resistance, could be
achieved through ecological pressures selecting for a new genetic trait.
Certainly it's known that cancer cells arriving by metastasis to a new
habitat (that is, new tissue) often establish populations with a different
genetic composition than that of the primary tumor. Shades of knotweed, *Phragmites,* and dozens of other invasive plants.

That both systems share many traits opens up at least one fortuitous research frontier stemming from the understanding of two malevolent but separate phenomena.

## Context-Dependent Context Dependence

"Maybe I should have my student look at some of the world's other big lakes to see how the morphospace thing works there."

Nick Mandrak is horizontal on a couch in a hotel room in Kelowna, British Columbia, feet extended over one armrest, hands behind his head on the other, thinking out loud. From all appearances, it's going well.

He'd traveled here for a gathering of the freshwater fishes section of the Committee on the Status of Endangered Wildlife in Canada (COSEWIC). The group met annually to work their way through various species-at-risk evaluations and make listing recommendations that would be passed to the federal minister of the environment for approval. Unfortunately, with an election under way, they'd be forwarding any such counsel into a limbo whose terminus was contingent on the speed at which the next government—old or new—could organize itself.

"And why, exactly, would that be," I prod, hoping to clarify Mandrak's morphospace train of thought without derailing it.

"Well, she's using the approach to look at context dependence for invasion potential between different Great Lakes basins, but now I'm thinking it would be cool to look at fish communities this way in several large freshwater lakes around the world for comparison. Like all those bizarre cyprinids in Lake Khanka. You'd predict that for an invasive species to do well there, it would have to have shape and growth-rate characteristics that could help it quickly exceed the gape size of the lake's two major predators—Amur pike and northern snakehead. Even Khanka's goldfish look like different ecomorphs; they have deeper bodies as well. I think there's something there."

What sparked this was yet another conversation about his long-ago visit to Lake Khanka with E. J. Crossman, the goldfish soup he'd daily been forced to eat, and the silver and grass carp he'd seen that may or may not have been indigenous. "If they *aren't* native, one reason for their success might be fast growth and deep bodies," he says.

As he'd mentioned to me previously, Mandrak felt strongly that morphospace work with invasive fishes needed to consider separately the effects of shape on the probability of establishment and the effects of shape on impact—that is, once an invading fish established, could morphospace predict whether it would compete directly or partially with indigenous species, and/or eat them? As a pioneer of trait-based models for invasive fishes, Mandrak's latest idea was to look at variable effects of the quality and quantity of data on predictive abilities. Presenting this at a spring 2014 conference in Spain, his talk concluded that various models developed for the Great Lakes using different trait and data sets all predicted invasion success at the same rate of about 80 percent—though each identified different traits as responsible.

That different traits were important in different lakes was of critical importance, pointing directly to the arbiter that Mandrak was fond of circling back to: context dependence. This time he was both more animated and more adamant as the idea took on greater import for him.

"I think context dependence is *the* nut that needs to be cracked in invasion biology," he asserts. "If you already think that probability of invasion is niche-based, then it *has* to be context dependent. The morphospace idea is inherently context dependent because it takes into consideration the presence of other fish. Trait-based models only look at successful and failed invasions in the context of the invader—ignoring the traits of the native species."

Along with Tony Ricciardi—who is equally bullish on context dependence—Mandrak co-supervised a doctoral student looking at invasive tench (*Tinca tinca*), a large cyprinid native to Europe and western Asia that had become sporadically established in the United States. Sometime in the early 1990s, it escaped a Quebec pond where it was being illegally farmed and swam into the Richelieu River, a tributary of the St. Lawrence. Somewhat miraculously, tench had yet to make it downstream into the St. Lawrence and thence westward to Lake Ontario or eastward toward the upper reaches of the Gulf of St. Lawrence (it can tolerate brackish water), but by the new millennium it had made its way to the Richelieu's source, Lake Champlain, which divides the northern reaches of Vermont and New York State. Tench prefer lakes and slow-moving waters with abundant vegetation and muddy bottoms from which they vacuum small mollusks and insect larvae in quantities that can lead to local turbidity and algal blooms, as well as competitive exclusion of native species. All that mucking around was fine with the tench—it could live in water from 0° to 24°C with oxygen levels lower even than those tolerated by carp. Currently there were invasion fronts in Lake Champlain, eastward toward Quebec City, and west to Montreal.

"What we'd like to know is how widely introduced species like tench inhabit different ecosystems around the globe—do they occupy the same ecological roles in different systems or do those roles vary? It won't surprise you if I say context dependence suggests the latter. And that's something we can look at, in this case through stable isotope analysis that might give you an idea of the trophic position or niche breadth of tench in different places it has invaded."

He offers the example of an April 2015 paper comparing stable isotope signatures in widely distributed and abundant round goby (*Neogobius melanostomus*) with those in more locally established, less abundant tubenose goby (*Proterorhinus semilunaris*) across their invaded ranges in the Great Lakes. The authors found that the broader, more plastic isotopic niche of round goby provided a basis for its wider establishment, and believed this might be applicable to many other invasive fish species. Mandrak also mentions a presentation he'd recently sat through that addressed the sacred cow of global niche conservatism in species, in which the researchers demonstrated variation—via plasticity in trophic position in a meta-analysis of fifteen species—suggesting these might not be so conservative after all.

Now the *Duh!* bell goes off in my brain. How conservative was the niche for a broadly distributed species across its *native* environment? And regardless, wasn't it logical to expect significantly greater niche variation for a species outside its native environment where various water qualities, chemistries, and substrates, as well as planktonic, macroinvertebrate, fish, and non-piscine vertebrate communities combine to create a de facto context-dependent context dependence?

"Absolutely," responds Mandrak. "After that presentation, a South African colleague stood up and said 'So what?' Basically what you're arguing. But looking at things through the morphospace lens may help us put a *reason* to what now merely sounds like stating the obvious."

Mandrak's core interest in morphospace was how it might one day be used in risk assessment. But that was a ways off, and, for the moment, risk assessment remained the biggest can of worms in the invasion game. Since it also seemed the most important, however, I wanted to know why. I mentally recap what I'd learned to date: the goal of risk assessment was to identify all those species likely to become invasive if they were introduced into a specific, exotic habitat or location; if you could do that beforehand, you could regulate the trade or movement of these species in those areas; and if you could regulate different pathways—such as ballast water, baitfish, the aquarium trade—you could prioritize which were riskiest (with new regulations, for instance, ballast discharge was no longer a huge risk; regression models run for ballast risk flat-lined while those for live-trade risk still showed an increase over time).

"You have to evaluate risk in order to make an informed decision about whether to do something or not, but the biggest problem with assessing risk is high uncertainty, so doing nothing might be OK, or it might be problematic. Our maximum entropy model shows that northern snakehead could survive quite far north in Canada, but another model doesn't show a climate match to the Great Lakes; climate-matching models are very sensitive to the type of distributional data you use. A model based on museum records alone, for instance, typically underestimates the degree of native distribution, and that, in turn, will lead to an underestimate of potential distribution *outside* its range. This is where you add in context dependence: with climate you only have coarse-level context dependence in terms of risk of *establishment*, whereas context dependence becomes more important when looking at risk of *impact*," says Mandrak, of a critical distinction not always made.

"Take our bighead and silver carp risk assessments for the Great Lakes. We look separately at the risk of establishment and impact for each lake, so that's already context dependent to a certain extent. And we just completed the risk assessment for grass carp, noting that each lake has various levels of remaining wetlands—the areas we expect them to impact. So with that, the spatio-temporal scope of the risk assessment becomes an issue. Do we mean *all* of the Great Lakes separately, or collectively? Or just the wetland areas? What if you need sixteen grass carp per hectare to have an impact and you have 100,000 hectares of wetland that would require 1.6 million grass carp to do so?"

I didn't know, and neither did Mandrak. But he was freeballing with passion and he was on a roll. "For most risk assessments uncertainty remains relatively high because there's still relatively little information available for most of the species we assess. This highlights another key point: risk assessments don't just need to be about a managerial decision. They can also be used to inform *research* directions that can ultimately reduce uncertainty. And what's that called? Adaptive management. In cases where uncertainty is high you make a management decision that you agree may change, that you'll revisit as more information becomes available; so it's key to fund the management decision *and* the research that will reduce uncertainty. And this is where we hit a wall. Politicians will of course say 'OK, we'll fund a managerial decision but, uh . . . there's no money for research.'"

In other words, with no specific suite of traits or methods that can be reliably employed to predict all invasive species establishments and impacts, no one upstairs wants to be held responsible. The most realistic approach, then, might involve circling back to something of universal utility—like morphospace, where we'd begun. But even that avenue, promising as it was, may already have been politicized. Not by governments, but by other scientists.

## Pride and Prejudice, Pattern and Process

At this point my discussion with Mandrak, always wandering, veers sharply, and we find ourselves addressing the flow of information on invasives, and the changing face of scientific communication in general—the kind of bitch session we might have shared with fellow grad students sitting around our office back in the Royal Ontario Museum days.

"If you look at all these high-impact papers coming out, you're mostly getting patterns, and post-hoc rationalization about why those patterns exist, but *process* is the detail you need to excavate," says Mandrak. "The scientific method involves observation, theory, prediction, and experiment, and these patterns are only first-order observations. I don't know if you remember, but when we were students there was a real knock on pattern papers—basically we were told you couldn't understand pattern until you knew something about the process behind it and then you'd publish on both."

Indeed I did remember—as well as railing against the overly reductionist approach in my own work.

"Take climate change and thermal tolerances," Mandrak continues. "The literature is based on 1950s experiments using juvenile fishes in test chambers. A colleague has fifty years of data on smallmouth bass from Algonquin Park that in no way match those reported in the literature. The point, again, is pattern versus process and thinking about details. So, thermal preferences some researchers use to predict climate change response in some species? All wrong. When I read pattern papers I immediately think 'fluff' because there's no process there."

Pattern papers, according to Mandrak, increasingly clogged high-impact journals while the papers that sought to understand process were seen as messy details that editors didn't want to trifle with because they weren't as clear. This problem, in part, was symptomatic of the digital age. Unlike the previous slow-motion dissemination of science via print periodicals—easy to follow but tedious to track—scientific discoveries are now, with the rapid, almost instant communications of today's online journals, tedious to follow but easy enough to track. Scientists are now so overwhelmed by information, with the number of peer-reviewed studies increasing exponentially along with the online vehicles they're published in, that they can never keep up. More so than before, journals and the societies or organizations publishing them compete for attention. In an age of big ideas, who'd publish the next one? A high-impact journal like *Ecology Letters* could run with the novel morphospace study and guarantee that it would generate readership and citations. Unfortunately, and often due to sheer volume, papers cited first in a subject area were the ones people tended to return to—so they continued being cited to the exclusion, Mandrak felt, of better studies that had been quickly buried because they weren't as rapidly communicated or headline sexy. In our current culture of digital "popularity," both attention and tactics figured heavily.

"For students getting PhDs right now it's all about 'impact factor.' They tweet out the papers they like, and then other Twitter-savvy scientists are more likely to cite those same papers. That's just how information flows these days. I was the last person I know to get on Google Scholar because I was basically forced to—but as a result I'm getting hundreds of download requests and those papers are going to be cited. So the easier it is to find a paper online the better; if it's not easy, or the search terms don't work well to find it, it won't be cited."

In other words, if all digital roads don't lead to your work, is it even there?

## The Aliens Among Us

Mention to anyone that you're writing a book on invasive species and both zeit and geist bubble up from the conversational woodwork: I saw one last summer . . . Oh, in our backyard . . . Yes, at the cottage . . . On our street . . . Along the highway . . . At the lake . . . In the news. Michael Colley, for instance, a herpetologist friend and avid duck hunter, responded by happily recalling for me his undergrad thesis at the University of Waterloo: a look at hunter-mediated seed dispersal in an Ontario marsh. He'd collected data by the simple method of combing the clothing of fellow hunters. And he'd found exactly what a raft of larger empirical studies on the detrimental effects of anthropochory (for example, among tourists visiting sub-Antarctic islands) had also found: hunters moved seeds of all sorts, and by dint of carrying the greatest density, cotton clothing had the greatest potential to increase dispersal rates of invasive plants. He'd recommended that hunters be educated to have a basic understanding of invasive species, as well as to properly clean clothing to reduce their impact. His was a de facto vote for synthetic camouflage.

Voyage anywhere and you'd also confront the issue. As I write this I've been traveling in Southeast Asia for two months and have yet to experience a single day—including a trek to Everest Base Camp at 5,400 meters in Nepal's high Himalaya—free of invasive species. While I expected introduced species in human-occupied landscapes, in particular those with high levels of international traffic, I didn't anticipate that they would be so obvious or pervasive. And yet there they were, *everywhere*: pigeons; English sparrows; Indian mynah birds; eastern gray squirrels; water buffaloes; house geckos; red-eared slider turtles; yellow crazy ants; feral dogs, cats, pigs, and cattle; Japanese rose; various eucalyptus; a dozen varieties of pine, kudzu, *Phragmites*—frickin' phrag no less, crowding out native shoreline vegetation along Myanmar's mighty Irrawaddy River—and endless tracts of tamarind dominating every quasi-arid landscape. (You may recall reporters catching up with President George W. Bush while he "relaxed" at his Texas ranch by hacking away at invasive tamarind with neoconservative determination.)

It took little digging to find a first-of-its-kind 2013 study out of Singapore, which concluded that invasive species were costing Southeast Asian countries some $33.5 billion USD each year, a figure that experts felt was only scratching the surface. Costs related to human health and the environment were $1.85 billion and $2.1 billion, respectively, but the agricultural sector was the one taking the biggest hit, accounting for $29.3 billion (90 percent of losses). While far less than the annual $120 billion USD loss to agriculture in the United States, it was enormous relative to the region's economy, meaning the impact of nonindigenous species in Asia was actually greater.

An invertebrate encountered with regularity in many Asian countries is the giant African land snail, an acknowledged problem in Central America and the United States that recently achieved something John F. Kennedy couldn't manage in 1961—an invasion of Cuba. These snails grow to twenty centimeters in length and come with the usual winning invasive character suite: ongoing reproduction, high fecundity, high hatch rate, few predators, and an indiscriminate palate for hundreds of plant species, including agricultural crops and fragile varieties indigenous to relatively small regions like islands. They tend to outcompete native snails and other small animals for fodder, leading to a drop in biodiversity. And while elsewhere they had been purposely introduced or were escapees of the pet trade, they had appeared "mysteriously" in Cuba—though it was really no mystery to invasion scientists. The snail easily hitchhikes on cargo ships in a range of goods, a proclivity shared with numerous other organisms and one that recently precipitated an interesting watershed for a cherished principle of ecology.

In a 2014 study in *Nature,* Matthew Helmus of Vriji Universiteit in Amsterdam led a team that tracked lizard movements around the Caribbean in a test of the revered MacArthur-Wilson theory of island biogeography. First described in the 1960s by Robert MacArthur and E. O. Wilson, the theory predicts that the richness of an island's flora and fauna is determined by two major factors—island location and island size. More isolated islands, regardless of size, are less likely to receive migrants from neighboring areas; nevertheless, larger islands offer a more diverse range of habitats for colonizing species, or those forms that subsequently speciate *in situ* from an immigrant stock (think finch diversity in the Galapagos; honeycreepers in Hawaii). Taken in concert, the two factors predict that large islands located close to shore will be home to the most species, while smaller, more distant islands will be home to the least. In 1969, Wilson, then at Harvard, and his then-student Daniel Simberloff (yes, *the* "tireless D. Simberloff"), set out to test the theory in what would become one of the most famous experiments in ecology. Fumigating a series of small mangrove islands in the Florida Keys to kill off their arthropods, they then monitored the species' return. As predicted, the largest islands closest to the mainland recovered their faunas most quickly, immortalizing the study in ecology textbooks and cementing island biogeography theory as one of the most common approaches to predicting species number in a given area. Elegant as this microcosm was, the theory long defied testing at larger scales. Helmus thought that a spike in successful colonization by nonnative species of *Anolis* lizard that paralleled an increase in international trade over the past few decades (analogous to experimental manipulation of a key variable in the island biogeography theory) could provide just such a test.

Prior to human occupation or ships, *Anolis* species moved around the Caribbean by hitchhiking on floating logs or vegetation mats, which molecular data suggest was accomplished once every 100,000 years or so; the rate at which nonnative *Anolis* now arrived on islands is about one *per year.* They accomplish this feat much like snails, ants, and other organisms do—by stowing away in plant and food cargo. According to a corollary of the island biogeography theory, a rise in migration rate (which is roughly equal to propagule pressure) should lead to the greatest increase in newly established species on those islands that bear the fewest varieties to begin with.

Working with Harvard evolutionary biologist Jonathan Losos, who has studied *Anolis* since the 1980s, Helmus mapped patterns of their recent spread by combining existing field data with those gathered from herpetologists and local lizard aficionados. The outcome matched the prediction—islands with the fewest natives gained the most exotics. Volume of shipping trade, however, was a better predictor of this diversity than was location, suggesting that island biogeography theory now only works if researchers also add a human factor to the equation—in this case replacing geographic isolation (distance between islands) with a measurement of economic isolation (proximity to shipping lanes).

Along with several previously mentioned post-hoc investigations—for instance, how fire ants spread around the globe—the study represents perhaps the keenest cutting edge of invasion biology: the integration of human autoecology into the broader synecology of a region. As the science matures, researchers will be able to focus on understanding ecological patterns *and* processes at broader and broader scales. In a 2015 paper, Belinda Gallardo and her colleagues set out to determine the global consistency of impacts of aquatic invasions across taxa and habitats. Their meta-analysis of 151 publications covered 733 individual cases and a wide range of invaders (primary producers, filter feeders, omnivores, predators), resident community components (macrophytes, phytoplankton, zooplankton, benthic invertebrates, fish), and habitats (rivers, lakes, estuaries). Unsurprisingly, the synthesis suggested a strong negative influence of invasive species on abundance in aquatic communities, particularly macrophytes, zooplankton, and fish. It also offered insight into the conundrum of time lags and the nature of impacts over time, finding these to be relatively consistent across habitats and experimental approaches. They concluded by proposing a framework of positive and negative links between invasive species at four trophic positions, and by identifying five different components of recipient communities (context dependence!) that might be used to anticipate the furthest-reaching consequences of invasion on the structure and functionality of aquatic ecosystems.

Ecosystem-level approaches are proving key to understanding and dealing with many problems, invasives included. Take the wolves of

Wyoming's Yellowstone National Park. Since the wolves' reintroduction in the mid-1990s, their effects have rippled across the entire structure of the food web that defines biodiversity in Northern Rockies ecosystems, demonstrating how ecosystems are structured neither by species nor diversity per se, but by species *interactions.* Competing ecological interpretations of these effects have generated a significant amount of debate over the relative strength of top-down versus bottom-up forces in determining herbivore and vegetation abundance (including irruptive scenarios), and, moreover, the role of natural enemies and climate as forces that structure food webs and modify ecosystem function.

Such reintroductions dwell under a branch of conservation biology known as "restoration ecology." In practice for over a century, reintroduction has been recognized as a bona fide tool only since 1988, when the International Union for the Conservation of Nature (IUCN) tabled its first guidelines on the subject. As reintroductions became more popular, however, so rose the risks to other animal populations. By 2009 it was clear that the IUCN guidelines needed to be revised to include our better understanding and practices around disease and invasive species threats. New guidelines dropped in 2013. One of the principals behind them was Axel Moehrenschlager, whom I'd interviewed on several occasions about the subject.

Co-chair of the Reintroduction Specialist Group under IUCN's Species Survival Commission, Moehrenschlager oversaw an active stable of graduate students and technicians between his posts at the University of Calgary and the Centre for Conservation Research that he'd founded at the Calgary Zoo, where he'd worked on everything from frogs to whooping cranes to Vancouver Island marmot to a host of African mammals, as well as local swift fox and black-footed ferret, both of which he'd wrested from the brink of extinction. As I'd gleaned from Moehrenschlager, conservation may be the only legitimate reason to move species around, but there were several different forms this could take: *reintroduction* is the intentional release of an organism inside an indigenous range from which it has disappeared (think Yellowstone's wolves); *reinforcement* is the release into an existing population of the same species (for example, reinforcing caribou to increase minimum functional herd size, or out-crossing cheetahs to avoid genetic bottlenecking); *assisted colonization* moves an organism outside its indigenous range to suitable foreign habitat to avoid extinction (see the upcoming lemur example); and *ecological replacement* is the release of an organism outside its current indigenous range in order to perform a lost ecological function (for instance, sending bison back to Russia after 30,000 years). There's an equal range of associated risk: to source populations and local ecosystems (disease, invasion, and gene escape or mixing are some of the potential problems), and on the human side, to the socioeconomic health of an area. At issue is which of these is relevant in any particular case.

"A primary consideration for species movements isn't necessarily the indigenous *range*, but whether release sites are suitable in terms of climate, habitat, and threats," noted Moehrenschlager. "On one hand these are proactive mitigation tools against extinction; on the other hand, some [Ricciardi and Simberloff fit here] construe them as opening a Pandora's box of moving nonnative species around with all the associated risks."

A high-risk example: British business mogul Sir Richard Branson purchased 120-acre Moskito, one of the Caribbean's British Virgin Islands, intending to create the world's most ecologically friendly island. He subsequently announced plans to introduce both ring-tailed and red-ruffed lemurs from their native Madagascar, an assisted colonization aimed at establishing wild populations that might guarantee these species' survival against deforestation threats in their homeland. The plan is benevolent and right-hearted, yes, but you can guarantee that should breeding populations establish they will be mercilessly poached as part of the sizzling global trade in endangered wildlife, with some lemurs doubtless finding their way into the wild once delivered to their illegal destinations.

A low-risk example: Moehrenschlager mentions the ecological replacement of tortoises in both the Seychelles and the Caribbean, where the extinction of these large, endemic grazing and browsing animals has had significant ecological consequences. Replacing these with exotic tortoise species that occupy analogous ecological niches in their indigenous ranges has successfully kickstarted a lost function in local ecosystems without any undue consequences. But as you've just learned, the global rampage of something like ranavirus could make tortoise introductions high risk *anywhere*. Other aspects of restoration ecology are also shifting: a changing climate means that habitats can quickly become suitable for species beyond their indigenous range, and vice versa—a primary thrust in the new guidelines. Invasion scenarios were possible, but, as they were always one of the first things accounted for, are less likely to emerge than other problems.

"We generally haven't had challenges with invasives through reintroductions," said Moehrenschlager, agreeing that some restoration cases still represent a fundamental tradeoff between the benefits of trying to prevent extinction and the potential risk of invasiveness. "The whole approach is a testament to the depths to which conservationists must reach to grapple with the extinction crisis—and to prepare for biodiversity challenges looming on the horizon. Over time these will no longer be exceptions but common dilemmas, so if we went down the road of *never* attempting reintroductions we'd be throwing in a lot of towels."

Not that Moehrenschlager didn't worry about invasions. He strongly believed, depending on the time frame (and he'd roll that back to the 1600s, the Age of Exploration), that invasive species, and not habitat loss, were *the* main reason for extinctions. And following the ideas outlined so far in this part of the book, he was looking at the bigger picture. "There's a whole scale beyond the salvation of individual species—like rewilding initiatives. I can't even tell you what that is though because the definition keeps shifting. It used to be this ludicrous thing about de-extinction—bringing saber-tooths and mammoths back to North America. The term de-extinction is weighted with religious connotations like resurrection, but that's not what rewilding aims for anyway. It's about ecosystem restoration. Or creating whole ecosystems, or ecological engineering—or something beyond. Maybe rewilding doesn't even know what it is."

Perhaps, but many who are convinced they do know are embracing it. Respected journalists and authors like J. B. MacKinnon and George Monbiot have both penned insightful, like-minded books on the subject. For MacKinnon, a realization that the prairie he grew up on wasn't the pristine grassland he'd always believed but rather the outcome of serious historical tinkering—that, along with other ecological transgressions, had vanquished the grizzly bear in favor of introduced cattle—started him on the path to *The Once and Future World: Nature as It Was, as It Is, as It Could Be.* "Picture the first place you thought of as nature. It is an illusion that has in many ways created our world," he writes, perfectly capturing my own sentiments inculcated over the decade that I've followed invasive species. In Monbiot's *Feral: Rewilding the Land, the Sea, and Human Life,* the author is bullish, among other initiatives, on the reintroduction of beaver, an extirpated keystone species in the United Kingdom, and isn't afraid to lay on the definitions, even when they keen to the complex: "Rewilding recognizes that nature consists not just of a collection of species but also of their ever-shifting relationships with each other and with the physical environment. It understands that to keep an ecosystem in a state of arrested development, to preserve it as if it were a jar of pickles, is to protect something which bears little relationship to the natural world . . . Rewilding, to me, is about resisting the urge to control nature and allowing it to find its own way. It involves reintroducing absent plants and animals (and in a few cases culling exotic species which cannot be contained by native wildlife), pulling down the fences, blocking the drainage ditches, but otherwise stepping back. At sea, it means excluding commercial fishing and other forms of exploitation. The ecosystems that result are best described not as wilderness, but as self-willed: governed not by human management but by their own processes. Rewilding has no end points, no view about what a 'right' ecosystem or a 'right' assemblage of species looks like."

Contrary to what apologists claimed, this was fully in line with how most invasion scientists saw things.

Both writers had encountered plenty of invasive species issues in their work and travels, with many cropping up in their respective volumes. And me? I was hard-pressed to work on any story assignment that *didn't* have an invasive angle—climate change in the Arctic (irruptive frogs), paleontological digs in Manitoba (bioturbation of fossil material by feral cattle), and disappearing insects (biodiversity impacts of introduced species). Even looks at beekeeping and winemaking had brought me face-to-face with trenchant invasive problems. One of these involved a trip to British Columbia's oldest continuously producing vineyard, Tantalus, the centerpiece of a growing regional hotbed of organic viticulture.

On a bluebird fall day in 2014, Jane Hatch, general operations and sales manager for Tantalus, met me on the back deck of an airy tasting room that offered sweeping views over vines sloping toward distant Okanagan Lake. I was ostensibly there to talk birds and bugs; Jane liked to chat vines and wines. Since those topics overlapped, we had a fruitful conversation.

In addition to hand-tended vines and no pesticide use, Tantalus was moving toward a fully biodynamic system (Rudolph Steiner's rediscovered ideal of ever-increasing ecological self-sufficiency that views any farm as a cohesive, interconnected living system), acknowledging the importance of local biodiversity. Cleaving the property was a natural dry-land forest, which served as habitat and forage for great horned owl, four species of hawk, and a range of other indigenous flora and fauna. The property's fifty-three beehives were a partnership with nearby Arlo's Honey Farm and, interestingly, weren't here in support of the wine (grapevines aren't reliant on bees for pollination, but other local plants and crops are). In addition to maintaining health and robustness in the natural ecosystem, the hives ensured that neighboring orchard, soft fruit, and vegetable farms remained productive in the face of viticulture. With various degrees of success (everything is an experiment), Tantalus also provided nesting boxes for useful avian insectivores like western bluebirds. Whatever arthropods the birds couldn't control were dealt with by the grape growers. On an organic farm, that could be a tractable problem. Leafhoppers, for instance, were common pests; Tantalus used pheromone cards to trap and kill them and, if an infestation got really testy, soap. But the biggest issue growers dealt with, Jane noted, was the invasive European starling.

In a now infamously hubristic error, the American Acclimatization Society released a hundred starlings into New York's Central Park in 1890–1891, part of a misguided project to introduce all the birds featured in Shakespeare's plays to North America. Smart, adaptable generalists, starlings spread quickly across the continent, establishing in the Okanagan by the early 1950s. A habit of taking over nesting cavities and driving native birds from their territory has landed the starlings on, you guessed it, the 100 Worst Alien Invasive Species list. Because some cry foul when you knock birds, however, it's important to educate the public on why starlings are dangerous pests: in addition to damaging all fruit crops, its dense populations spread diseases like salmonellosis, chlamydiosis, avian tuberculosis, and histoplasmosis. Starlings cause an estimated $800 million USD damage annually to agricultural crops in North America, with conservative estimates in the Okanagan-Similkameen region of over $4 million CAD annually.

Paralleling its growth as one of North America's top wine regions, the number of starlings plaguing the Okanagan was now astronomical, their sky-darkening flocks a familiar late-summer scourge. More fruit being grown means more food. More food equals more happy starlings having more babies than normal (they were already prolific as birds go, with up to three broods a year from which each breeding pair averages eight surviving fledglings). Ergo, wine success has equaled starling success. The dry-land forest at Tantalus offers perches for raptors that keep starlings nervous— but it also offers perches for starlings. To keep the population even at a level termed "obnoxious" required more than a natural fear of the property's raptors. Thus Tantalus employed protective nets over vines, propane cannons, and audio playback of recorded bird distress calls—standard fare throughout the industry. In addition, and more broadly effective, the British Columbia Grapegrowers' Association ran a collective starling control program in the Okanagan to which wineries voluntarily contributed: starlings were netted and gassed up and down the Okanagan in numbers that ran into five figures annually (the high was 88,290 in 2010).

Starlings show up in early fall when grapes are just starting to build up their Brix Degrees (the universal measurement of sugar content employed by oenologists), and, drunk on the prospect of a free-for-all fruit bacchanalia, quickly become aggressive. That explained the cannon explosions I heard every few minutes.

"Robins and flickers can be problems to a lesser extent and they're actually cheekier—you look at them when they're sneaking a grape and they look back like they've been caught in the act. So there's an affinity with them, and we pluck them out of the nets when they get caught and turn them loose," said Jane. "Should we value one species over another? Well, we certainly don't let starlings flounder to death if they get caught in netting but . . . we do pay the fee for eradication efforts. I guess eradication isn't a good word—*control.*"

Checking herself on that recalled the freighted nature of *all* language when it came to the aliens among us.

## Time Traveler

A sampling of daily news and email notifications from autumn 2016: the Zika epidemic continues unabated; zebra mussel are spreading north through Manitoba; the U.S. Army Corps of Engineers is focusing attention on new potential passages into the Great Lakes for Asian carps; in a comic demonstration of Matthew Helmus's biogeography-shipping theory, a Caribbean *Anolis* hatches out of a potted plant in a Walmart in Nova Scotia; on the tragedy ledger, a three-year-old child dies in Victoria, British Columbia, from eating a death-cap mushroom, an increasingly common invasive transplanted on European trees that resembles native edible straw mushroom; in eastern North America, feathers of the northern flicker, whose typical yellow pigmentation derives from the bird's diet during its molt, are turning red as a result of the birds' feeding on the short-lived berry crop of an invasive plant, with implications for sexual selection, fitness, and population stability; and in Melbourne, Australia, four people die and hundreds are sent to hospital after a wild thunderstorm causes the rain-sodden pollen of invasive ryegrass to explode and disperse over the city, sparking widespread asthma attacks—even in many who've never before suffered from it.

With similar litanies now a daily feature of global news, what does the future hold in the absence of money or political will to deal with them? Are the Everglades, Great Lakes, and other unique habitats already lost? Are we destined, as many predict, to witness the complete invasional meltdown of ecosystems we know and love? It's possible, because once an ecological wobble starts, it's often hard to stop, and many invasions show all the signs of such entropy: similar to more familiar atmospheric-, geologic-, and technology-based "natural" disasters, invasions can become self-replicating, their incidence and impacts involving chain reactions and non-linear phenomena.

This we know: with a changing climate come shifting ranges and measurable effects on the distribution and abundance of many native species—whether this means lizards in southern California disappear-

ing because skyrocketing springtime temperatures and plant dieback
are paving the way for invasives to push them out, or an egg-laying lizard from southern Europe establishing itself on a northern Canadian
island where the climate, until recently, wasn't conducive to incubation.
Such scenarios have scientists eyeing climate-driven "invasion sequenc-
es": for example, an aggressive Argentine ant that invaded the southern
United States fifty years ago to wreak havoc and flummox all control efforts is now being displaced by an invading Asian ant that becomes active earlier in spring, constructing colonies and exploiting resources
before the sleepy Argentines awake from their winter siesta.

    With warming oceans, tropical marine species are increasingly be-
coming established in subtropical marine ecosystems. The venomous Pacific red lionfish (*Pterois volitans*), for instance, is out of control in the
waters of Florida, the Bahamas, Cuba, and soon (having arrived in 2015),
the Mediterranean. A danger to unwitting swimmers, this fish, a top predator, is also negatively influencing (well, destroying) fish communities on
host reefs. Yet despite even this stark example, industry-minded governments continue to ignore the lessons of transoceanic introductions—as
seen in the increasing number of diseases and parasites spread to native
fishes by the open-pen farming of Atlantic salmon, which have been
translocated from Norway to British Columbia, Chile, and New Zealand.

    And Florida's Burmese pythons? Still doing spectacularly well,
thanks. An admirable if not exquisite demonstration of evolution's capacity for adaptation—so much so that snake hunters from the Irula
tribe of south India's Tamil Nadu region have been recruited to find the
increasingly wary serpents. Though we've peeked in on a range of representative circum-global invasive issues—many of these fortuitously
discovered in my own unlikely backyard—theoretically we haven't
strayed far from the slow susurrations of those pythons through the
Everglades' sawgrass. And we can add this: climate and habitat models
demonstrate that environmental conditions are currently suitable for
the Burmese python to occupy a much broader geographic range in the
southeast United States; furthermore, models that account for predicted
near-future changes in average temperature and precipitation show that
potential range to be even greater. Global warming could facilitate the
spread of not only Burmese pythons, but also related giants-in-waiting

already established in Florida—reticulated pythons, anacondas, and boa constrictors. Full-time employment for the Irula.

It's a lot to keep an eye on, both today and in the future. But what we've learned thus far from Florida's python crisis will, if nothing else, aid in preventing similar contretemps elsewhere—or at least force us to anticipate their inevitability.

During the herpetology conference in Saint John, running along a waterfront trail, I brushed by virtual walls of invasive tansy. As I paralleled the chest-high monoculture I thought: *Ah—tansy must be the local curse, the invasive everyone here wants to be rid of.* A few steps later, however, rounding a corner under a highway overpass where the trail wound closer to the water, I realized it wasn't true, and what I saw stopped me in my tracks.

A palisade of *Phragmites* lined the water's edge; atop adjacent riprap stood a dense bank of knotweed; giant hogweed ringed a nearby playground. Scattered amidst this was a mix of Japanese rose, blue weed, knapweed, and Himalayan blackberry, the same European and Asian colonizers I'd spent the summer hanging out with some 4,500 km away on the opposite coast. I thought again of what Elton observed in 1958, as prescient then, in the year of my first birthday, as today: "Instead of six continental realms of life, with all their minor components of mountain tops, islands and fresh waters, separated by barriers to dispersal, there will be only one world, with the remaining wild species dispersed up to the limits set by their genetic characteristics, not to the narrower limits set by mechanical barriers as well."

Saint John was Canada's first incorporated city. A British colony. And nothing had changed. It was still being colonized. One world. One love. Or not.

Farther along I came to world-famous Reversing Rapids, so named because of a phenomenon linked to the Bay of Fundy's record fifteen-meter tides. Twice a day, the Saint John River—here tumbling over a deep underwater ledge to cause a downstream series of waves and whirlpools—appears to reverse its flow, rising 4.5 meters to create similarly dynamic riffles and eddies running upstream. Reversing Rapids' most famous tourist was Darwin's geologist buddy Charles Lyell, who visited in 1852 after digging the world's earliest-known reptile out of sea cliffs at the head of Fundy near Joggins, Nova Scotia (a title that 315-million-year-old *Hylonomus lyelli* retains to this day). Pausing, quite literally, in Lyell's footsteps, I considered the rapids' compelling geologic backstory, unknown to him at the time.

The gorge constricting the Saint John River here is a shear zone between two stray continental fragments, or terranes. Although both originated in the Southern Hemisphere, they differ in age by some half-billion years, coming together here as the Iapetus Ocean closed during the final assembly of Pangaea more than 300 million years ago. Conjointly welded to the North American plate, a line of weakness nevertheless remained between the two allochthonous terranes; during later crustal movements, a fault developed along this boundary in which the river now flowed. Finally and most significantly, this very spot was where modern-day Europe and North America said their long goodbye as the Eurasian and North American plates pulled apart during Pangaea's final breakup some 80 to 100 million years ago. The separation precipitated the glorious isolation by which diverging floras and faunas on those two continents would track different vectors through evolutionary time. So how ironic it was that I'd just seen how these two tectonic plates, so distant in time and space, were actually meeting again. As my surroundings attested, you needn't move rocks to rejoin continents. You could do it by proxy—with plants.

## Love, Hate, and Solastalgia

It was a November day after a cold rainstorm, and I was enjoying a last run along Whistler's Valley Trail before a predicted snowfall would close it for the season. In the few places where sunlight sought out fissures in the clouds, the ground steamed lightly. As I ran, deep breathing delivered a post-rain petrichor layered with autumn's metallic tang—dead leaves, pigment breakdown, hints of decay. Dodging busy cross-trail slug traffic—some native, most not—I absently logged other familiar plant and animal denizens of the Whistler Valley, not just a few of them new arrivals as well. Unlike the young adults who flocked here from Australia, Europe, and the United States to work for a ski season or two, these organisms had been displaced through no conscious acts of their own, yet similarly flourished by adapting to whatever parameters their adoptive community imposed. Rounding a corner, the canopy opened and a wall of vegetation crowded each trailside. Here stalks, blades, and tendrils of mostly nonnative plants reached with silent intention toward the intervening gap in hopes of being brushed and stripped of a seed pod or burr. I admired their mute pluck and the evolutionary system that had served them for millennia, one that neither differentiated a worn deer trail from an asphalt vein, nor myself from a black bear. But now I was wise to their ploy and inclined to mind the entreaties. I kept to the center and sprinted on.

Awareness had changed my relationship with invasives. Like Tony Ricciardi, I didn't hate them, but I also didn't enjoy seeing them occupy vast areas when, more pointedly, I knew what *should* be there. I had become actively vigilant, posting problematic sightings to the SSISC, commiserating with frog-finders and broom-bashers, preaching the Gospel of Indigenous in the name of Rehabilitation. Sure it remained edifying to run through a forest or meadow whether I knew what was native or not, but it was no longer possible to ignore it when I did. This was, after all, an untangling bank.

It turns out that there's a word to describe my discomfort, this fleeting syndrome of human distress reflecting a syndrome of ecosystem distress: *solastalgia*. As coined by the Australian environmental philosopher Glenn Albrecht using the landscape modifications of prolonged drought and strip-mining as examples, solastalgia spoke directly to impacts of both local and global environmental change on an individual's sense of place, negative effects that were exacerbated by feelings of powerlessness or lack of control over an unfolding shift. As opposed to the more familiar *nostalgia*—melancholia or homesickness experienced when separated from a cherished locale—solastalgia is the distress of environmental change on those still occupying their home environment.

The unease, however, ran even deeper, as I now found myself questioning how I *felt* on every run. Eventually, as movement is wont to do (*solvitur ambulando*—it is solved by walking—Diogenes said), I was delivered of an answer: despite a strong belief in human responsibility when it came to addressing invasive species, in reality there was nowhere for my mind to settle comfortably on the issue. Like everyone I'd met who was given to thought or study concerning invasive species, I was forever doomed to oscillate between its essential schism of subjectivity and objectivity: they were alien, I was alien; they were nature, I was nature.

And none of us, it seemed, was going anywhere.

# Literature and Other Sources

## Literature Sources

Albrecht, Glenn, G. M. Sartore, L. Connor, N. Higginbotham, S. Freeman, B. Kelly, H. Stain, A. Tonna, and G. Pollard. "Solastalgia: The Distress Caused by Environmental Change." *Australas Psychiatry* 15 (2007): S95–98. doi: 10.1080/10398560701701288.

Allombert, Sylvain, and J.-L. Martin. "The Effects of Deer on Invertebrate Abundance and Diversity." Pp. 87–92 in A. J. Gaston, T. E. Golumbia, J.-L. Martin, and S. T. Sharpe, eds., *Lessons from the Islands: Introduced Species and What They Tell Us about How Ecosystems Work.* Proceedings from the Research Group on Introduced Species, 2002 Symposium, Queen Charlotte City, B.C. Ottawa: Canadian Wildlife Service/Environment Canada, 2008.

Ashworth, William. *The Late, Great Lakes.* Toronto: Collins, 1986.

Azzurro, Ernesto, V. M. Tuset, A. Lombarte, F. Maynou, D. Simberloff, A. Rodriguez-Perez, and R. V. Sole. "External Morphology Explains the Success of Biological Invasions." *Ecology Letters* (2014). doi: 10.1111/ele.12351.

Bailey, John P., and A. P. Conolly. "Prize-Winners to Pariahs—A History of Japanese Knotweed *s.l.* (Polygonaceae) in the British Isles." *Watsonia* 23 (2000): 93–110.

Bassett, Carol Ann. *Galapagos at the Crossroads: Pirates, Biologists, Tourists, and Creationists Battle for Darwin's Cradle of Evolution.* Washington, DC: National Geographic, 2009.

Bogoch, Isaac, O. J. Brady, M. U. G. Kraemer, M. German, M. I. Creatore, M. A. Kulkarni, J. S. Brownstein, S. R. Mekaru, S. I. Hay, E. Groot, A. Watts, and K. Khan. "Anticipating the International Spread of Zika Virus from Brazil." *Lancet* 387 (2016): 335–336. doi: 10.1016/ S0140-6736(16)00080-5.

Bradley, Bethany A., R. Early, and C. J. B. Sorte. "Space to Invade? Comparative Range Infilling and Potential Range of Invasive and Native Plants." *Global Ecology and Biogeography* 24 (2014): 348–359. doi: 10.1111/geb.12275.

Buck, Eugene H., H. F. Upton, C. V. Stern, and J. E. Nicols. "Asian Carp and the Great
    Lakes Region." *Congressional Research Service Reports,* August 6, 2010.
Burdick, Alan. *Out of Eden: An Odyssey of Ecological Invasion.* New York: Farrar, Strauss,
    and Giroux, 2005.
Cera, George M. *Save Florida Eat an Iguana: The Iguana Cookbook.* N.p.: George Cera,
    2009.
Chown, Stephen L., A. H. L. Huiskes, N. J. M. Gremmen, J. E. Lee, A. Terauds, K. Crosbie, Y.
    Frenot, K. A. Hughes, S. Imura, K. Kiefer, M. Lebouvier, B. Raymond, M. Tsujimoto,
    C. Ware, B. Van de Vijver, and D. M. Bergstrom. "Continent-Wide Risk Assessment
    for the Establishment of Nonindigenous Species in Antarctica." *Proceedings of the
    National Academy of Sciences of the U.S.A.* 109 (2012): 4938–4943. doi: 10.1073
    /pnas.1119787109.
Christy, Bryan. *The Lizard King: The True Crimes and Passions of the World's Greatest
    Reptile Smugglers.* New York: Hatchette, 2008.
Coates, Peter. *American Perceptions of Immigrant and Invasive Species: Strangers on the
    Land.* Berkeley: University of California Press, 2006.
Cockburn, Andrew. "Weed Whackers: Monsanto, Glyphosate, and the War on Invasive
    Species." *Harper's,* September 2015.
Conolly, Ann P. "The Distribution and History in the British Isles of Some Alien Species
    of *Polygonum* and *Reynoutria.*" *Watsonia* 11 (1977): 291–311.
Côté, Steve D. "Extirpation of a Large Black Bear Population by Introduced White-
    Tailed Deer." *Conservation Biology* 19 (2005): 1668–1671.
Darwin, Charles. *The Formation of Vegetable Mould Through the Action of Worms: With
    Observations on Their Habits.* London: Murray, 1881.
———. *On the Origin of Species by Means of Natural Selection; or, The Preservation of
    Favoured Races in the Struggle for Life.* London: Murray, 1859.
Dobson, Andy P. "Yellowstone Wolves and the Forces That Structure Natural Ecosys-
    tems." *PLoS Biology* 12 (2014): e1002025. doi: 10.1371/journal.pbio.1002025.
Dorcas, Michael E., and J. D. Willson. *Invasive Pythons in the United States: Ecology of an
    Introduced Predator.* Athens: University of Georgia Press, 2011.
Dorcas, Michael E., J. D. Willson, R. N. Reed, R. W. Snow, M. R. Rochford, M. A. Miller,
    W. E. Meshaka, Jr., P. T. Andreadis, F. J. Mazzotti, C. M. Romagosa, and K. M. Hart.
    "Severe Mammal Declines Coincide with Proliferation of Invasive Burmese
    Pythons in Everglades National Park." *Proceedings of the National Academy of
    Sciences of the U.S.A.* 109 (2012): 2418–2422. doi: 10.1073/pnas.1115226109.
Elton, Charles S. *The Ecology of Invasions by Animals and Plants.* Chicago: University of
    Chicago Press, 1958.
Frelich, Lee E., and P. B. Reich. "Will Environmental Changes Reinforce the Impact of
    Global Warming on the Prairie-Forest Border of Central North America?" *Fron-
    tiers in Ecology and Environment* 8 (2010): 371–378.
Galil, Bella S., F. Boero, M. L. Campbell, J. T. Carlton, E. Cook, S. Fraschetti, S. Gollasch,
    C. L. Hewitt, A. Jelmert, E. Macpherson, A. Marchini, C. McKenzie, D. Minchin,
    A. Occhipinti-Ambrogi, H. Ojaveer, S. Olenin, S. Piraino, and G. M. Ruiz. " 'Double

Trouble': The Expansion of the Suez Canal and Marine Bioinvasions in the Mediterranean Sea." *Biological Invasions* 17 (2015): 973–976.

Gallardo, Belinda, M. Clavero, M. I. Sánchez, and M. Vilà. "Global Ecological Impacts of Invasive Species in Aquatic Ecosystems." *Global Change Biology* 22 (2015): 151–163. doi: 10.1111/gcb.13004.

Gladieux, Pierre, A. Feurtey, M. E. Hood, A. Snirc, J. Clavel, C. Dutech, M. Roy, and T. Giraud. "The Population Biology of Fungal Invasions." *Molecular Ecology* 24 (2015): 1969–1986. doi: 10.1111/mec.13028.

Gotzek, Dietrich, H. J. Axen, A. V. Suarez, S. H. Cahan, and D. Shoemaker. "Global Invasion History of the Tropical Fire Ant: A Stowaway on the First Global Trade Routes." *Molecular Ecology* 24 (2015): 374–388.

Grady, Wayne. *The Great Lakes: The Natural History of a Changing Region.* Vancouver: Greystone, 2007.

Green, Jared. "The Brave New World of Ecological Restoration." *The Dirt,* April 4, 2011. http://dirt.asla.org/2011/04/06/the-rise-of-novel-ecosystems.

Greenberg, Daniel A., and D. M. Green. "Effects of an Invasive Plant on Population Dynamics in Toads." *Conservation Biology* 27 (2013): 1049–1057. doi: 10.1111/cobi.12078.

Grosholz, Edwin D., R. E. Crafton, R. E. Fontana, J. R. Pasari, S. L. Williams, and C. J. Zabin. "Aquaculture as a Vector for Marine Invasions in California." *Biological Invasions* 17 (2015): 1471–1484. doi:10.1007/s10530–014-0808-9.

Groves, Jason. "The Ecology of Invasions: Reflections from a Damaged Planet." *The Global South* 3 (2009): 30–41.

Helmus, Matthew, D. L. Mahler, and J. B. Losos. "Island Biogeography in the Anthropocene." *Nature* 513 (2014): 543–546. doi: 10.1038/nature13739.

Hendrix, Paul F., M. A. Callaham, Jr., J. M. Drake, C. Huang, S. W. James, B. A. Snyder, and W. Zhang. "Pandora's Box Contained Bait: The Global Problem of Introduced Earthworms." *Annual Review of Ecology, Evolution, and Systematics* 39 (2008): 593–613. doi: 10.1146/annurev.ecolsys.39.110707.173426.

Hughes, Kevin A., and P. Convey. "Determining the Native/Non-Native Status of Newly Discovered Terrestrial and Freshwater Species in Antarctica—Current Knowledge, Methodology and Management Action." *Journal of Environmental Management* 93 (2012): 52–66.

Huiskes, Ad H. L., N. J. M. Gremmen, D. M. Bergstrom, Y. Frenot, K. A. Hughes, S. Imura, K. Kiefer, M. Lebouvier, J. E. Lee, M. Tsujimoto, C. Ware, B. Van de Vijver, and S. L. Chown. "Aliens in Antarctica: Assessing Transfer of Plant Propagules by Human Visitors to Reduce Invasion Risk." *Biological Conservation* 171 (2014): 278–284.

Hutchings, Jeffrey A., C. Walters, and R. L. Haedrich. "Is Scientific Inquiry Incompatible with Government Information Control?" *Canadian Journal of Fisheries and Aquatic Science* 54 (1997): 1198–1210.

Jaffe, Mark. *And No Birds Sing: The Story of an Ecological Disaster in a Tropical Paradise.* New York: Simon & Shuster, 1994.

Jeschke, Jonathan M., L. G. Aparicio, S. Haider, T. Heger, G. J. Lortie, P. Pyšek, and D. L. Strayer. "Support for Major Hypotheses in Invasion Biology Is Uneven and Declining." *Neobiota* 14 (2012): 1–20. doi: 10.3897/neobiota.14.3435.

Kahn, Kamran, I. Bogoch, J. S. Brownstein, J. Miniota, A. Nicolucci, W. Hu, E. O. Nsoesie, M. Cetron, M. I. Creatore, M. German, and A. Wilder-Smith. "Assessing the Origin of and Potential for International Spread of Chikungunya Virus from the Caribbean." *PLoS Currents* (2014). doi: 10.1371/currents.outbreaks.2134a0a7bf37fd 8d388181539fea2da5.

Ketcham, Christopher. "What the Deer Are Telling Us." *Nautilus,* December 11, 2014. http://nautil.us/issue/101/in-our-nature/what-the-deer-are-telling-us-rp.

Kilpatrick, Howard J., A. M. LaBonte, and K. C. Stafford. "The Relationship Between Deer Density, Tick Abundance, and Human Cases of Lyme Disease in a Residential Community." *Journal of Medical Entomology* 51 (2014): 777–784.

Kolar, Cynthia S., and D. M. Lodge. "Ecological Predictions and Risk Assessment for Alien Fishes in North America." *Science* 298 (2002): 1233–1236. doi: 10.1126/science.1075753.

———. "Progress in Invasion Biology: Predicting Invaders." *Trends in Ecology and Evolution* 16 (2001): 199–204.

Kraus, Fred. *Alien Reptiles and Amphibians: A Scientific Compendium and Analysis.* Berlin: Springer, 2009.

Landers, Jackson. *Eating Aliens: One Man's Adventures Hunting Invasive Animal Species.* North Adams: Storey, 2012.

Lodge, David M., M. A. Lewis, J. F. Shogren, and R. P. Keller. "Introduction to Biological Invasions: Biological, Economic and Social Perspectives." Pp. 1–24 in R. P. Keller, D. M. Lodge, M. A. Lewis, and J. F. Shogren, eds., *Bioeconomics of Invasive Species: Integrating Ecology, Economics, Policy, and Management.* New York: Oxford University Press, 2009.

Loss, Scott R., T. Will, and P. P. Marra. "The Impact of Free-Ranging Domestic Cats on Wildlife of the United States." *Nature Communications* 4 (2013). doi: 10.1038/ncomms2380.

Luhring, Thomas, and M. Wagner. "Chemical Risk Information Guides Migratory Movements of Semelparous Sea Lamprey: Implications for Control and Conservation." Presented at a meeting of the American Society of Ichthyologists and Herpetologists, Chattanooga, Tennessee, July 2014.

MacKinnon, James B. *The Once and Future World: Nature as It Was, as It Is, as It Could Be.* Toronto: Random House, 2013.

Mandrak, Nicholas E. "Potential Invasion of the Great Lakes by Fish Species Associated with Climatic Warming." *Journal of Great Lakes Research* 15 (1989): 306–316.

Mandrak, Nicholas E., and B. Cudmore. "The Fall of Native Fishes and the Rise of Non-Native Fishes in the Great Lakes Basin." *Aquatic Ecosystem Health & Management* 13 (2010): 255–268. doi: 10.1080/14634988.2010.507150.

———. "Risk Assessment: Cornerstone of an Aquatic Invasive Species Program." *Aquatic Ecosystem Health & Management* 18 (2015): 312–320. doi: 10.1080/14634988.2015.1046357.

Marco, Diana E., S. A. Cannas, M. A. Montemurro, B. Hu, and S. Cheng. "Comparable Ecological Dynamics Underlie Early Cancer Invasion and Species Dispersal, Involving Self-Organizing Processes." *Journal of Theoretical Biology* 256 (2009): 65–75. doi: 10.1016/j.jtbi.2008.09.011.

McCleery, Robert A., A. Sovie, R. N. Reed, M. W. Cunningham, M. E. Hunter, and K. M. Hart. "Marsh Rabbit Mortalities Tie Pythons to the Precipitous Decline of Mammals in the Everglades." *Proceedings of the Royal Society B* 282 (2015). doi: 10.1098/rspb.2015.0120.

Molina-Montenegro, Marco A., C-U Fernando; C. Rodrigo, P. Convey, F. Valladares, and E. Gianoli. "Occurrence of the Non-Native Annual Bluegrass on the Antarctic Mainland and Its Negative Effects on Native Plants." *Conservation Biology* 26 (2012): 717–723. doi: 10.1111/j.1523-1739.2012.01865.

Monbiot, George. *Feral: Rewilding the Land, the Sea and Human Life.* Toronto: Allen Lane, 2013.

Mooney, Harold A., and Richard J. Hobbs., eds. *Invasive Species in a Changing World.* Washington, DC: Island Press, 2000.

Muirhead, Jim R., M. S. Minton, W. A. Miller, and G. M. Ruiz. "Projected Effects of the Panama Canal Expansion on Shipping Traffic and Biological Invasions." *Diversity and Distributions* 21 (2015): 75–87.

Murray, Rylee G., V. D. Popescu, W. J. Palen, and P. Govindarajulu. "Relative Performance of Ecological Niche and Occupancy Models for Predicting Invasions by Patchily-Distributed Species." *Biological Invasions* 17 (2014): 2691–2706. doi: 10.1007/s10530-015-0906-3.

Nagy, Kelsi, and P. D. Johnson II, eds. *Trash Animals: How We Live with Nature's Filthy, Feral, Invasive and Unwanted Species.* Minneapolis: University of Minnesota Press, 2013.

Nghiem Le T. P., T. Soliman, D. C. J. Yeo, H. T. W. Tan, T. A. Evans, J. D. Mumford, R. P. Keller, R. H. A. Baker, R. T. Corlett, and L. R. Carrasco. "Economic and Environmental Impacts of Harmful Non-Indigenous Species in Southeast Asia." *PLoS ONE* 8 (2013): e71255. doi: 10.1371/journal.pone.0071255.

Orion, Tao. *Beyond the War on Invasive Species: A Permaculture Approach to Ecosystem Restoration.* White River Junction, VT: Chelsea Green, 2015.

Parepa, Madalin, M. Fischer, C. Krebs, and Oliver Bossdorf. "Hybridization Increases Invasive Knotweed Success." *Evolutionary Applications* 7 (2013): 413–420. doi: 10.1111/eva.12139.

Pearce, Fred. *The New Wild: Why Invasive Species Will Be Nature's Salvation.* Boston: Beacon Press, 2015.

Perez, Larry. *Snake in the Grass: An Everglades Invasion.* Sarasota, FL: Pineapple Press, 2012.

Pettitt-Wade, Harri, K. W. Wellband, Daniel D. Heath, and Aaron T. Fisk. "Niche Plasticity in Invasive Fishes in the Great Lakes." *Biological Invasions* 17 (2015): 2565–2580. doi: 10.1007/s10530-015-0894-3.

Quammen, David. "Planet of Weeds." *Harper's,* October 1998, 57–69.

Ricciardi, Anthony, and H. J. MacIsaac. "The Book That Began Invasion Ecology." *Nature* 452 (2008): 34.

Ricciardi, Anthony, M. E. Palmer, and N. D. Yan. "Should Biological Invasions Be Managed as Natural Disasters?" *BioScience* 61 (2011): 312–317.

Ricciardi, Anthony, and D. Simberloff. "Fauna in Decline: First Do No Harm." *Science* 345 (2014): 884.

Richardson, David M., and A. Ricciardi. "Misleading Criticisms of Invasion Science: A Field Guide." *Diversity & Distributions* 19 (2013): 1461–1467. doi: 10.1111/ddi .12150.

Riley, John L. *The Once and Future Great Lakes Country: An Ecological History.* Montreal: McGill-Queen's University Press, 2013.

Roberts, Philip D., H. Diaz-Soltero, D. J. Hemming, M. J. Parr, R. H. Shaw, N. Wakefield, H. J. Wright, and A. B. R. Witt. "What Is the Evidence That Invasive Species Are a Significant Contributor to the Decline or Loss of Threatened Species?" *CABI Invasive Species Systematic Review* (2015). cabi.org/Uploads/isc/IAS%20Systematic%20Review.pdf.

Santaniello, Neil. "Walking Catfish Let Off the Hook." *Miami Sun Sentinel,* January 9, 2005.

Schulz, Kathryn. "The Really Big One." *New Yorker,* July 20, 2015.

Scott, Timothy Lee. *Invasive Plant Medicine: The Ecological Benefits and Healing Abilities of Invasives.* Rochester, VT: Healing Arts Press, 2010.

Scott, William B., and E. J. Crossman. *The Freshwater Fishes of Canada.* Ottawa: Fisheries Research Board of Canada, 1973.

Simberloff, Daniel. "How Much Information on Population Biology Is Needed to Manage Introduced Species?" *Conservation Biology* 17 (2003): 83–92.

———. *Invasive Species: What Everyone Needs to Know.* New York: Oxford University Press, 2013.

Simberloff, Daniel, M. Souza, M. A. Nuñez, M. N. Barrios-Garcia, and W. Bunn. "The Natives Are Restless, But Not Often and Mostly When Disturbed." *Ecology* 93 (2012): 598–607.

Simberloff, Daniel, and B. Von Holle. "Positive Interactions of Nonindigenous Species: Invasional Meltdown?" *Biological Invasions* 1 (1999): 21–32.

Smith, Jennie Erin. *Stolen World: A Tale of Reptiles, Smuggling, and Skullduggery.* New York: Crown, 2011.

Stern, Charles V., H. F. Upton, and C. Brougher. "Asian Carp and the Great Lakes Region." *Congressional Research Service Reports,* January 23, 2014.

Stevenson, Robert Louis. *The Silverado Squatters.* London: Chatto & Windus, 1883.

Stolzenburg, William. *Rat Island: Predators in Paradise and the World's Greatest Wildlife Rescue.* London: Bloomsbury, 2011.

Thompson, Ken. *Where Do Camels Belong? Why Invasive Species Aren't All Bad.* Vancouver: Greystone, 2014.

Turner, Chris. *The War on Science: Muzzled Scientists and Wilful Blindness in Stephen Harper's Canada.* Vancouver: Greystone, 2013.

Ware, Chris, D. M. Bergstrom, E. Müller, and I. G. Alsos. "Humans Introduce Viable Seeds to the Arctic on Footwear." *Biological Invasions* 14 (2012): 567–577.

Warnecke, Lisa, J. M. Turner, T. K. Bollinger, J. M. Lorch, V. Misra, P. M. Cryan, G. Wibbelt, D. S. Blehert, and C. K. R. Willis. "Inoculation of Bats with European *Geomyces destructans* Supports the Novel Pathogen Hypothesis for the Origin of White-Nose Syndrome." *Proceedings of the National Academy of Sciences of the U.S.A.* 109 (2012): 6999–7003.

Weber, Karl, ed. *Cane Toads and Other Rogue Species.* New York: PublicAffairs, 2010.

## Additional Sources

All websites current as of March 1, 2017.

### ANTARCTIC/SUB-ANTARCTIC

- Bouvetøya, vimeo.com/ondemand/bouvetoya.
- British Antarctic Survey, bas.ac.uk.
- South Georgia Heritage Trust, sght.org.

### BLACK SPINY-TAILED IGUANA

- Predatory nature, youtube.com/watch?v=pXr5DJUiCFQ.
- Presence in Boca Grande, Florida, youtube.com/watch?v=cy5ClYqZDeU.

### CARP

- Asian Carp Canada, asiancarp.ca (webinar series is available under the Resources tab).
- "Australia Commits $15m in Bid to Eradicate Carp Using Herpes Virus," theguardian.com/environment/2016/may/01/australia-budget-2016–15m-eradicate-carp-herpes-virus-murray-darling.
- GLMRIS, glmris.anl.gov/glmris-report.
- Illinois-Indiana Sea Grant Be a Hero—Transport Zero™ program, transportzero.org.
- Lock closing study, ilchamber.org/lockclosingstudy.html.
- Michigan lawsuit, michigan.gov/documents/ag/1-Appendix-Renewed_Motion_310133_7.pdf.
- Office of the Assistant Secretary of the Army, FY 2011 Civil Works Budget for the U.S. Army Corps of Engineers, Washington DC, February 2010, p. LRD-132.
- Silver carp jumping, youtube.com/watch?v=tLmJjRqXDCo.

### CATS

- Invasive Animals Cooperative Research Centre, *Review of Cat Ecology and Management Strategies in Australia,* pestsmart.org.au/wp-content/uploads/2010/03/CatReport_web.pdf.

• "Threatened Species Given Lifeline by New Bait Developed to Kill Feral Cats" (Eradicat), theguardian.com/environment/2015/jan/30/new-baits-that-appeal-to-feral-cats-could-aid-recovery-of-threatened-species.

## FINLAND

• Invasive species portal, vieraslajit.fi.
• National strategy on invasive species, vieraslajit.fi/sites/default/files/Finlands_national_strategy_on_invasive_alien_species.pdf#overlay-context=fi/content/national-strategy-invasive-alien-species.

## HAWAII INSECTS

• Kahului Airport Pest Risk Assessment, Plant Quarantine Branch, Hawaii Department of Agriculture, November 2002. hawaiiag.org/PQ/KARA20Report20Final.pdf.

## HIV

• Origins, avert.org/origin-hiv-aids.htm.
• Statistics, amfar.org/worldwide-aids-stats.

## INTERNATIONAL ORGANIZATIONS

• Alien Invasive Species Inventories for Europe, europe-aliens.org.
• Australia Invasive Species Council, invasives.org.au.
• Centre for Agriculture and Biosciences International, cabi.org.
• European Network on Invasive Species, nobanis.org.
• Finland Invasive Species, vieraslajit.fi/fi/content/welcome-invasive-alien-species-portal.
• Great Britain Non-Native Species Secretariat, nonnativespecies.org.
• IUCN Global Invasive Species Database, iucngisd.org/gisd.
• IUCN Invasive Species Specialist Group, issg.org.
• IUCN 100 worst invasives, issg.org/worst100_species.html.
• Ministry for Primary Industries New Zealand, biosecurity.govt.nz.

## NORTH AMERICAN ORGANIZATIONS

• Canadian Aquatic Invasive Species Network, caisn.ca.
• EDRR Ontario, edrrontario.ca.
• Fisheries and Oceans Canada, dfo-mpo.gc.ca/science/environmental-environnement/ais-eae/index-eng.htm.
• Forest Invasives Canada, forestinvasives.ca.
• Government of Canada, invasivespecies.gc.ca.
• Invasive Species Centre Canada, invasivespeciescentre.ca.
• North American Invasive Species Network, naisn.org.
• Save Our People Action Committee, Newfoundland moose reporting, sopacnl.com.

• University of Georgia Center for Invasive Species and Ecosystem Health, invasive.org.
• U.S. Department of Agriculture, invasivespeciesinfo.gov.
• U.S. National Invasive Species Council, invasivespecies.gov.
• U.S. Fish and Wildlife Service, fws.gov/invasives.

### ZIKA

• Monitoring, cdc.gov/zika/index.html.
• Origins, who.int/emergencies/zika-virus/timeline/en.

# Index